Student's Solutions Manual

Differential Equations & Linear Algebra

Fourth Edition

C. Henry Edwards

David E. Penney
The University of Georgia

David T. Calvis
Baldwin Wallace University

4

ISBN-13: 978-0-13-449814-0
ISBN-10: 0-13-449814-3

CONTENTS

PREFACE

This is a solutions manual to accompany the textbook **DIFFERENTIAL EQUATIONS &
LINEAR ALGEBRA** (4th edition, 2018) by C. Henry Edwards, David E. Penney, and David T.
Calvis. We include solutions to most of the odd-numbered problems in the text.

Our goal is to support teaching of the subject of differential equations with linear algebra in every
way that we can. We therefore invite comments and suggested improvements for future printings of
this manual, as well as advice regarding features that might be added to increase its usefulness in
subsequent editions. Additional supplementary material can be found at the Expanded Applications
web site listed below.

Henry Edwards
David Calvis

`h.edwards@mindspring.com`
`dcalvis@bw.edu`

`http://goo.gl/UYnW2g`

CHAPTER 1

FIRST-ORDER DIFFERENTIAL EQUATIONS

SECTION 1.1

DIFFERENTIAL EQUATIONS AND MATHEMATICAL MODELS

The main purpose of Section 1.1 is simply to introduce the basic notation and terminology of differential equations, and to show the student what is meant by a solution of a differential equation. Also, the use of differential equations in the mathematical modeling of real-world phenomena is outlined.

Problems 1-12 are routine verifications by direct substitution of the suggested solutions into the given differential equations. We include here just some typical examples of such verifications.

3. If $y_1 = \cos 2x$ and $y_2 = \sin 2x$, then $y_1' = -2\sin 2x$ $y_2' = 2\cos 2x$, so
$y_1'' = -4\cos 2x = -4y_1$ and $y_2'' = -4\sin 2x = -4y_2$. Thus $y_1'' + 4y_1 = 0$ and $y_2'' + 4y_2 = 0$.

5. If $y = e^x - e^{-x}$, then $y' = e^x + e^{-x}$, so $y' - y = \left(e^x + e^{-x}\right) - \left(e^x - e^{-x}\right) = 2e^{-x}$. Thus
$y' = y + 2e^{-x}$.

11. If $y = y_1 = x^{-2}$, then $y' = -2x^{-3}$ and $y'' = 6x^{-4}$, so
$$x^2 y'' + 5x y' + 4y = x^2 \left(6x^{-4}\right) + 5x\left(-2x^{-3}\right) + 4\left(x^{-2}\right) = 0.$$
If $y = y_2 = x^{-2}\ln x$, then $y' = x^{-3} - 2x^{-3}\ln x$ and $y'' = -5x^{-4} + 6x^{-4}\ln x$, so
$$x^2 y'' + 5x y' + 4y = x^2\left(-5x^{-4} + 6x^{-4}\ln x\right) + 5x\left(x^{-3} - 2x^{-3}\ln x\right) + 4\left(x^{-2}\ln x\right)$$
$$= \left(-5x^{-2} + 5x^{-2}\right) + \left(6x^{-2} - 10x^{-2} + 4x^{-2}\right)\ln x = 0.$$

13. Substitution of $y = e^{rx}$ into $3y' = 2y$ gives the equation $3r e^{rx} = 2e^{rx}$, which simplifies to $3r = 2$. Thus $r = 2/3$.

15. Substitution of $y = e^{rx}$ into $y'' + y' - 2y = 0$ gives the equation $r^2 e^{rx} + r e^{rx} - 2e^{rx} = 0$, which simplifies to $r^2 + r - 2 = (r+2)(r-1) = 0$. Thus $r = -2$ or $r = 1$.

The verifications of the suggested solutions in Problems 17-26 are similar to those in Problems 1-12. We illustrate the determination of the value of C only in some typical cases. However, we illustrate typical solution curves for each of these problems.

17. $C = 2$

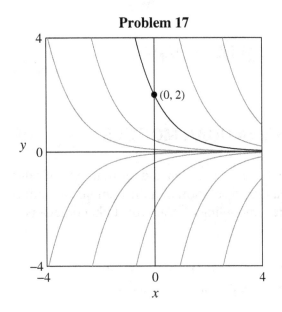

Problem 17

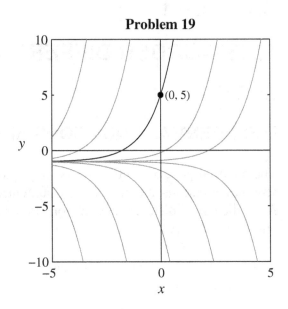

Problem 19

19.   If $y(x) = Ce^x - 1$, then $y(0) = 5$ gives $C - 1 = 5$, so $C = 6$.

21. $C = 7$.

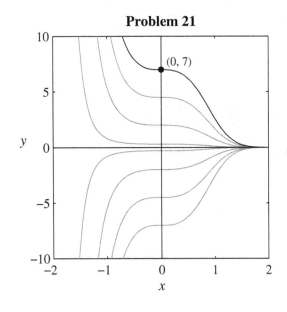

Problem 21

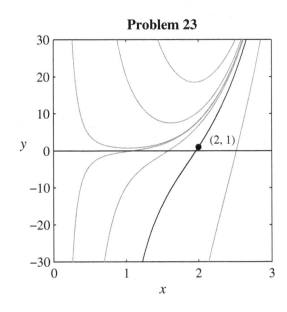

Problem 23

23.   If $y(x) = \frac{1}{4}x^5 + Cx^{-2}$, then $y(2) = 1$ gives $\frac{1}{4} \cdot 32 + C \cdot \frac{1}{8} = 1$, or $C = -56$.

25.   If $y = \tan(x^3 + C)$, then $y(0) = 1$ gives the equation $\tan C = 1$. Hence one value of C is $C = \pi / 4$, as is this value plus any integral multiple of π.

Problem 25

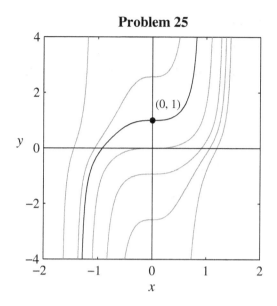

27. $y' = x + y$

29. If $m = y'$ is the slope of the tangent line and m' is the slope of the normal line at (x, y), then the relation $mm' = -1$ yields $m' = -1/y' = (y - 1)/(x - 0)$. Solving for y' then gives the differential equation $(1 - y)y' = x$.

31. The slope of the line through (x, y) and $(-y, x)$ is $y' = (x - y)/(-y - x)$, so the differential equation is $(x + y)y' = y - x$.

In Problems 32-36 we get the desired differential equation when we replace the "time rate of change" of the dependent variable with its derivative with respect to time t, the word "is" with the $=$ sign, the phrase "proportional to" with k, and finally translate the remainder of the given sentence into symbols.

33. $dv/dt = kv^2$ 35. $dN/dt = k(P - N)$

37. The second derivative of any linear function is zero, so we spot the two solutions $y(x) \equiv 1$ and $y(x) = x$ of the differential equation $y'' = 0$.

39. We reason that if $y = kx^2$, then each term in the differential equation is a multiple of x^2. The choice $k = 1$ balances the equation and provides the solution $y(x) = x^2$.

41. We reason that if $y = ke^x$, then each term in the differential equation is a multiple of e^x. The choice $k = \frac{1}{2}$ balances the equation and provides the solution $y(x) = \frac{1}{2}e^x$.

43.   **(a)** We need only substitute $x(t) = 1/(C - kt)$ in both sides of the differential equation $x' = kx^2$ for a routine verification.

(b) The zero-valued function $x(t) \equiv 0$ obviously satisfies the initial value problem $x' = kx^2$, $x(0) = 0$.

45.   Substitution of $P' = 1$ and $P = 10$ into the differential equation $P' = kP^2$ gives $k = \frac{1}{100}$, so Problem 43(a) yields a solution of the form $P(t) = 1/\left(C - \frac{1}{100}t\right)$. The initial condition $P(0) = 2$ now yields $C = \frac{1}{2}$, so we get the solution

$$P(t) = \frac{1}{\dfrac{1}{2} - \dfrac{t}{100}} = \frac{100}{50 - t}.$$

We now find readily that $P = 100$ when $t = 49$ and that $P = 1000$ when $t = 49.9$. It appears that P grows without bound (and thus "explodes") as t approaches 50.

47.   **(a)** $y(10) = 10$ yields $10 = 1/(C - 10)$, so $C = 101/10$.

(b) There is no such value of C, but the constant function $y(x) \equiv 0$ satisfies the conditions $y' = y^2$ and $y(0) = 0$.

(c) It is obvious visually (in Fig. 1.1.8 of the text) that one and only one solution curve passes through each point (a, b) of the xy-plane, so it follows that there exists a unique solution to the initial value problem $y' = y^2$, $y(a) = b$.

SECTION 1.2

INTEGRALS AS GENERAL AND PARTICULAR SOLUTIONS

This section introduces **general solutions** and **particular solutions** in the very simplest situation — a differential equation of the form $y' = f(x)$ — where only direct integration and evaluation of the constant of integration are involved. Students should review carefully the elementary concepts of velocity and acceleration, as well as the fps and mks unit systems.

1.   Integration of $y' = 2x + 1$ yields $y(x) = \int (2x + 1)\, dx = x^2 + x + C$. Then substitution of $x = 0$, $y = 3$ gives $3 = 0 + 0 + C = C$, so $y(x) = x^2 + x + 3$.

3.   Integration of $y' = \sqrt{x}$ yields $y(x) = \int \sqrt{x}\, dx = \frac{2}{3}x^{3/2} + C$. Then substitution of $x = 4$, $y = 0$ gives $0 = \frac{16}{3} + C$, so $y(x) = \frac{2}{3}\left(x^{3/2} - 8\right)$.

5.   Integration of $y' = (x+2)^{-1/2}$ yields $y(x) = \int (x+2)^{-1/2} dx = 2\sqrt{x+2} + C$. Then substitution of $x = 2$, $y = -1$ gives $-1 = 2 \cdot 2 + C$, so $y(x) = 2\sqrt{x+2} - 5$.

7.   Integration of $y' = \dfrac{10}{x^2+1}$ yields $y(x) = \int \dfrac{10}{x^2+1} dx = 10 \tan^{-1} x + C$. Then substitution of $x = 0$, $y = 0$ gives $0 = 10 \cdot 0 + C$, so $y(x) = 10 \tan^{-1} x$.

9.   Integration of $y' = \dfrac{1}{\sqrt{1-x^2}}$ yields $y(x) = \int \dfrac{1}{\sqrt{1-x^2}} dx = \sin^{-1} x + C$. Then substitution of $x = 0$, $y = 0$ gives $0 = 0 + C$, so $y(x) = \sin^{-1} x$.

11.   If $a(t) = 50$, then $v(t) = \int 50 dt = 50t + v_0 = 50t + 10$. Hence
$$x(t) = \int (50t + 10) dt = 25t^2 + 10t + x_0 = 25t^2 + 10t + 20.$$

13.   If $a(t) = 3t$, then $v(t) = \int 3t\, dt = \tfrac{3}{2}t^2 + v_0 = \tfrac{3}{2}t^2 + 5$. Hence
$$x(t) = \int \left(\tfrac{3}{2}t^2 + 5\right) dt = \tfrac{1}{2}t^3 + 5t + x_0 = \tfrac{1}{2}t^3 + 5t.$$

15.   If $a(t) = 4(t+3)^2$, then $v(t) = \int 4(t+3)^2 dt = \tfrac{4}{3}(t+3)^3 + C = \tfrac{4}{3}(t+3)^3 - 37$ (taking $C = -37$ so that $v(0) = -1$). Hence
$$x(t) = \int \tfrac{4}{3}(t+3)^3 - 37\, dt = \tfrac{1}{3}(t+3)^4 - 37t + C = \tfrac{1}{3}(t+3)^4 - 37t - 26.$$

17.   If $a(t) = (t+1)^{-3}$, then $v(t) = \int (t+1)^{-3} dt = -\tfrac{1}{2}(t+1)^{-2} + C = -\tfrac{1}{2}(t+1)^{-2} + \tfrac{1}{2}$ (taking $C = \tfrac{1}{2}$ so that $v(0) = 0$). Hence
$$x(t) = \int -\tfrac{1}{2}(t+1)^{-2} + \tfrac{1}{2}\, dt = \tfrac{1}{2}(t+1)^{-1} + \tfrac{1}{2}t + C = \tfrac{1}{2}\left[(t+1)^{-1} + t - 1\right].$$
(taking $C = -\tfrac{1}{2}$ so that $x(0) = 0$).

Students should understand that Problems 19-22, though different at first glance, are solved in the same way as the preceding ones, that is, by means of the fundamental theorem of calculus in the form $x(t) = x(t_0) + \int_{t_0}^{t} v(s)\,ds$ cited in the text. Actually in these problems

$x(t) = \int_0^t v(s)\,ds$, since t_0 and $x(t_0)$ are each given to be zero.

19.   The graph of $v(t)$ shows that $v(t) = \begin{cases} 5 & \text{if } 0 \le t \le 5 \\ 10-t & \text{if } 5 \le t \le 10 \end{cases}$, so that

$x(t) = \begin{cases} 5t + C_1 & \text{if } 0 \le t \le 5 \\ 10t - \frac{1}{2}t^2 + C_2 & \text{if } 5 \le t \le 10 \end{cases}$. Now $C_1 = 0$ because $x(0) = 0$, and continuity of

$x(t)$ requires that $x(t) = 5t$ and $x(t) = 10t - \frac{1}{2}t^2 + C_2$ agree when $t = 5$. This implies

that $C_2 = -\frac{25}{2}$, leading to the graph of $x(t)$ shown.

Alternate solution for Problem 19 (and similar for 20-22): The graph of $v(t)$ shows

that $v(t) = \begin{cases} 5 & \text{if } 0 \le t \le 5 \\ 10-t & \text{if } 5 \le t \le 10 \end{cases}$. Thus for $0 \le t \le 5$, $x(t) = \int_0^t v(s)\,ds$ is given by

$\int_0^t 5\,ds = 5t$, whereas for $5 \le t \le 10$ we have

$$x(t) = \int_0^t v(s)\,ds = \int_0^5 5\,ds + \int_5^t 10 - s\,ds$$

$$= 25 + \left(10s - \frac{s^2}{2}\Big|_{s=5}^{s=t}\right) = 25 + 10t - \frac{t^2}{2} - \frac{75}{2} = 10t - \frac{t^2}{2} - \frac{25}{2}.$$

The graph of $x(t)$ is shown.

Problem 19

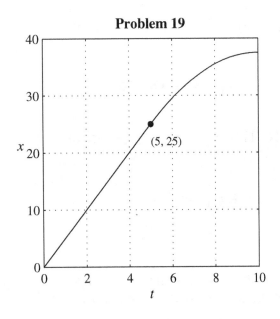

Problem 21

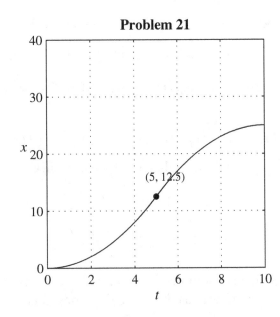

21.   The graph of $v(t)$ shows that $v(t) = \begin{cases} t & \text{if } 0 \le t \le 5 \\ 10-t & \text{if } 5 \le t \le 10 \end{cases}$, so that

$x(t) = \begin{cases} \frac{1}{2}t^2 + C_1 & \text{if } 0 \le t \le 5 \\ 10t - \frac{1}{2}t^2 + C_2 & \text{if } 5 \le t \le 10 \end{cases}$. Now $C_1 = 0$ because $x(0) = 0$, and continuity of

$x(t)$ requires that $x(t) = \frac{1}{2}t^2$ and $x(t) = 10t - \frac{1}{2}t^2 + C_2$ agree when $t = 5$. This implies that $C_2 = -25$, leading to the graph of $x(t)$ shown.

23. $v(t) = -9.8t + 49$, so the ball reaches its maximum height ($v = 0$) after $t = 5$ seconds. Its maximum height then is $y(5) = -4.9(5)^2 + 49(5) = 122.5$ meters.

25. $a = -10$ m/s^2 and $v_0 = 100$ km/h ≈ 27.78 m/s, so $v = -10t + 27.78$, and hence $x(t) = -5t^2 + 27.78t$. The car stops when $v = 0$, that is $t \approx 2.78$ s, and thus the distance traveled before stopping is $x(2.78) \approx 38.59$ meters.

27. $a = -9.8$ m/s^2, so $v = -9.8t - 10$ and $y = -4.9t^2 - 10t + y_0$. The ball hits the ground when $y = 0$ and $v = -9.8t - 10 = -60$ m/s, so $t \approx 5.10$ s. Hence the height of the building is
$$y_0 = 4.9(5.10)^2 + 10(5.10) \approx 178.57 \text{ m}.$$

29.   Integration of $dv/dt = 0.12t^2 + 0.6t$ with $v(0) = 0$ gives $v(t) = 0.04t^3 + 0.3t^2$. Hence $v(10) = 70$ ft/s. Then integration of $dx/dt = 0.04t^3 + 0.3t^2$ with $x(0) = 0$ gives $x(t) = 0.01t^4 + 0.1t^3$, so $x(10) = 200$ ft. Thus after 10 seconds the car has gone 200 ft and is traveling at 70 ft/s.

31.   If $a = -20$ m/s^2 and $x_0 = 0$, then the car's velocity and position at time t are given by $v = -20t + v_0$ and $x = -10t^2 + v_0t$. It stops when $v = 0$ (so $v_0 = 20t$), and hence when $x = 75 = -10t^2 + (20t)t = 10t^2$. Thus $t = \sqrt{7.5}$ s, so
$$v_0 = 20\sqrt{7.5} \approx 54.77 \text{ m/s} \approx 197 \text{ km/hr}.$$

33.   If $v_0 = 0$ and $y_0 = 20$, then $v = -at$ and $y = -\frac{1}{2}at^2 + 20$. Substitution of $t = 2$, $y = 0$ yields $a = 10$ ft/s^2. If $v_0 = 0$ and $y_0 = 200$, then $v = -10t$ and $y = -5t^2 + 200$. Hence $y = 0$ when $t = \sqrt{40} = 2\sqrt{10}$ s and $v = -20\sqrt{10} \approx -63.25$ ft/s.

35.   If $v_0 = 0$ and $y_0 = h$, then the stone's velocity and height are given by $v = -gt$ and $y = -0.5gt^2 + h$, respectively. Hence $y = 0$ when $t = \sqrt{2h/g}$, so $v = -g\sqrt{2h/g} = -\sqrt{2gh}$.

37. We use units of miles and hours. If $x_0 = v_0 = 0$, then the car's velocity and position after t hours are given by $v = at$ and $x = \frac{1}{2}at^2$, respectively. Since $v = 60$ when $t = 5/6$, the velocity equation yields . Hence the distance traveled by 12:50 pm is $x = \frac{1}{2} \cdot 72 \cdot (5/6)^2 = 25$ miles .

39. Integration of $y' = (9/v_S)(1 - 4x^2)$ yields $y = (3/v_S)(3x - 4x^3) + C$, and the initial condition $y(-1/2) = 0$ gives $C = 3/v_S$. Hence the swimmer's trajectory is $y(x) = (3/v_S)(3x - 4x^3 + 1)$. Substitution of $y(1/2) = 1$ now gives $v_S = 6\,\text{mph}$.

41. The bomb equations are $a = -32$, $v = -32t$, and $s_B = s = -16t^2 + 800$ with $t = 0$ at the instant the bomb is dropped. The projectile is fired at time $t = 2$, so its corresponding equations are $a = -32$, $v = -32(t-2) + v_0$, and $s_P = s = -16(t-2)^2 + v_0(t-2)$ for $t \geq 2$ (the arbitrary constant vanishing because $s_P(2) = 0$). Now the condition $s_B(t) = -16t^2 + 800 = 400$ gives $t = 5$, and then the further requirement that $s_P(5) = 400$ yields $v_0 = 544/3 \approx 181.33$ ft/s for the projectile's needed initial velocity.

43. The velocity and position functions for the spacecraft are $v_S(t) = 0.0098t$ and $x_S(t) = 0.0049t^2$, and the corresponding functions for the projectile are $v_P(t) = \frac{1}{10}c = 3\times10^7$ and $x_P(t) = 3\times10^7 t$. The condition that $x_S = x_P$ when the spacecraft overtakes the projectile gives $0.0049t^2 = 3\times10^7 t$, whence

$$t = \frac{3\times10^7}{0.0049} \approx 6.12245\times10^9\,\text{s} \approx \frac{6.12245\times10^9}{(3600)(24)(365.25)} \approx 194\,\text{years} .$$

Since the projectile is traveling at $\frac{1}{10}$ the speed of light, it has then traveled a distance of about 19.4 light years, which is about 1.8367×10^{17} meters.

45. Equation (10) gives

$$v(t)^2 - v_0^2 = (at + v_0)^2 - v_0^2 = a^2t^2 + 2atv_0 + \cancel{v_0^2} - \cancel{v_0^2} = a^2t^2 + 2atv_0,$$

whereas by Eq. (11),

$$2a[x(t) - x_0] = 2a\left(\frac{1}{2}at^2 + v_0t + \cancel{x_0} - \cancel{x_0}\right) = a^2t^2 + 2av_0t,$$

proving the formula.

To apply this formula to Example 2, let x_0 denote (as in the example) the height of the lander above the lunar surface at the moment when the retrorockets should be activated. Thus $v_0 = -450$. We further take $x(t) = 0$ and $v(t) = 0$, corresponding to the lander's touch down on the planet's surface. Because $a = +2.5$, our formula gives

$(-450)^2 - 0^2 = 2 \cdot 2.5 \cdot (x_0 - 0)$, or $x_0 = \dfrac{(-450)^2}{2 \cdot 2.5} = 40,500\,\text{m}$, in agreement with the example.

SECTION 1.3

SLOPE FIELDS AND SOLUTION CURVES

The instructor may choose to delay covering Section 1.3 until later in Chapter 1. However, before proceeding to Chapter 2, it is important that students come to grips at some point with the question of the existence of a unique solution of a differential equation — and realize that it makes no sense to look for the solution without knowing in advance that it exists. It may help some students to simplify the statement of the existence-uniqueness theorem as follows:

Suppose that the function $f(x, y)$ and the partial derivative $\partial f / \partial y$ are both continuous in some neighborhood of the point (a, b). Then the initial value problem

$$\frac{dy}{dx} = f(x, y),\ y(a) = b$$

has a unique solution in some neighborhood of the point a.

Slope fields and geometrical solution curves are introduced in this section as a concrete aid in visualizing solutions and existence-uniqueness questions. Instead, we provide some details of the construction of the figure for the Problem 1 answer, and then include without further comment the similarly constructed figures for Problems 2 through 9.

1.   The following sequence of *Mathematica* 7 commands generates the slope field and the solution curves through the given points. Begin with the differential equation $dy / dx = f(x, y)$, where

```
f[x_, y_] := -y - Sin[x]
```

Then set up the viewing window

```
a = -3; b = 3; c = -3; d = 3;
```

The slope field is then constructed by the command

```
dfield = VectorPlot[{1, f[x, y]}, {x, a, b}, {y, c, d},
   PlotRange -> {{a, b}, {c, d}}, Axes -> True, Frame -> True,
   FrameLabel -> {TraditionalForm[x], TraditionalForm[y]},
   AspectRatio -> 1, VectorStyle -> {Gray, "Segment"},
   VectorScale -> {0.02, Small, None},
   FrameStyle -> (FontSize -> 12), VectorPoints -> 21,
   RotateLabel -> False]
```

The original curve shown in Fig. 1.3.15 of the text (and its initial point not shown there) are plotted by the commands

```
x0 = -1.9; y0 = 0;
point0 = Graphics[{PointSize[0.025], Point[{x0, y0}]}];
soln = NDSolve[{y'[x] == f[x, y[x]], y[x0] == y0}, y[x],
    {x, a, b}];
curve0 = Plot[soln[[1, 1, 2]], {x, a, b}, PlotStyle ->
    {Thickness[0.0065], Blue}];
Show[curve0, point0]
```

(The *Mathematica* **NDSolve** command carries out an approximate numerical solution of the given differential equation. Numerical solution techniques are discussed in Sections 2.4–2.6 of the textbook.)

The coordinates of the 12 points are marked in Fig. 1.3.15 in the textbook. For instance the 7th point is $(-2.5, 1)$. It and the corresponding solution curve are plotted by the commands

```
x0 = -2.5; y0 = 1;
point7 = Graphics[{PointSize[0.025], Point[{x0, y0}]}];
soln = NDSolve[{y'[x] == f[x, y[x]], y[x0] == y0}, y[x],
    {x, a, b}];
curve7 = Plot[soln[[1, 1, 2]], {x, a, b},
    PlotStyle -> {Thickness[0.0065], Blue}];
Show[curve7, point7]
```

The following command superimposes the two solution curves and starting points found so far upon the slope field:

```
Show[dfield, point0, curve0, point7, curve7]
```

We could continue in this way to build up the entire graphic called for in the problem. Here is an alternative looping approach, variations of which were used to generate the graphics below for Problems 1-10:

```
points = {{-2.5,2}, {-1.5,2}, {-0.5,2}, {0.5,2}, {1.5,2},
    {2.5,2}, {-2,-2}, {-1,-2}, {0,-2}, {1,-2}, {2,-2}, {-2.5,1}};
curves = {}; (* start with null lists *)
dots = {};
Do [
    x0 = points[[i, 1]];
    y0 = points[[i, 2]];
    newdot = Graphics[{PointSize[0.025],Point[{x0, y0}]}];
    dots = AppendTo[dots, newdot];
    soln = NDSolve[{y'[x] == f[x, y[x]],y[x0] == y0}, y[x],
        {x, a, b}];
    newcurve = Plot[soln[[1, 1, 2]], {x, a, b},
        PlotStyle -> {Thickness[0.0065], Black}];
    AppendTo[curves, newcurve],
    {i, 1, Length[points]}];
Show[dfield, curves, dots, PlotLabel -> Style["Problem 1", Bold,
    11]]
```

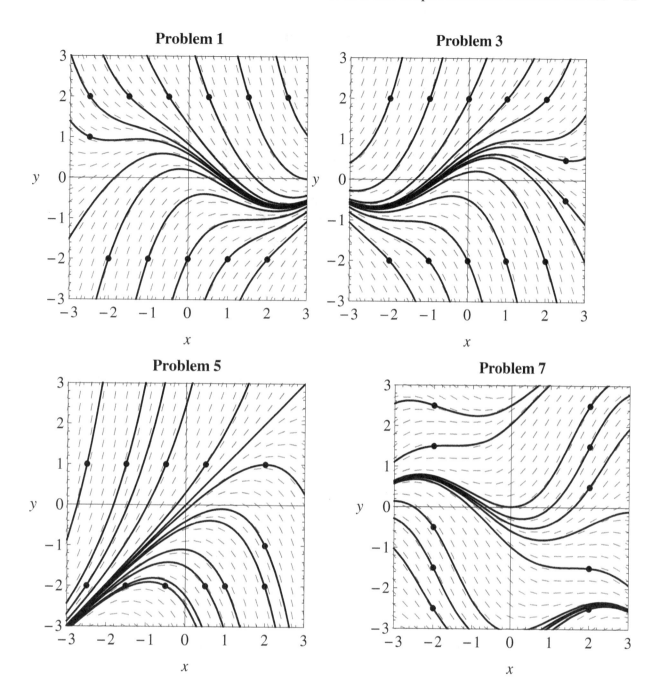

Problem 9

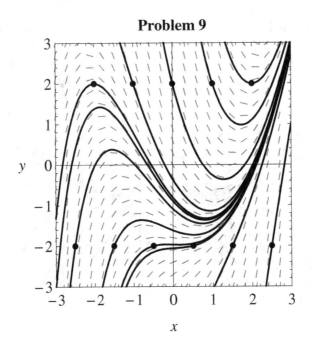

11. Because both $f(x,y) = 2x^2y^2$ and $D_y f(x,y) = 4x^2 y$ are continuous everywhere, the existence-uniqueness theorem of Section 1.3 in the textbook guarantees the existence of a unique solution in some neighborhood of $x = 1$.

13. Both $f(x,y) = y^{1/3}$ and $\partial f/\partial y = \frac{1}{3} y^{-2/3}$ are continuous near $(0,1)$, so the theorem guarantees the existence of a unique solution in some neighborhood of $x = 0$.

15. The function $f(x,y) = (x-y)^{1/2}$ is not continuous at $(2,2)$ because it is not even defined if $y > x$. Hence the theorem guarantees neither existence nor uniqueness in any neighborhood of the point $x = 2$.

17. Both $f(x,y) = (x-1)/y$ and $\partial f/\partial y = -(x-1)/y^2$ are continuous near $(0,1)$, so the theorem guarantees both existence and uniqueness of a solution in some neighborhood of $x = 0$.

19. Both $f(x,y) = \ln(1+y^2)$ and $\partial f/\partial y = 2y/(1+y^2)$ are continuous near $(0,0)$, so the theorem guarantees the existence of a unique solution near $x = 0$.

21. The figure shown can be constructed using commands similar to those in Problem 1, above. Tracing this solution curve, we see that $y(-4) \approx 3$. (An exact solution of the differential equation yields the more accurate approximation $y(-4) = 3 + e^{-4} \approx 3.0183$.)

Problem 21

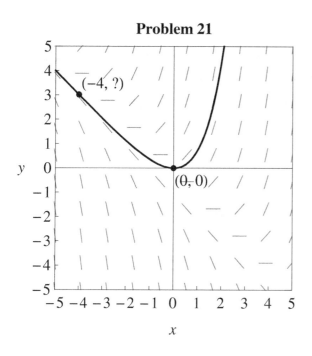

Problem 23

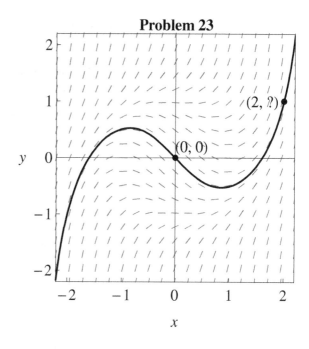

23.   Tracing the curve in the figure shown, we see that $y(2) \approx 1$.   A more accurate approximation is $y(2) \approx 1.0044$.

25.   The figure indicates a limiting velocity of 20 ft/sec — about the same as jumping off a $6\frac{1}{4}$-foot wall, and hence quite survivable.   Tracing the curve suggests that $v(t) = 19$ ft/sec when t is a bit less than 2 seconds.   An exact solution gives $t \approx 1.8723$ then.

Problem 25

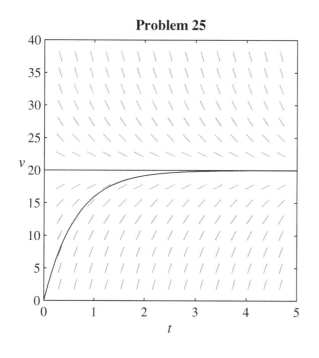

Problem 27a

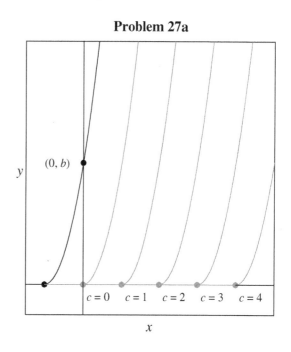

27.   **a)** It is clear that $y(x)$ satisfies the differential equation at each x with $x < c$ or $x > c$, and by examining left- and right-hand derivatives we see that the same is true at $x = c$. Thus $y(x)$ not only satisfies the differential equation for all x, it also satisfies the given initial value problem whenever $c \geq 0$. The infinitely many solutions of the initial value problem are illustrated in the figure. Note that $f(x, y) = 2\sqrt{y}$ is not continuous in any neighborhood of the origin, and so Theorem 1 guarantees neither existence nor uniqueness of solution to the given initial value problem. As it happens, existence occurs, but not uniqueness.

b) If $b < 0$, then the initial value problem $y' = 2\sqrt{y}$, $y(0) = b$ has <u>no</u> solution, because the square root of a negative number would be involved. If $b > 0$, then we get a unique solution curve through $(0, b)$ defined for all x by following a parabola (as in the figure, in black) — down (and leftward) to the x-axis and then following the x-axis to the left. Finally if $b = 0$, then starting at $(0, 0)$ we can follow the positive x-axis to the point $(c, 0)$ and then branch off on the parabola $y = (x - c)^2$, as shown in gray. Thus there are infinitely many solutions in this case.

29.   As with Problem 27, it is clear that $y(x)$ satisfies the differential equation at each x with $x < c$ or $x > c$, and by examining left- and right-hand derivatives we see that the same is true at $x = c$. Looking at the figure on the left below, we see that if, for instance, $b > 0$, then we can start at the point (a, b) and follow a branch of a cubic down to the x-axis, then follow the x-axis an arbitrary distance before branching down on another cubic. This gives infinitely many solutions of the initial value problem $y' = 3y^{2/3}$, $y(a) = b$ that are defined for all x. However, if $b \neq 0$, then there is only a single cubic $y = (x - c)^3$ passing through (a, b), so the solution is unique near $x = a$ (as Theorem 1 would predict).

Problem 29

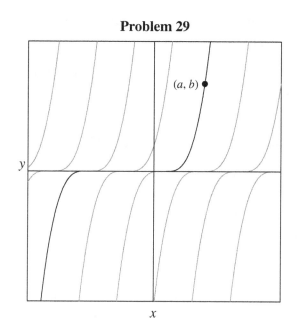

Problem 31

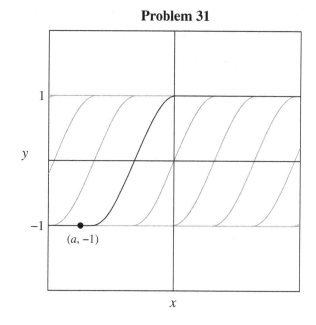

31. The function $y(x) = \begin{cases} -1 & \text{if } x < c - \pi/2 \\ \sin(x-c) & \text{if } c - \pi/2 < x < c + \pi/2 \\ +1 & \text{if } x < c + \pi/2 \end{cases}$ satisfies the given differential

equation on the interval $c - \dfrac{\pi}{2} < x < c + \dfrac{\pi}{2}$, since $y'(x) = \cos(x-c) > 0$ there and thus

$$\sqrt{1-y^2} = \sqrt{1-\sin^2(x-c)} = \sqrt{\cos^2(x-c)} = \cos(x-c) = y'.$$

Moreover, the same is true for $x < \dfrac{\pi}{2}$ and $x > c + \dfrac{\pi}{2}$ (since $y^2 \equiv 1$ and $y' \equiv 0$ there), and

at $x = \dfrac{\pi}{2}, c + \dfrac{\pi}{2}$ by examining one-sided derivatives. Thus $y(x)$ satisfies the given differential equation for all x.

If $|b| > 1$, then the initial value problem $y' = \sqrt{1-y^2}$, $y(a) = b$ has no solution because the square root of a negative number would be involved. If $|b| < 1$, then there is only one curve of the form $y = \sin(x-c)$ through the point (a,b); this gives a unique solution. But if $b = \pm 1$, then we can combine a left ray of the line $y = -1$, a sine curve from the line $y = -1$ to the line $y = +1$, and then a right ray of the line $y = +1$. Looking at the figure, we see that this gives infinitely many solutions (defined for all x) through any point of the form $(a, \pm 1)$.

33.   Looking at the figure provided in the answers section of the textbook, it suffices to observe that, among the pictured curves $y = x/(cx-1)$ for all possible values of c,

- there is a unique one of these curves through any point not on either coordinate axis;
- there is no such curve through any point on the y-axis other than the origin; and
- there are infinitely many such curves through the origin $(0,0)$.

But in addition we have the constant-valued solution $y(x) \equiv 0$ that "covers" the x-axis. It follows that the given differential equation has near (a,b)

- a unique solution if $a \ne 0$;
- no solution if $a = 0$ but $b \ne 0$;
- infinitely many different solutions if $a = b = 0$.

Once again these findings are consistent with Theorem 1.

35.   **(a)** With a computer algebra system we find that the solution of the initial value problem $y' = x - y + 1$, $y(-3) = -0.2$ is $y(x) = x + 2.8e^{-x-3}$, whence $y(2) \approx 2.0189$. With the same differential equation but with initial condition $y(-3) = +0.2$ the solution is $y(x) = x + 3.2e^{-x-3}$, whence $y(2) \approx 2.0216$.

(b) Similarly, the solution of the initial value problem $y' = x - y + 1$, $y(-3) = -0.5$ is $y(x) = x + 2.5e^{-x-3}$, whence $y(2) \approx 2.0168$. With the same differential equation but with initial condition $y(-3) = +0.5$ the solution is $y(x) = x + 3.5e^{-x-3}$, whence $y(2) \approx 2.0236$. Thus the initial values $y(-3) = \pm 0.5$ that are not close both yield $y(2) \approx 2.02$.

SECTION 1.4

SEPARABLE EQUATIONS AND APPLICATIONS

Of course it should be emphasized to students that the possibility of separating the variables is the first one you look for. The general concept of natural growth and decay is important for all differential equations students, but the particular applications in this section are optional. Torricelli's law in the form of Equation (24) in the text leads to some nice concrete examples and problems.

Also, in the solutions below, we make free use of the fact that if C is an arbitrary constant, then so is $5 - 3C$, for example, which we can (and usually do) replace simply with C itself. In the same way we typically replace e^C by C, with the understanding that C is then an arbitrary nonzero constant.

1.   For $y \neq 0$ separating variables gives $\int \dfrac{dy}{y} = -\int 2x\, dx$, so that $\ln|y| = -x^2 + C$, or

$y(x) = \pm e^{-x^2 + C} = Ce^{-x^2}$, where C is an arbitrary nonzero constant. (The equation also has the singular solution $y \equiv 0$.)

3.   For $y \neq 0$ separating variables gives $\int \dfrac{dy}{y} = \int \sin x\, dx$, so that $\ln|y| = -\cos x + C$, or

$y(x) = \pm e^{-\cos x + C} = Ce^{-\cos x}$, where C is an arbitrary nonzero constant. (The equation also has the singular solution $y \equiv 0$.)

5.   For $-1 < y < 1$ and $x > 0$ separating variables gives $\int \dfrac{dy}{\sqrt{1 - y^2}} = \int \dfrac{1}{2\sqrt{x}}\, dx$, so that

$\sin^{-1} y = \sqrt{x} + C$, or $y(x) = \sin\left(\sqrt{x} + C\right)$. (The equation also has the singular solutions $y \equiv 1$ and $y \equiv -1$.)

7.   For $y \neq 0$ separating variables gives $\int \dfrac{dy}{y^{1/3}} = \int 4x^{1/3}\, dx$, so that $\frac{3}{2} y^{2/3} = 3x^{4/3} + C$, or

$y(x) = \left(2x^{4/3} + C\right)^{3/2}$. (The equation also has the singular solution $y \equiv 0$.)

9.   For $y \neq 0$ separating variables and decomposing into partial fractions give

$\int \dfrac{dy}{y} = \int \dfrac{2}{1 - x^2}\, dx = \int \dfrac{1}{1 + x} + \dfrac{1}{1 - x}\, dx$, so that $\ln|y| = \ln|1 + x| - \ln|1 - x| + C$, or

$|y| = C\left|\dfrac{1 + x}{1 - x}\right|$, where C is an arbitrary positive constant, or $y(x) = C\dfrac{1 + x}{1 - x}$, where C is an arbitrary nonzero constant. (The equation also has the singular solution $y \equiv 0$.)

11.   For $y > 0$ separating variables gives $\int \dfrac{dy}{y^3} = \int x\, dx$, so that $-\dfrac{1}{2y^2} = \dfrac{x^2}{2} + C$, or

$y(x) = \left(C - x^2\right)^{-1/2}$, where C is an arbitrary constant. Likewise $y(x) = -\left(C - x^2\right)^{-1/2}$ for $y < 0$. (The equation also has the singular solution $y \equiv 0$.)

13.   Separating variables gives $\int \dfrac{y^3}{y^4 + 1}\, dy = \int \cos x\, dx$, so that $\dfrac{1}{4}\ln\left(y^4 + 1\right) = \sin x + C$, where C is an arbitrary constant.

15.   For $x \neq 0$ and $y \neq 0$, $\dfrac{\sqrt{2}}{2}$ separating variables gives $\displaystyle\int \dfrac{2}{y^2} - \dfrac{1}{y^4}\, dy = \int \dfrac{1}{x} - \dfrac{1}{x^2}\, dx$, so that

$-\dfrac{2}{y} + \dfrac{1}{3y^3} = \ln|x| + \dfrac{1}{x} + C$, where C is an arbitrary constant.

17.   Factoring gives $y' = 1 + x + y + xy = (1+x)(1+y)$, and then for $y \neq -1$ separating variables gives $\displaystyle\int \dfrac{1}{1+y}\, dy = \int 1 + x\, dx$, so that $\ln|1+y| = x + \dfrac{1}{2}x^2 + C$, where C is an arbitrary constant. (The equation also has the singular solution $y \equiv -1$.)

19.   For $y \neq 0$ separating variables gives $\displaystyle\int \dfrac{1}{y}\, dy = \int e^x\, dx$, so that $\ln|y| = e^x + C$, or

$|y| = C\exp(e^x)$, where C is an arbitrary positive constant, or finally $y = C\exp(e^x)$, where C is an arbitrary nonzero constant. The initial condition $y(0) = 2e$ implies that $C \cdot \exp(e^0) = 2e$, or $C = 2$, leading to the particular solution $y(x) = 2\exp(e^x)$.

21.   For $|x| > 4$ separating variables gives $\displaystyle\int 2y\, dy = \int \dfrac{x}{\sqrt{x^2 - 16}}\, dx$, so that

$y^2 = \sqrt{x^2 - 16} + C$. The initial condition $y(5) = 2$ implies that $C = 1$, leading to the particular solution $y(x) = \sqrt{1 + \sqrt{x^2 - 16}}$.

23.   Rewriting the differential equation as $\dfrac{dy}{dx} = 2y - 1$, we see that for $y \neq \dfrac{1}{2}$ separating variables gives $\displaystyle\int \dfrac{1}{2y - 1}\, dy = \int dx$, so that $\dfrac{1}{2}\ln|2y - 1| = x + C$, or $|2y - 1| = Ce^{2x}$, where C is an arbitrary positive constant, or finally $2y - 1 = Ce^{2x}$, which is to say $y = \dfrac{1}{2}(Ce^{2x} + 1)$, where C is an arbitrary nonzero constant. The initial condition $y(1) = 1$ implies that $C = \dfrac{1}{e^2}$, leading to the particular solution $y(x) = \dfrac{1}{2}\left(\dfrac{1}{e^2}e^{2x} + 1\right) = \dfrac{1}{2}(e^{2x-2} + 1)$.

25.   Rewriting the differential equation as $x\dfrac{dy}{dx} = 2x^2 y + y$, we see that for $x, y \neq 0$ separating variables gives $\displaystyle\int \dfrac{1}{y}\, dy = \int 2x + \dfrac{1}{x}\, dx$, so that $\ln|y| = x^2 + \ln|x| + C$, or $|y| = C|x|e^{x^2}$, where C is an arbitrary positive constant, or $y = Cxe^{x^2}$, where C is an arbitrary nonzero constant. The initial condition $y(1) = 1$ implies that $C = \dfrac{1}{e}$, leading to the particular solution

$$y(x) = xe^{x^2-1}.$$

27.   Separating variables gives $\displaystyle\int e^y dy = \int 6e^{2x}dx$, so that $e^y = 3e^{2x} + C$, or $y = \ln\left(3e^{2x} + C\right)$. The initial condition $y(0) = 0$ implies that $C = -2$, leading to the particular solution $y(x) = \ln\left(3e^{2x} - 2\right)$.

29.   **(a)** For $y \neq 0$ separation of variables gives the general solution $\displaystyle\int \frac{1}{y^2} dy = \int dx$, so that

$$-\frac{1}{y} = x + C, \text{ or } y(x) = \frac{1}{C-x}.$$

(b) Inspection yields the singular solution $y(x) \equiv 0$ that corresponds to *no* value of the constant C.

(c) The figure illustrates that there is a unique solution curve through every point in the xy-plane.

(c) Finally, if $b > 0$, then near (a, b) there are exactly *two* solution curves through this point, corresponding to the two indicated parabolas through (a, b), one ascending, and one descending, with increasing x. (Again, see Problem 31.)

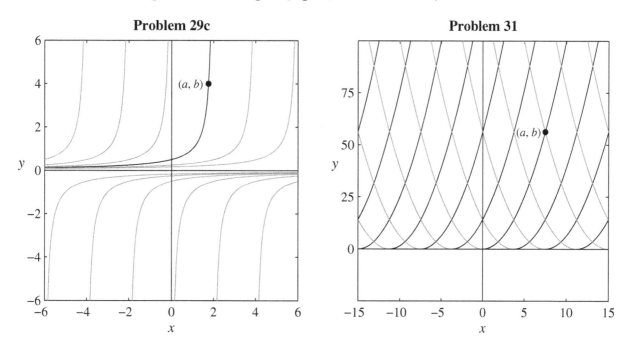

Problem 29c   **Problem 31**

31.   As noted in Problem 30, the solutions of the differential equation $(dy/dx)^2 = 4y$ consist of the solutions of $dy/dx = 2\sqrt{y}$ together with those of $dy/dx = -2\sqrt{y}$, and again we must have $y \geq 0$. Imposing the initial condition $y(a) = b$, where $b > 0$, upon the general solution $y(x) = (x - C)^2$ found in Problem 30 gives $b = (a - C)^2$, which leads to the two values $C = a \pm \sqrt{b}$, and thus to the two particular solutions $y(x) = \left(x - a \pm \sqrt{b}\right)^2$. For these two particular solutions we have $y'(a) = \pm 2\sqrt{b}$, where $(+)$ corresponds to $dy/dx = 2\sqrt{y}$ and $(-)$ corresponds to $dy/dx = -2\sqrt{y}$. It follows that whereas the solutions of $(dy/dx)^2 = 4y$ through (a,b) contain two parabolic segments, one ascending and one descending from left to right, the solutions of $dy/dx = 2\sqrt{y}$ through (a,b) (the black curves in the figure) contain only ascending parabolic segments, whereas for $dy/dx = -2\sqrt{y}$ the (gray) parabolic segments are strictly descending. Thus the answer to the question is "no", because the descending parabolic segments represent solutions of $(dy/dx)^2 = 4y$ but not of $dy/dx = 2\sqrt{y}$. From all this we arrive at the following answers to parts (a)-(c):

(a) No solution curve if $b < 0$;

(b) A unique solution curve if $b > 0$;

(c) Infinitely many solution curves if $b = 0$, because in this case (as noted in the solution for Problem 30) we can pick any $c > a$ and define the solution

$$y(x) = \begin{cases} 0 & \text{if } x < c \\ (x-c)^2 & \text{if } x \geq c \end{cases}.$$

33.   The population growth rate is $k = \ln(30000/25000)/10 \approx 0.01823$, so the population of the city t years after 1960 is given by $P(t) = 25000e^{0.01823t}$. The expected year 2000 population is then $P(40) = 25000e^{0.01823 \times 40} \approx 51840$.

35.   As in the textbook discussion of radioactive decay, the number of ^{14}C atoms after t years is given by $N(t) = N_0 e^{-0.0001216t}$. Hence we need only solve the equation

$\frac{1}{6}N_0 = N_0 e^{-0.0001216t}$ for the age t of the skull, finding $t = \dfrac{\ln 6}{0.0001216} \approx 14735$ years.

37.   The amount in the account after t years is given by $A(t) = 5000e^{0.08t}$. Hence the amount in the account after 18 years is given by $A(18) = 5000e^{0.08 \times 18} \approx 21{,}103.48$ dollars.

39. To find the decay rate of this drug in the dog's blood stream, we solve the equation $\frac{1}{2} = e^{-5k}$ (half-life 5 hours) for k, finding $k = (\ln 2)/5 \approx 0.13863$. Thus the amount in the dog's bloodstream after t hours is given by $A(t) = A_0 e^{-0.13863t}$. We therefore solve the equation $A(1) = A_0 e^{-0.13863} = 50 \times 45 = 2250$ for A_0, finding $A_0 \approx 2585$ mg, the amount to anesthetize the dog properly.

41. Taking $t = 0$ when the body was formed and $t = T$ now, we see that the amount $Q(t)$ of ^{238}U in the body at time t (in years) is given by $Q(t) = Q_0 e^{-kt}$, where $k = (\ln 2)/(4.51 \times 10^9)$. The given information implies that $\dfrac{Q(T)}{Q_0 - Q(T)} = 0.9$. Upon substituting $Q(t) = Q_0 e^{-kt}$ we solve readily for $e^{kT} = 19/9$, so that $T = (1/k)\ln(19/9) \approx 4.86 \times 10^9$. Thus the body was formed approximately 4.86 billion years ago.

43. Because $A = 0$ in Newton's law of cooling, the differential equation reduces to $T' = -kT$, and the given initial temperature then leads to $T(t) = 25e^{-kt}$. The fact that $T(20) = 15$ yields $k = (1/20)\ln(5/3)$, and finally we solve the equation $5 = 25e^{-kt}$ for t to find $t = \ln 5/k \approx 63$ min.

45. **(a)** The light intensity at a depth of x meters is given by $I(x) = I_0 e^{-1.4x}$. We solve the equation $I(x) = I_0 e^{-1.4x} = \frac{1}{2}I_0$ for x, finding $x = (\ln 2)/1.4 \approx 0.495$ meters.

(b) At depth 10 meters the intensity is $I(10) = I_0 e^{-1.4 \times 10} \approx (8.32 \times 10^{-7})I_0$, that is, 0.832 of one millionth of the light intensity I_0 at the surface.

(c) We solve the equation $I(x) = I_0 e^{-1.4x} = 0.01I_0$ for x, finding $x = (\ln 100)/1.4 \approx 3.29$ meters.

47. If $N(t)$ denotes the number of people (in thousands) who have heard the rumor after t days, then the initial value problem is $N' = k(100 - N)$, $N(0) = 0$. Separating variables leads to $\ln(100 - N) = -kt + C$, and the initial condition $N(0) = 0$ gives $C = \ln 100$. Then $100 - N = 100e^{-kt}$, so $N(t) = 100(1 - e^{-kt})$. Substituting $N(7) = 10$ and solving for k gives $k = \ln(100/90)/7 \approx 0.01505$. Finally, 50,000 people have heard the rumor after $t = (\ln 2)/k \approx 46.05$ days, by solving the equation $100(1 - e^{-kt}) = 50$ for t.

49.   Newton's law of cooling gives $\dfrac{dT}{dt} = k(70 - T)$, and separating variables and integrating

lead to $\ln(T - 70) = -kt + C$. The initial condition $T(0) = 210$ gives $C = \ln 140$, and

then $T(30) = 140$ gives $\ln 70 = -30k + \ln 140$, or $k = (\ln 2)/30$, so that

$T(t) = e^{-kt+C} + 70 = 140e^{-kt} + 70$. Finally, setting $T(t) = 100$ gives $140e^{-kt} + 70 = 100$, or

$t = \left[\ln(14/3)\right]/k \approx 66.67$ minutes, or 66 minutes and 40 seconds.

51.   **(a)** The initial condition gives $A(t) = 15e^{-kt}$, and then $A(5) = 10$ implies that $15e^{-kt} = 10$,

or $e^{kt} = \dfrac{3}{2}$, or $k = \dfrac{1}{5}\ln\dfrac{3}{2}$. Thus

$$A(t) = 15\exp\left(-\frac{t}{5}\ln\frac{3}{2}\right) = 15\left(\frac{3}{2}\right)^{-t/5} = 15\left(\frac{2}{3}\right)^{t/5}.$$

(b) After 8 months we have $A(8) = 15\left(\dfrac{2}{3}\right)^{8/5} \approx 7.84$ su .

(c) $A(t) = 1$ when $A(t) = 15\left(\dfrac{2}{3}\right)^{t/5} = 1$, that is $t = 5\dfrac{\ln(\frac{1}{15})}{\ln(\frac{2}{3})} \approx 33.3944$. Thus it will be safe

to return after about 33.4 months.

53.   As in Problem 52, if $L(t)$ denotes the number of Native American language families at

time t (in years), then $L(t) = e^{kt}$ for some constant k, and the condition that

$L(6000) = e^{6000k} = 1.5$ gives $k = \dfrac{1}{6000}\ln\dfrac{3}{2}$. If "now" corresponds to time $t = T$, then we

are given that $L(T) = e^{kT} = 150$, so $T = \dfrac{1}{k}\ln 150 = 6000\dfrac{\ln 150}{\ln(3/2)} \approx 74146.48$. This result

suggests that the ancestors of today's Native Americans first arrived in the western hemi-
sphere about 74 thousand years ago.

55.   With $A = \pi \cdot 3^2$ and $a = \pi(1/12)^2$, and taking $g = 32$ ft/sec^2, Equation (30) reduces to

$162y' = -\sqrt{y}$, which we solve to find $324\sqrt{y} = -t + C$. The initial condition $y(0) = 9$

leads to $C = 972$, and so $y = 0$ when $t = 972$ sec , that is 16 min 12 sec.

57.   The solution of $y' = -k\sqrt{y}$ is given by $2\sqrt{y} = -kt + C$. The initial condition $y(0) = h$

(the height of the cylinder) yields $C = 2\sqrt{h}$. Then substituting $t = T$ and $y = 0$ gives

$k = 2\sqrt{h}/T$. It follows that $y = h\left(1 - \dfrac{t}{T}\right)^2$. If r denotes the radius of the cylinder, then

$$V(y) = \pi r^2 y = \pi r^2 h \left(1 - \frac{t}{T}\right)^2 = V_0 \left(1 - \frac{t}{T}\right)^2.$$

59. **(a)** Since $x^2 = by$, the cross-sectional area is $A(y) = \pi x^2 = \pi b y$. Hence equation (30) becomes $y^{1/2} y' = -k = -(a/\pi b)\sqrt{2g}$, with general solution $\frac{2}{3} y^{3/2} = -kt + C$. The initial condition $y(0) = 4$ gives $C = 16/3$, and then $y(1) = 1$ yields $k = 14/3$. It follows that the depth at time t is $y(t) = (8 - 7t)^{2/3}$.

(b) The tank is empty after $t = 8/7$ hr , that is, at 1:08:34 p.m.

(c) We see above that $k = \frac{a}{\pi b}\sqrt{2g} = \frac{14}{3}$. Substitution of $a = \pi r^2$ and $b = 1$ and $g = 32 \cdot 3600^2$ ft/hr² yields $r = \frac{1}{60}\sqrt{\frac{7}{12}}$ ft ≈ 0.15 in for the radius of the bottom hole.

61. $A(y) = \pi\left(8y - y^2\right)$ as in Example 8 in the text, but now $a = \frac{\pi}{144}$ in Equation (30), so that the initial value problem is $18\left(8y - y^2\right)y' = -\sqrt{y}$, $y(0) = 8$. Separating variables gives $\int 18\left(y^{3/2} - 8y^{1/2}\right) dy = \int dt$, or $18\left(\frac{2}{5}y^{5/2} - \frac{16}{3}y^{3/2}\right) = t + C$, and the initial condition gives $C = 18\left(\frac{2}{5}8^{5/2} - \frac{16}{3}8^{3/2}\right)$. We seek the value of t when $y = 0$, which is given by $-C \approx 869$ sec $= 14$ min 29 sec .

63. **(a)** As in Example 6, the initial value problem is $\pi\left(8y - y^2\right)\frac{dy}{dt} = -\pi k\sqrt{y}$, $y(0) = 4$, where $k = 0.6r^2\sqrt{2g} = 4.8r^2$. Separating variables and applying the initial condition just as in the Example 6 solution, we find that $\frac{16}{3}y^{3/2} - \frac{2}{5}y^{5/2} = -kt + \frac{448}{15}$. When we substitute $y = 2$ (ft) and $t = 1800$ sec (that is, 30 min) we find that $k \approx 0.009469$. Finally, $y = 0$ when $t = \frac{448}{15k} \approx 3154$ sec $= 53$ min 34 sec . Thus the tank is empty at 1:53:34 p.m.

(b) The radius of the bottom hole is $r = \sqrt{\frac{k}{4.8}} \approx 0.04442$ ft ≈ 0.53 in , thus about half an inch.

65.   The temperature $T(t)$ of the body satisfies the differential equation $\dfrac{dT}{dt} = k(70 - T)$.

Separating variables gives $\displaystyle\int \dfrac{1}{70 - T}\, dT = \int k\, dt$, or (since $T(t) > 70$ for all t)

$\ln(T - 70) = -kt + C$. If we take $t = 0$ at the (unknown) time of death, then applying the initial condition $T(0) = 98.6$ gives $C = \ln 28.6$, and so $T(t) = 70 + 28.6e^{-kt}$. Now suppose that 12 noon corresponds to $t = a$. This gives the two equations

$$T(a) = 70 + 28.6e^{-ka} = 80$$
$$T(a+1) = 70 + 28.6e^{-k(a+1)} = 75,$$

which simplify to

$$28.6e^{-ka} = 10$$
$$28.6e^{-ka}e^{-k} = 5.$$

These latter equations imply that $e^{-k} = 5/10 = 1/2$, so that $k = \ln 2$. Finally, we can substitute this value of k into the first of the previous two equations to find that

$a = \dfrac{\ln 2.86}{\ln 2} \approx 1.516 \text{ hr} \approx 1 \text{ hr } 31 \text{ min}$, so the death occurred at 10:29 a.m.

67.   We still have $t = t_0 e^{kx}$, but now the given information yields the conditions

$$t_0 + 1 = t_0 e^{4k}$$
$$t_0 + 2 = t_0 e^{7k}$$

at 8 a.m. and 9 a.m., respectively. Elimination of t_0 gives the equation $2e^{4k} - e^{7k} - 1 = 0$, which cannot be easily factored, unlike the corresponding equation in Problem 66. Letting $u = e^k$ gives $2u^4 - u^7 - 1 = 0$, and solving this equation using MATLAB or other technology leads to three real and four complex roots. Of the three real roots, only $u \approx 1.086286$ satisfies $u > 1$, and thus represents the desired solution. This means that $k \approx \ln 1.086286 \approx 0.08276$. Using this value, we finally solve either of the preceding pair of equations for $t_0 \approx 2.5483 \text{ hr} \approx 2 \text{ hr } 33 \text{ min}$. Thus it began to snow at 4:27 a.m.

69.   Substitution of $v = dy/dx$ in the differential equation for $y = y(x)$ gives $a\dfrac{dv}{dx} = \sqrt{1 + v^2}$,

and separation of variables then yields $\displaystyle\int \dfrac{1}{\sqrt{1 + v^2}}\, dv = \int \dfrac{1}{a}\, dx$, or $\sinh^{-1} v = \dfrac{x}{a} + C_1$, or

$\dfrac{dy}{dx} = \sinh\left(\dfrac{x}{a} + C_1\right)$. The fact that $y'(0) = 0$ implies that $C_1 = 0$, so it follows that

$\dfrac{dy}{dx} = \sinh\left(\dfrac{x}{a}\right)$, or $y(x) = a\cosh\left(\dfrac{x}{a}\right) + C$. Of course the (vertical) position of the x-axis can be adjusted so that $C = 0$, and the units in which T and ρ are measured may be adjusted so that $a = 1$. In essence, then, the shape of the hanging cable is the hyperbolic cosine graph $y = \cosh x$.

SECTION 1.5

LINEAR FIRST-ORDER EQUATIONS

1. An integrating factor is given by $\rho = \exp\left(\int 1\,dx\right) = e^x$, and multiplying the differential equation by ρ gives $e^x y' + e^x y = 2e^x$, or $D_x\left(e^x \cdot y\right) = 2e^x$. Integrating then leads to $e^x \cdot y = \int 2e^x\,dx = 2e^x + C$, and thus to the general solution $y = 2 + Ce^{-x}$. Finally, the initial condition $y(0) = 0$ implies that $C = -2$, so the corresponding particular solution is $y(x) = 2 - 2e^{-x}$.

3. An integrating factor is given by $\rho = \exp\left(\int 3\,dx\right) = e^{3x}$, and multiplying the differential equation by ρ gives $D_x\left(y \cdot e^{3x}\right) = 2x$. Integrating then leads to $y \cdot e^{3x} = x^2 + C$, and thus to the general solution $y(x) = \left(x^2 + C\right)e^{-3x}$.

5. We first rewrite the differential equation for $x > 0$ as $y' + \dfrac{2}{x} y = 3$. An integrating factor is given by $\rho = \exp\left(\int \dfrac{2}{x}\,dx\right) = e^{2\ln x} = x^2$, and multiplying the equation by ρ gives $x^2 \cdot y' + 2xy = 3$, or $D_x\left(y \cdot x^2\right) = 3x^2$. Integrating then leads to $y \cdot x^2 = x^3 + C$, and thus to the general solution $y(x) = x + \dfrac{C}{x^2}$. Finally, the initial condition $y(1) = 5$ implies that $C = 4$, so the corresponding particular solution is $y(x) = x + \dfrac{4}{x^2}$.

7. We first rewrite the differential equation for $x > 0$ as $y' + \dfrac{1}{2x} y = \dfrac{5}{\sqrt{x}}$. An integrating factor is given by $\rho = \exp\left(\int \dfrac{1}{2x}\,dx\right) = e^{(\ln x)/2} = \sqrt{x}$, and multiplying the equation by ρ gives $\sqrt{x} \cdot y' + \dfrac{1}{2\sqrt{x}} y = 5$, or $D_x\left(y \cdot \sqrt{x}\right) = 5$. Integrating then leads to $y \cdot \sqrt{x} = 5x + C$, and thus to the general solution $y(x) = 5\sqrt{x} + \dfrac{C}{\sqrt{x}}$.

9.   We first rewrite the differential equation for $x > 0$ as $y' - \dfrac{1}{x}y = 1$. An integrating factor

is given by $\rho = \exp\left(\int -\dfrac{1}{x}dx\right) = \dfrac{1}{x}$, and multiplying the equation by ρ gives

$\dfrac{1}{x}y' - \dfrac{1}{x^2}y = \dfrac{1}{x}$, or $D_x\left(y \cdot \dfrac{1}{x}\right) = \dfrac{1}{x}$. Integrating then leads to $y \cdot \dfrac{1}{x} = \ln x + C$, and thus to

the general solution $y(x) = x \ln x + Cx$. Finally, the initial condition $y(1) = 7$ implies

that $C = 7$, so the corresponding particular solution is $y(x) = x \ln x + 7x$.

11.   We first collect terms and rewrite the differential equation for $x > 0$ as

$y' + \left(\dfrac{1}{x} - 3\right)y = 0$. An integrating factor is given by

$$\rho = \exp\left[\int\left(\dfrac{1}{x} - 3\right)dx\right] = e^{\ln x - 3x} = xe^{-3x},$$

and multiplying by ρ gives $xe^{-3x} \cdot y' + \left(e^{-3x} - 3xe^{-3x}\right)y = 0$, or $D_x\left(y \cdot xe^{-3x}\right) = 0$. Inte-

grating then leads to $y \cdot xe^{-3x} = C$, and thus to the genral solution $y(x) = Cx^{-1}e^{3x}$. Final-

ly, the initial condition $y(1) = 0$ implies that $C = 0$, so the corresponding particular solu-

tion is $y(x) \equiv 0$, that is, the solution is the zero function.

13.   An integrating factor is given by $\rho = \exp\left(\int 1\,dx\right) = e^x$, and multiplying by ρ gives

$e^x \cdot y' + e^x y = e^{2x}$, or $D_x\left(y \cdot e^x\right) = e^{2x}$. Integrating then leads to $y \cdot e^x = \dfrac{1}{2}e^{2x} + C$, and

thus to the general solution $y(x) = \dfrac{1}{2}e^x + Ce^{-x}$. Finally, the initial condition $y(0) = 1$

implies that $C = \dfrac{1}{2}$, so the corresponding particular solution is $y(x) = \dfrac{1}{2}e^x + \dfrac{1}{2}e^{-x}$, that

is, $y = \cosh x$.

15.   An integrating factor is given by $\rho = \exp\left(\int 2x\,dx\right) = e^{x^2}$, and multiplying by ρ gives

$e^{x^2} \cdot y' + 2xe^{x^2}y = xe^{x^2}$, or $D_x\left(y \cdot e^{x^2}\right) = xe^{x^2}$. Integrating then leads to $y \cdot e^{x^2} = \dfrac{1}{2}e^{x^2} + C$,

and thus to the general solution $y(x) = \dfrac{1}{2} + Ce^{-x^2}$. Finally, the initial condition

$y(0) = -2$ implies that $C = -\dfrac{5}{2}$, so the corresponding particular solution is

$y(x) = \dfrac{1}{2} - \dfrac{5}{2}e^{-x^2}$.

17.   We first rewrite the differential equation for $x > -1$ as $y' + \dfrac{1}{1+x} y = \dfrac{\cos x}{1+x}$. An integrat-

ing factor is given by $\rho = \exp\left(\displaystyle\int \dfrac{1}{1+x} dx\right) = 1 + x$, and multiplying by ρ gives

$(1+x) y' + y = \cos x$ (which happens to be the original differential equation), or

$D_x\left[y \cdot (1+x)\right] = \cos x$. Integrating then leads to $y \cdot (1+x) = \sin x + C$, and thus to the

general solution $y(x) = \dfrac{\sin x + C}{1+x}$. Finally, the initial condition $y(0) = 1$ implies that

$C = 1$, so the corresponding particular solution is $y(x) = \dfrac{1 + \sin x}{1+x}$.

19.   For $x > 0$ an integrating factor is given by $\rho = \exp\left(\displaystyle\int \cot x \, dx\right) = e^{\ln(\sin x)} = \sin x$, and mul-

tiplying by ρ gives $(\sin x) \cdot y' + (\cos x) y = \sin x \cos x$, or $D_x(y \cdot \sin x) = \sin x \cos x$. In-

tegrating then leads to $y \cdot \sin x = \dfrac{1}{2} \sin^2 x + C$, and thus to the general solution

$y(x) = \dfrac{1}{2} \sin x + C \csc x$.

21.   We first rewrite the differential equation for $x > 0$ as $y' - \dfrac{3}{x} y = x^3 \cos x$. An integrating

factor is given by $\rho = \exp\left(\displaystyle\int -\dfrac{3}{x} dx\right) = e^{-3\ln x} = x^{-3}$, and multiplying by ρ gives

$x^{-3} \cdot y' - 3x^{-4} y = \cos x$, or $D_x(y \cdot x^{-3}) = \cos x$. Integrating then leads to

$y \cdot x^{-3} = \sin x + C$, and thus to the general solution $y(x) = x^3 \sin x + Cx^3$. Finally, the ini-

tial condition $y(2\pi) = 0$, so the corresponding particular solution is $y(x) = x^3 \sin x$.

23.   We first rewrite the differential equation for $x > 0$ as $y' + \left(2 - \dfrac{3}{x}\right) y = 4x^3$. An integrat-

ing factor is given by $\rho = \exp\left(\displaystyle\int 2 - \dfrac{3}{x} dx\right) = \exp(2x - 3\ln x) = x^{-3} e^{2x}$, and multiplying by

ρ gives $x^{-3} e^{2x} \cdot y' + (2x^{-3} - 3x^{-4}) e^{2x} y = 4e^{2x}$, or $D_x(y \cdot x^{-3} e^{2x}) = 4e^{2x}$. Integrating then

leads to $y \cdot x^{-3} e^{2x} = 2e^{2x} + C$, and thus to the general solution $y(x) = 2x^3 + Cx^3 e^{-2x}$.

25.   We first rewrite the differential equation as $y' + \dfrac{3x^3}{x^2+1}y = \dfrac{6x}{x^2+1}e^{-\frac{3}{2}x^2}$. An integrating

factor is given by $\rho = \exp\left(\displaystyle\int \frac{3x^3}{x^2+1}dx\right)$. Long division of polynomials shows that

$\dfrac{3x^3}{x^2+1} = 3x - \dfrac{3x}{x^2+1}$, and so

$$\rho = \exp\left(\int 3x - \frac{3x}{x^2+1}dx\right) = \exp\left[\frac{3}{2}x^2 - \frac{3}{2}\ln(x^2+1)\right] = (x^2+1)^{-3/2}e^{\frac{3}{2}x^2}.$$

Multiplying by ρ gives

$$(x^2+1)^{-3/2}e^{\frac{3}{2}x^2}\cdot y' + 3x^3(x^2+1)^{-5/2}e^{\frac{3}{2}x^2}y = 6x(x^2+1)^{-5/2},$$

or (as can be verified using the product rule twice, together with some algebra)

$D_x\left[y\cdot(x^2+1)^{-3/2}e^{\frac{3}{2}x^2}\right] = 6x(x^2+1)^{-5/2}$. Integrating then leads to

$$y\cdot(x^2+1)^{-3/2}e^{\frac{3}{2}x^2} = \int 6x(x^2+1)^{-5/2}dx = -2(x^2+1)^{-3/2}+C,$$

and thus to the general solution $y = \left[-2 + C(x^2+1)^{3/2}\right]e^{-\frac{3}{2}x^2}$. Finally, the initial condi-

tion $y(0)=1$ implies that $C=3$, so the corresponding particular solution is

$$y = \left[-2 + 3(x^2+1)^{3/2}\right]e^{-\frac{3}{2}x^2}.$$

The strategy in each of Problems 26-28 is to use the inverse function theorem to conclude that at
points (x,y) where $\dfrac{dy}{dx} \neq 0$, x is locally a function of y with $\dfrac{dx}{dy}\cdot\dfrac{dy}{dx}=1$. Thus the given differential equation is equivalent to one in which x is the dependent variable and y as the independent variable, and this latter equation may be easier to solve than the one originally given. It may not be feasible, however, to solve the resulting solution for the original dependent variable y.

27.   At points (x,y) with $x+ye^y \neq 0$, rewriting the differential equation as $\dfrac{dy}{dx} = \dfrac{1}{x+ye^y}$

shows that $\dfrac{dx}{dy} = x+ye^y$, or (putting x' for $\dfrac{dx}{dy}$) $x'-x = ye^y$, a linear equation for the

dependent variable x as a function of the independent variable y. An integrating factor is
given by $\rho = \exp\left(\int -1\,dy\right) = e^{-y}$, and multiplying by ρ gives $e^{-y}\cdot x' - e^{-y}x = y$, or

$D_y\left(x \cdot e^{-y}\right) = y$. Integrating then leads to $x \cdot e^{-y} = \dfrac{1}{2}y^2 + C$, and thus to the general (implicit) solution $x(y) = \left(\dfrac{1}{2}y^2 + C\right)e^y$.

29.   We first rewrite the differential equation as $y' - 2xy = 1$. An integrating factor is given by $\rho = \exp\left(\int -2x\,dx\right) = e^{-x^2}$, and multiplying by ρ gives $e^{-x^2} \cdot y' - 2xe^{-x^2}y = e^{-x^2}$, or $D_x\left(y \cdot e^{-x^2}\right) = e^{-x^2}$. Integrating then leads to $y \cdot e^{-x^2} = \int e^{-x^2}\,dx$. Any antiderivative of e^{-x^2} differs by a constant (call it C) from the definite integral $\displaystyle\int_0^x e^{-t^2}\,dt$, and so we can write $y \cdot e^{-x^2} = \displaystyle\int_0^x e^{-t^2}\,dt + C$. The definition of $\operatorname{erf}(x)$ then gives

$$y \cdot e^{-x^2} = \frac{\sqrt{\pi}}{2}\operatorname{erf}(x) + C,$$ and thus the general solution $y(x) = e^{x^2}\left[\dfrac{\sqrt{\pi}}{2}\operatorname{erf}(x) + C\right]$.

31.   **(a)** The fundamental theorem of calculus implies, for any value of C, that

$$y_c'(x) = Ce^{-\int P(x)\,dx}\left[-P(x)\right] = -P(x)y_c(x),$$

and thus that $y_c'(x) + P(x)y_c(x) = 0$. Therefore y_c is a general solution of $\dfrac{dy}{dx} + P(x)y = 0$.

(b) The product rule and the fundamental theorem of calculus imply that

$$y_p'(x) = e^{-\int P(x)\,dx} \cdot Q(x)e^{\int P(x)\,dx} + e^{-\int P(x)\,dx}\left[-P(x)\right] \cdot \int\left(Q(x)e^{\int P(x)\,dx}\right)dx$$

$$= Q(x) - P(x)e^{-\int P(x)\,dx}\int\left(Q(x)e^{\int P(x)\,dx}\right)dx$$

$$= Q(x) - P(x)y_p(x),$$

and thus that $y_p'(x) + P(x)y_p(x) = Q(x)$. Therefore y_p is a particular solution of $\dfrac{dy}{dx} + P(x)y = Q(x)$.

(c) The stated assumptions imply that

$$y'(x) + P(x)y = y_c'(x) + y_p'(x) + P(x)\left[y_c(x) + y_p(x)\right]$$

$$= \left[y_c'(x) + P(x)y_c(x)\right] + \left[y_p'(x) + P(x)y_p(x)\right]$$

$$= 0 + Q(x)$$

$$= Q(x),$$

proving that $y(x)$ is a general solution of $\dfrac{dy}{dx} + P(x)y = Q(x)$.

33. Let $x(t)$ denote the amount of salt (in kg) in the tank after t seconds. We want to know when $x(t) = 10$. In the notation of Equation (18) of the text, the differential equation for $x(t)$ is

$$\frac{dx}{dt} = r_i c_i - \frac{r_o}{V} x = (5\,\text{L/s})(0\,\text{kg/L}) - \frac{5\,\text{L/s}}{1000\,\text{L}} \cdot x\,\text{kg},$$

or $\dfrac{dx}{dt} = -\dfrac{x}{200}$. Separating variables gives the general solution $x(t) = Ce^{-t/200}$, and the initial condition $x(0) = 100$ implies that $C = 100$, and so $x(t) = 100e^{-t/200}$. Setting $x(t) = 10$ gives $10 = 100e^{-t/200}$, or $t = 200\ln 10 \approx 461$ sec, that is, about 7 min 41 sec.

35. The only difference from the Example 4 solution in the textbook is that $V = 1640\,\text{km}^3$ and $r = 410\,\text{km}^3/\text{yr}$ for Lake Ontario, so the time required is

$$t = \frac{V}{r}\ln 4 = 4\ln 4 \approx 5.5452 \text{ years.}$$

37. Let $x(t)$ denote the amount of salt (in lb) after t seconds. Because the volume of liquid in the tank is increasing by 2 gallon each minute, the volume after t sec is $100 + 2t$ gallons. Thus in the notation of Equation (18) of the text, the differential equation for $x(t)$ is

$$\frac{dx}{dt} = r_i c_i - \frac{r_o}{V} x = (5\,\text{gal/s})(1\,\text{lb/gal}) - \frac{3\,\text{gal/s}}{(100 + 2t)\,\text{gal}} \cdot x\,\text{lb},$$

or $\dfrac{dx}{dt} + \dfrac{3}{100 + 2t} x = 5$. An integrating factor is given by

$$\rho = \exp\left(\int \frac{3}{100 + 2t}\,dt\right) = (100 + 2t)^{3/2},$$ and multiplying the differential equation by ρ gives

$$(100 + 2t)^{3/2} \cdot \frac{dx}{dt} + 3(100 + 2t)^{1/2} x = 5(100 + 2t)^{3/2},$$

or $D_t\left[(100 + 2t)^{3/2} \cdot x\right] = 5(100 + 2t)^{3/2}$. Integrating then leads to

$$(100 + 2t)^{3/2} \cdot x = \int 5(100 + 2t)^{3/2}\,dt = (100 + 2t)^{5/2} + C,$$

and thus to the general solution $x(t) = 100 + 2t + C(100 + 2t)^{-3/2}$. The initial condition $x(0) = 50$ implies that $50 = 100 + C \cdot 100^{-3/2}$, or $C = -50000$, and so the desired particular solution is $x(t) = 100 + 2t - \dfrac{50000}{(100 + 2t)^{3/2}}$. Finally, because the tank starts out with

300 gallons of excess capacity and the volume of its contents increases at $2\,\text{gal/s}$, the tank is full when $t = \dfrac{300\,\text{gal}}{2\,\text{gal/s}} = 150\,\text{s}$. At this time the tank contains

$$x(150) = 400 - \frac{50000}{(400)^{3/2}} = 393.75\,\text{lb} \text{ of salt.}$$

39.   **(a)** In the notation of Equation (18) of the text, the differential equation for $x(t)$ is

$$\frac{dx}{dt} = r_i c_i - \frac{r_o}{V}x = \left(10\,\text{gal/min}\right)(0) - \left(10\,\text{gal/min}\right)\left(\frac{x}{100}\right),$$

or $\dfrac{dx}{dt} + \dfrac{1}{10}x = 0$. Separating variables leads to the general solution $x(t) = Ce^{-t/10}$, and the initial condition $x(0) = 100$ implies that $C = 100$. Thus $x(t) = 100e^{-t/10}$. In the same way, the differential equation for $y(t)$ is

$$\frac{dy}{dt} = \left(10\,\text{gal/min}\right)\left(\frac{x}{100}\right) - \left(10\,\text{gal/min}\right)\left(\frac{y}{100}\right),$$

because the volume of liquid in each tank remains constant at 2 gal. Substituting the result of part (a) gives $\dfrac{dy}{dt} + \dfrac{1}{10}y = 10e^{-t/10}$. An integrating factor is given by

$$\rho = \exp\left(\int \frac{1}{10}\,dt\right) = e^{t/10}, \text{ and multiplying the differential equation by } \rho \text{ gives}$$

$e^{t/10} \cdot \dfrac{dy}{dt} + \dfrac{1}{10}e^{t/10}y = 10$, or $D_t\left(e^{t/10} \cdot y\right) = 10$. Integrating then leads to $e^{t/10} \cdot y = 10t + C$, and thus to the general solution $y(t) = \left(10t + C\right)e^{-t/10}$. The initial condition $y(0) = 0$ implies that $C = 0$, so that $y(t) = 10te^{-t/10}$.

(b) By Part (a), $y'(t) = 10\left(-\dfrac{t}{10}e^{-t/10} + e^{-t/10}\right) = e^{-t/10}\left(10 - t\right)$, which is zero for $t = 10$. Furthermore, $y'(t) > 0$ for $0 < t < 10$, and $y'(t) < 0$ for $t > 10$, which implies that $y(t)$ reaches its absolute maximum at $t = 10$ min. The maximum amount of ethanol in tank 2 is therefore $y(10) = 100e^{-1} \approx 36.79$ gal.

41.   **(a)** Between time t and time $t + \Delta t$, the amount $A(t)$ (in thousands of dollars) increases by a deposit of $0.12S(t)\Delta t$ (12% per year of annual salary) as well as interest earnings of $0.06A(t)\Delta t$ (6% per year of current balance). It follows that

$$\Delta A \approx 0.12S(t)\Delta t + 0.06A(t)\Delta t,$$

leading to the linear differential equation $\dfrac{dA}{dt} = 0.12S + 0.06A = 3.6e^{t/20} + 0.06A$, or

$$\frac{dA}{dt} - 0.06A = 3.6e^{t/20}.$$

(b) An integrating factor is given by $\rho = \exp\left(\int -0.06\,dt\right) = e^{-0.06t}$, and multiplying the

differential equation by ρ gives $e^{-0.06t} \cdot \dfrac{dA}{dt} - 0.06e^{-0.06t} A = 3.6e^{t/20}e^{-0.06t} = 3.6e^{-0.01t}$, or

$D_t\left(e^{-0.06t} \cdot A\right) = 3.6e^{-0.01t}$. Integrating then leads to $e^{-0.06t} \cdot A = -360e^{-0.01t} + C$, and thus to

the general solution $A(t) = -360e^{0.05t} + Ce^{0.06t}$. The initial condition $A(0) = 0$ implies

that $C = 360$, so that $A(t) = 360\left(e^{0.06t} - e^{0.05t}\right)$. At age 70 she will have

$A(40) \approx 1308.283$ thousand dollars, that is, \$1,308,283.

43.   **(a)** First we rewrite the differential equation as $y' + y = x$. An integrating factor is given

by $\rho = \exp\left(\int 1\,dx\right) = e^x$, and multiplying the differential equation by ρ gives

$e^x \cdot y' + e^x y = xe^x$, or $D_x\left(e^x \cdot y\right) = xe^x$. Integrating (by parts) then leads to

$e^x \cdot y = \int xe^x\,dx = xe^x - e^x + C$, and thus to the general solution $y(x) = x - 1 + Ce^{-x}$.

Then the fact that $\lim\limits_{x\to\infty} e^{-x} = 0$ implies that every solution curve approaches the straight

line $y = x - 1$ as $x \to \infty$.

(b) The initial condition $y(-5) = y_0$ imposed upon the general solution in part (a) im-

plies that $y_0 = -5 - 1 + Ce^5$, and thus that $C = e^{-5}(y_0 + 6)$. Hence the solution of the ini-

tial value problem $y' = x - y$, $y(-5) = y_0$ is $y(x) = x - 1 + (y_0 + 6)e^{-x-5}$. Substituting

$x = 5$, we therefore solve the equation $4 + (y_0 + 6)e^{-10} = y_1$ with

$$y_1 = 3.998, 3.999, 4, 4.001, 4.002$$

for the desired initial values

$$y_0 = -50.0529, -28.0265, -6.0000, 16.0265, 38.0529,$$

respectively.

45.   The volume of the reservoir (in millions of cubic meters, denoted m-m^3) is 2. In the nota-

tion of Equation (18) of the text, the differential equation for $x(t)$ is

$$\frac{dx}{dt} = r_i c_i - \frac{r_o}{V}x = \left(0.2\,\text{m-m}^3/\text{month}\right)\left(10\,\text{L/m}^3\right) - \left(0.2\,\text{m-m}^3/\text{month}\right)\left(\frac{x}{2}\,\text{L/m}^3\right),$$

or $\dfrac{dx}{dt}+\dfrac{1}{10}x=2$. An integrating factor is given by $\rho=e^{t/10}$, and multiplying the differ-

ential equation by ρ gives $e^{t/10}\cdot\dfrac{dx}{dt}+\dfrac{1}{10}e^{t/10}x=2e^{t/10}$, or $D_t\left(e^{t/10}\cdot x\right)=2e^{t/10}$. Integrating

then leads to $e^{t/10}\cdot x=20e^{t/10}+C$, and thus to the general solution $x(t)=20+Ce^{-t/10}$.

The initial condition $x(0)=0$ implies that $C=-20$, and so $x(t)=20\left(1-e^{-t/10}\right)$, which

shows that indeed $\lim\limits_{t\to\infty}x(t)=20$ (million liters). This was to be expected because the

reservoir's pollutant concentration should ultimately match that of the incoming water,

namely $10\,\text{L/m}^3$. Finally, since the volume of reservoir remains constant at 2 m-m^3, a

pollutant concentration of $5\,\text{L/m}^3$ is reached when $\dfrac{x(t)}{2}=5$, that is, when

$10=20\left(1-e^{-t/10}\right)$, or $t=10\ln 2\approx 6.93\,\text{months}$.

SECTION 1.6

SUBSTITUTION METHODS AND EXACT EQUATIONS

It is traditional for every elementary differential equations text to include the particular types of
equation that are found in this section. However, no one of them is vitally important solely in its
own right. Their main purpose (at this point in the course) is to familiarize students with the
technique of transforming a differential equation by substitution. The subsection on airplane
flight trajectories (together with Problems 56–59) is included as an application, but is optional
material and may be omitted if the instructor desires.

The differential equations in Problems 1-15 are homogeneous, and so we solve by means of the
substitution $v=y/x$ indicated in Equation (8) of the text. In some cases we present solutions by
other means, as well.

1. For $x\neq 0$ and $x+y\neq 0$ we rewrite the differential equation as $\dfrac{dy}{dx}=\dfrac{x-y}{x+y}=\dfrac{1-\dfrac{y}{x}}{1+\dfrac{y}{x}}$.

Substituting $v=\dfrac{y}{x}$ then gives $v+x\dfrac{dv}{dx}=\dfrac{1-v}{1+v}$, or $x\dfrac{dv}{dx}=\dfrac{1-v}{1+v}-v=\dfrac{1-2v-v^2}{1+v}$. Sepa-

rating variables leads to $\displaystyle\int\dfrac{v+1}{v^2+2v-1}dv=-\int\dfrac{1}{x}dx$, or $\dfrac{1}{2}\ln\left|v^2+2v-1\right|=-\ln|x|+C$, or

$\left|v^2+2v-1\right|=Cx^{-2}$, where C is an arbitrary positive constant, or finally

$v^2 + 2v - 1 = Cx^{-2}$, where C is an arbitrary nonzero constant. Back-substituting $\dfrac{y}{x}$ for v

then gives the solution $\left(\dfrac{y}{x}\right)^2 + 2\dfrac{y}{x} - 1 = Cx^{-2}$, or $y^2 + 2xy - x^2 = C$.

3.   For x, y with $xy > 0$ we rewrite the differential equation as $\dfrac{dy}{dx} = \dfrac{y}{x} + 2\sqrt{\dfrac{y}{x}}$. Substituting

$v = \dfrac{y}{x}$ then gives $v + x\dfrac{dv}{dx} = v + 2\sqrt{v}$, or $x\dfrac{dv}{dx} = 2\sqrt{v}$. Separating variables leads to

$\displaystyle\int \dfrac{1}{\sqrt{v}}\,dv = \int \dfrac{2}{x}\,dx$, or $2\sqrt{v} = 2\ln|x| + C$, or $v = \left(\ln|x| + C\right)^2$. Back-substituting $\dfrac{y}{x}$ for v

then gives the solution $y = x\left(\ln|x| + C\right)^2$.

5.   For $x \ne 0$ and $x + y \ne 0$ we rewrite the differential equation as

$\dfrac{dy}{dx} = \dfrac{y}{x} \cdot \dfrac{x - y}{x + y} = \dfrac{y}{x} \cdot \dfrac{1 - \dfrac{y}{x}}{1 + \dfrac{y}{x}}$. Substituting $v = \dfrac{y}{x}$ then gives $v + x\dfrac{dv}{dx} = v \cdot \dfrac{1 - v}{1 + v}$, or

$x\dfrac{dv}{dx} = v \cdot \dfrac{1 - v}{1 + v} - v = \dfrac{-2v^2}{1 + v}$. Separating variables leads to $\displaystyle\int \dfrac{1 + v}{v^2}\,dv = -2\int \dfrac{1}{x}\,dx$, or

$-\dfrac{1}{v} + \ln|v| = -2\ln|x| + C$. Back-substituting $\dfrac{y}{x}$ for v then gives

$-\dfrac{x}{y} + \ln\left|\dfrac{y}{x}\right| = -2\ln|x| + C$, or $\ln\left|\dfrac{y}{x}\right| + 2\ln|x| = \dfrac{x}{y} + C$, or $\ln|xy| = \dfrac{x}{y} + C$.

7.   For $x, y \ne 0$ we rewrite the differential equation as $\dfrac{dy}{dx} = \left(\dfrac{x}{y}\right)^2 + \dfrac{y}{x}$. Substituting $v = \dfrac{y}{x}$

then gives $v + x\dfrac{dv}{dx} = \left(\dfrac{1}{v}\right)^2 + v$, or $x\dfrac{dv}{dx} = \left(\dfrac{1}{v}\right)^2$. Separating variables leads to

$\displaystyle\int v^2\,dv = \int \dfrac{1}{x}\,dx$, or $v^3 = 3\ln|x| + C$. Back-substituting $\dfrac{y}{x}$ for v then gives

$\left(\dfrac{y}{x}\right)^3 = 3\ln|x| + C$, or $y^3 = x^3\left(3\ln|x| + C\right)$.

Alternatively, the substitution $v = y^3$, which implies that $v' = 3y^2 y'$, gives $\dfrac{1}{3}xv' = x^3 + v$,

or $v' - \dfrac{3}{x}v = 3x^2$, a linear equation in v as a function of x. An integrating factor is given by $\rho = \exp\left(-\int \dfrac{3}{x}dx\right) = x^{-3}$, and multiplying the differential equation by ρ gives

$x^{-3} \cdot v' - 3x^{-4}v = 3x^{-1}$, or $D_x\left(x^{-3} \cdot v\right) = 3x^{-1}$. Integrating then gives $x^{-3} \cdot v = 3\ln|x| + C$,

and finally back-substituting y^3 for v yields $y^3 = x^3\left(3\ln|x| + C\right)$, as determined above.

9. For $x \neq 0$ we rewrite the differential equation as $\dfrac{dy}{dx} = \dfrac{y}{x} + \left(\dfrac{y}{x}\right)^2$. Substituting $v = \dfrac{y}{x}$

then gives $v + x\dfrac{dv}{dx} = v + v^2$, or $x\dfrac{dv}{dx} = v^2$. Separating variables leads to

$\displaystyle\int \dfrac{1}{v^2}dv = \int \dfrac{1}{x}dx$, or $-\dfrac{1}{v} = \ln|x| + C$. Back-substituting $\dfrac{y}{x}$ for v then gives the solution

$y = \dfrac{x}{C - \ln|x|}$.

11. For $x^2 - y^2 \neq 0$ and $x \neq 0$ we rewrite the differential equation as

$\dfrac{dy}{dx} = \dfrac{2xy}{x^2 - y^2} = \dfrac{2\dfrac{y}{x}}{1 - \left(\dfrac{y}{x}\right)^2}$. Substituting $v = \dfrac{y}{x}$ then gives $v + x\dfrac{dv}{dx} = \dfrac{2v}{1 - v^2}$, or

$x\dfrac{dv}{dx} = \dfrac{2v}{1 - v^2} - v = \dfrac{v + v^3}{1 - v^2}$. Separating variables leads to $\displaystyle\int \dfrac{1 - v^2}{v + v^3}dv = \int \dfrac{1}{x}dx$, or (after

decomposing into partial fractions) $\displaystyle\int \dfrac{1}{v} - \dfrac{2v}{v^2 + 1}dv = \int \dfrac{1}{x}dx$, or

$\ln|v| - \ln\left(v^2 + 1\right) = \ln|x| + C$, or $\dfrac{v}{v^2 + 1} = Cx$. Back-substituting $\dfrac{y}{x}$ for v then gives the

solution $\dfrac{y}{x} = Cx\left[\left(\dfrac{y}{x}\right)^2 + 1\right]$, or finally $y = C\left(x^2 + y^2\right)$.

13. For $x > 0$ we rewrite the differential equation as

$$\dfrac{dy}{dx} = \dfrac{y}{x} + \dfrac{\sqrt{x^2 + y^2}}{x} = \dfrac{y}{x} + \sqrt{1 + \left(\dfrac{y}{x}\right)^2}\,.$$

Substituting $v = \dfrac{y}{x}$ then gives $v + x\dfrac{dv}{dx} = v + \sqrt{1 + v^2}$, or $x\dfrac{dv}{dx} = \sqrt{1 + v^2}$. Separating vari-

ables leads to $\displaystyle\int \dfrac{1}{\sqrt{1 + v^2}}dv = \int \dfrac{1}{x}dx$, or (by means of either the substitution $v = \tan\theta$ or an integral table) $\ln\left(v+\sqrt{v^2+1}\right)=\ln|x|+C$, or finally $v+\sqrt{v^2+1}=Cx$. Back-substituting $\dfrac{y}{x}$ for v then gives the solution $y+\sqrt{y^2+x^2}=Cx^2$.

15. For $x\neq 0$ and $x+y\neq 0$ we rewrite the differential equation as

$$\frac{dy}{dx}=\frac{-y(3x+y)}{x(x+y)}=-\frac{y}{x}\cdot\frac{3+\dfrac{y}{x}}{1+\dfrac{y}{x}}.$$

Substituting $v=\dfrac{y}{x}$ then gives $v+x\dfrac{dv}{dx}=-v\dfrac{3+v}{1+v}=-\dfrac{3v+v^2}{1+v}$, or

$$x\frac{dv}{dx}=-\frac{3v+v^2}{1+v}-v=\frac{-4v-2v^2}{1+v}.$$

Separating variables leads to $\displaystyle\int\frac{1+v}{4v+2v^2}\,dv=-\int\frac{1}{x}\,dx$, or $\dfrac{1}{4}\ln\left|4v+2v^2\right|=-\ln|x|+C$, or

$x^4\left(4v+2v^2\right)=C$, or simply $x^4\left(2v+v^2\right)=C$. Back-substituting $\dfrac{y}{x}$ for v then gives the solution $x^2\left(2xy+y^2\right)=C$.

17. The expression $4x+y$ suggests the substitution $v=4x+y$, which implies that $y=v-4x$, and thus that $y'=v'-4$. Substituting gives $v'-4=v^2$, or $v'=v^2+4$, a separable equation for v as a function of x. Separating variables gives $\displaystyle\int\frac{1}{v^2+4}\,dv=\int dx$, or $\dfrac{1}{2}\tan^{-1}\dfrac{v}{2}=x+C$, or $v=2\tan(2x+C)$. Finally, back-substituting $4x+y$ for v leads to the solution $y=2\tan(2x+C)-4x$.

The differential equations in Problems 19-25 are Bernoulli equations, and so we solve by means of the substitution $v=y^{1-n}$ indicated in Equation (10) of the text. (Problem 25 also admits of another solution.)

19. We first rewrite the differential equation for $x,y>0$ as $y'+\dfrac{2}{x}y=\dfrac{5}{x^2}y^3$, a Bernoulli equation with $n=3$. The substitution $v=y^{1-3}=y^{-2}$ implies that $y=v^{-1/2}$ and thus that $y'=-\dfrac{1}{2}v^{-3/2}v'$. Substituting gives $-\dfrac{1}{2}v^{-3/2}v'+\dfrac{2}{x}v^{-1/2}=\dfrac{5}{x^2}v^{-3/2}$, or $v'-\dfrac{4}{x}v=-\dfrac{10}{x^2}$, a linear equation for v as a function of x. An integrating factor is given by

$\rho = \exp\left(\int -\dfrac{4}{x}\,dx\right) = x^{-4}$, and multiplying the differential equation by ρ gives

$\dfrac{1}{x^4}v' - \dfrac{4}{x^5}v = -\dfrac{10}{x^6}$, or $D_x\left(\dfrac{1}{x^4}\cdot v\right) = -\dfrac{10}{x^6}$. Integrating then leads to $\dfrac{1}{x^4}\cdot v = \dfrac{2}{x^5}+C$, or

$v = \dfrac{2}{x}+Cx^4 = \dfrac{2+Cx^5}{x}$. Finally, back-substituting y^{-2} for v gives the general solution

$y^{-2} = \dfrac{2+Cx^5}{x}$, or $y^2 = \dfrac{x}{2+Cx^5}$.

21. We first rewrite the differential equation as $y'-y = y^3$, a Bernoulli equation with $n = 3$.

The substitution $v = y^{1-3} = y^{-2}$ implies that $y = v^{-1/2}$ and thus that $y' = -\dfrac{1}{2}v^{-3/2}v'$. Sub-

stituting gives $-\dfrac{1}{2}v^{-3/2}v' - v^{-1/2} = v^{-3/2}$, or $v'+2v = -2$, a linear equation for v as a func-

tion of x. An integrating factor is given by $\rho = \exp\left(\int 2\,dx\right) = e^{2x}$, and multiplying the

differential equation by ρ gives $e^{2x}\cdot v' + 2e^{2x}v = -2e^{2x}$, or $D_x\left(e^{2x}\cdot v\right) = -2e^{2x}$. Integrat-

ing then leads to $e^{2x}\cdot v = -e^{2x}+C$, or $v = -1+Ce^{-2x}$. Finally, back-substituting y^{-2} for

v gives the general solution $y^{-2} = -1+Ce^{-2x}$, or $y^2 = \dfrac{1}{Ce^{-2x}-1}$.

23. We first rewrite the differential equation for $x > 0$ as $y'+\dfrac{6}{x}y = 3y^{4/3}$, a Bernoulli equa-

tion with $n = 4/3$. The substitution $v = y^{1-(4/3)} = y^{-1/3}$ implies that $y = v^{-3}$ and thus that

$y' = -3v^{-4}v'$. Substituting gives $-3v^{-4}v' + \dfrac{6}{x}v^{-3} = 3v^{-4}$, or $v'-\dfrac{2}{x}v = -1$, a linear equation

for v as a function of x. An integrating factor is given by $\rho = \exp\left(\int -\dfrac{2}{x}\,dx\right) = \dfrac{1}{x^2}$, and

multiplying the differential equation by ρ gives $\dfrac{1}{x^2}v' - \dfrac{2}{x^3}v = -\dfrac{1}{x^2}$, or

$D_x\left(\dfrac{1}{x^2}v\right) = -\dfrac{1}{x^2}$. Integrating then leads to $\dfrac{1}{x^2}v = \dfrac{1}{x}+C$, or $v = x+Cx^2$. Finally, back-

substituting $y^{-1/3}$ for v gives the general solution $y^{-1/3} = x+Cx^2$, or $y = \left(x+Cx^2\right)^{-3}$.

25. We first rewrite the differential equation for $x, y > 0$ as $y'+\dfrac{1}{x}y = \dfrac{1}{\left(1+x^4\right)^{1/2}}y^{-2}$, a Ber-

noulli equation with $n = -2$. The substitution $v = y^{1-(-2)} = y^3$ implies that $y = v^{1/3}$ and thus that $y' = \frac{1}{3}v^{-2/3}v'$. Substituting gives $\frac{1}{3}v^{-2/3}v' + \frac{1}{x}v^{1/3} = \frac{1}{\left(1+x^4\right)^{1/2}}v^{-2/3}$, or

$v' + \frac{3}{x}v = \frac{3}{\left(1+x^4\right)^{1/2}}$, a linear equation for v as a function of x. An integrating factor is

given by $\rho = \exp\left(\int \frac{3}{x}dx\right) = x^3$, and multiplying the differential equation by ρ gives

$x^3 \cdot v' + 3x^2 v = \frac{3x^3}{\left(1+x^4\right)^{1/2}}$, or $D_x\left(x^3 \cdot v\right) = \frac{3x^4}{\left(1+x^4\right)^{1/2}}$. Integrating then leads to

$x^3 \cdot v = \frac{3}{2}\left(1+x^4\right)^{1/2} + C$, or $v = \frac{3\left(1+x^4\right)^{1/2} + C}{2x^3}$. Finally, back-substituting y^3 for v gives

the general solution $y^3 = \frac{3\left(1+x^4\right)^{1/2} + C}{2x^3}$.

Alternatively, for $x \neq 0$, the substitution $v = xy$, which implies that $v' = xy' + y$ and that

$y = \frac{v}{x}$, gives $\frac{v^2}{x^2}v'\left(1+x^4\right)^{1/2} = x$. Separating variables leads to $\int v^2\,dv = \int \frac{x^3}{\left(1+x^4\right)^{1/2}}dx$,

or $\frac{1}{3}v^3 = \frac{1}{2}\left(1+x^4\right)^{1/2} + C$, or $v^3 = \frac{3\left(1+x^4\right)^{1/2} + C}{2}$. Back-substituting xy for v then

gives the solution $y^3 = \frac{3\left(1+x^4\right)^{1/2} + C}{2x^3}$, as determined above.

As with Problems 16-18, the differential equations in Problems 27-29 rely upon substitutions that are generally suggested by the equations themselves. Two of these equations are also Bernoulli equations.

27.   The substitution $v = y^3$, which implies that $v' = 3y^2 y'$, gives $xv' - v = 3x^4$, or (for $x > 0$)

$v' - \frac{1}{x}v = 3x^3$, a linear equation for v as a function of x. An integrating factor is given by

$\rho = \exp\left(\int -\frac{1}{x}dx\right) = \frac{1}{x}$, and multiplying the differential equation by ρ gives

$\frac{1}{x} \cdot v' - \frac{1}{x^2}v = 3x^2$, or $D_x\left(\frac{1}{x} \cdot v\right) = 3x^2$. Integrating then leads to $\frac{1}{x} \cdot v = x^3 + C$, or

$v = x^4 + Cx$. Finally, back-substituting y^3 for v gives the general solution $y^3 = x^4 + Cx$,

or $y = \left(x^4 + Cx\right)^{1/3}$.

Alternatively, for $x, y > 0$ we can first rewrite the differential equation as

$$y' - \frac{1}{3x} y = x^3 y^{-2},$$ a Bernoulli equation with $n = -2$. This leads to the substitution

$v = y^{1-(-2)} = y^3$ used above.

29. The substitution $v = \sin^2 y$, which implies that $v' = (2 \sin y \cos y) y'$, gives $xv' = 4x^2 + v$,

or (for $x > 0$) $v' - \frac{1}{x} v = 4x$, a linear equation for v as a function of x. An integrating fac-

tor is given by $\rho = \exp\left(\int -\frac{1}{x} dx \right) = \frac{1}{x}$, and multiplying the differential equation by ρ

gives $\frac{1}{x} \cdot v' - \frac{1}{x^2} v = 4$, or $D_x \left(\frac{1}{x} \cdot v \right) = 4$. Integrating then leads to $\frac{1}{x} \cdot v = 4x + C$, or

$v = 4x^2 + Cx$. Finally, back-substituting $\sin^2 y$ for v gives the general solution
$\sin^2 y = 4x^2 + Cx$.

Each of the differential equations in Problems 31–41 is of the form $M\,dx + N\,dy = 0$, and the
exactness condition $\partial M / \partial y = \partial N / \partial x$ is routine to verify. For each problem we give the principal
steps in the calculation corresponding to the method of Example 9 in this section.

31. The condition $F_x = M$ implies that $F(x, y) = \int 2x + 3y\, dx = x^2 + 3xy + g(y)$, and then

the condition $F_y = N$ implies that $3x + g'(y) = 3x + 2y$, or $g'(y) = 2y$, or $g(y) = y^2$.
Thus the solution is given by $x^2 + 3xy + y^2 = C$.

33. The condition $F_x = M$ implies that $F(x, y) = \int 3x^2 + 2y^2\, dx = x^3 + xy^2 + g(y)$, and then

the condition $F_y = N$ implies that $4xy + g'(y) = 4xy + 6y^2$, or $g'(y) = 6y^2$, or

$g(y) = 2y^3$. Thus the solution is given by $x^3 + 2xy^2 + 2y^3 = C$.

35. The condition $F_x = M$ implies that $F(x, y) = \int x^3 + \frac{y}{x} dx = \frac{1}{4} x^4 + y \ln x + g(y)$, and

then the condition $F_y = N$ implies that $\ln x + g'(y) = y^2 + \ln x$, or $g'(y) = y^2$, or

$g(y) = \frac{1}{3} y^3$. Thus the solution is given by $\frac{1}{4} x^4 + \frac{1}{3} y^3 + y \ln x = C$.

37. The condition $F_x = M$ implies that $F(x, y) = \int \cos x + \ln y\, dx = \sin x + x \ln y + g(y)$,

and then the condition $F_y = N$ implies that $\frac{x}{y} + g'(y) = \frac{x}{y} + e^y$, or $g'(y) = e^y$, or

$g(y) = e^y$. Thus the solution is given by $\sin x + x \ln y + e^y = C$.

39.   The condition $F_x = M$ implies that $F(x,y) = \int 3x^2 y^3 + y^4 \, dx = x^3 y^3 + xy^4 + g(y)$, and

then the condition $F_y = N$ implies that $3x^3 y^2 + 4xy^3 + g'(y) = 3x^3 y^2 + y^4 + 4xy^3$, or

$g'(y) = y^4$, or $g(y) = \dfrac{1}{5} y^5$. Thus the solution is given by $x^3 y^3 + xy^4 + \dfrac{1}{5} y^5 = C$.

41.   The condition $F_x = M$ implies that $F(x,y) = \int \dfrac{2x}{y} - \dfrac{3y^2}{x^4} \, dx = \dfrac{x^2}{y} + \dfrac{y^2}{x^3} + g(y)$, and then

the condition $F_y = N$ implies that $-\dfrac{x^2}{y^2} + \dfrac{2y}{x^3} + g'(y) = -\dfrac{x^2}{y^2} + \dfrac{2y}{x^3} + \dfrac{1}{\sqrt{y}}$, or

$g'(y) = \dfrac{1}{\sqrt{y}}$, or $g(y) = 2\sqrt{y}$. Thus the solution is given by $\dfrac{x^2}{y} + \dfrac{y^2}{x^3} + 2\sqrt{y} = C$.

In Problems 43-47 either the dependent variable y or the independent variable x (or both) is missing, and so we use the substitutions in equations (34) and/or (36) of the text to reduce the given differential equation to a first-order equation for $p = y'$.

43.   Since the dependent variable y is missing, we can substitute $y' = p$ and $y'' = p'$ as in Equation (34) of the text. This leads to $xp' = p$, a separable equation for p as a function of x. Separating variables gives $\int \dfrac{dp}{p} = \int \dfrac{dx}{x}$, or $\ln p = \ln x + \ln C$, or $p = Cx$, that is,

$y' = Cx$. Finally, integrating gives the solution $y(x) = \dfrac{1}{2} Cx^2 + B$, which we rewrite as

$y(x) = Ax^2 + B$.

45.   Since the independent variable x is missing, we can substitute $y' = p$ and $y'' = p\dfrac{dp}{dy}$ as in

Equation (36) of the text. This leads to $p\dfrac{dp}{dy} + 4y = 0$, or $\int p \, dp = -\int 4y \, dy$, or

$\dfrac{1}{2} p^2 = -2y^2 + C$, or $p = \sqrt{2C - 4y^2} = 2\sqrt{C - y^2}$ (replacing $\dfrac{C}{2}$ simply with C in the last

step). Thus $\dfrac{dy}{dx} = 2\sqrt{C - y^2}$. Separating variables once again yields $\int \dfrac{dy}{2\sqrt{C - y^2}} = \int dx$,

or $\int \dfrac{dy}{2\sqrt{k^2 - y^2}} = \int dx$, upon replacing C with k^2. Integrating gives

$x = \int \dfrac{dy}{2\sqrt{k^2 - y^2}} = \dfrac{1}{2} \sin^{-1} \dfrac{y}{k} + D$; solving for y leads to the solution

$y(x) = k \sin(2x - 2D) = k(\sin 2x \cos 2D - \cos 2x \sin 2D),$

or simply $y(x) = A\cos 2x + B\sin 2x$. (A much easier method of solution for this equation will be introduced in Chapter 3.)

47.   Since the dependent variable y is missing, we can substitute $y' = p$ and $y'' = p'$ as in Equation (34) of the text. This leads to $p' = p^2$, a separable equation for p as a function of x. Separating variables gives $\int \dfrac{dp}{p^2} = \int x\,dx$, or $-\dfrac{1}{p} = x + B$, or $p = -\dfrac{1}{x+B}$, that is, $\dfrac{dy}{dx} = -\dfrac{1}{x+B}$. Finally, integrating gives the solution $y(x) = A - \ln|x+B|$.

Alternatively, since the independent variable x is also missing, we can instead substitute $y' = p$ and $y'' = p\dfrac{dp}{dy}$ as in Equation (36) of the text. This leads to $p\dfrac{dp}{dy} = p^2$, or $\int \dfrac{dp}{p} = \int dy$, or $\ln p = y + C$, or $p = Ce^{y}$, that is, $\dfrac{dy}{dx} = Ce^{y}$. Separating variables once again leads to $\int e^{-y}\,dy = C\int dx$, or $-e^{-y} = Cx + D$, or

$$y = -\ln(Cx + D) = -\ln\left[C\left(x + \frac{D}{C}\right)\right] = -\ln C - \ln\left(x + \frac{D}{C}\right).$$

Putting $A = -\ln C$ and $B = \dfrac{D}{C}$ gives the same solution as found above.

49.   Since the independent variable x is missing, we can substitute $y' = p$ and $y'' = p\dfrac{dp}{dy}$ as in Equation (36) of the text. This leads to $py\dfrac{dp}{dy} + 4p^2 = yp$, or $y\dfrac{dp}{dy} + p = y$, a linear equation for p as a function as a function of y which we can rewrite as $D_y(y \cdot p) = y$, or $y \cdot p = \dfrac{1}{2}y^2 + C$, or $p = \dfrac{y^2 + C}{2y}$, that is, $\dfrac{dy}{dx} = \dfrac{y^2 + C}{2y}$. Separating variables leads to $\int \dfrac{2y}{y^2 + C}\,dy = \int dx$, or $x = \int \dfrac{2y\,dy}{y^2 + C} = \ln(y^2 + C) + B$. Solving for y leads to the solution $y^2 + C = e^{x+B} = Be^{x}$, or finally $y(x) = \pm\sqrt{A + Be^{x}}$.

51.   Since the independent variable x is missing, we can substitute $y' = p$ and $y'' = p\dfrac{dp}{dy}$ as in Equation (36) of the text. This leads to $pp' = 2yp^3$, or $p' = 2yp^2$, or $\int \dfrac{1}{p^2}\,dp = \int 2y\,dy$, or $-\dfrac{1}{p} = y^2 + C$, or $p = -\dfrac{1}{y^2 + C}$, that is, $\dfrac{dy}{dx} = -\dfrac{1}{y^2 + C}$. Separating variables once again leads to $\int y^2 + C\,dy = -\int dx$, or $\frac{1}{3}y^3 + Cy = -x + D$, or finally the solution

$y^3 + 3x + Ay + B = 0$.

53.   Since the independent variable x is missing, we can substitute $y' = p$ and $y'' = p\dfrac{dp}{dy}$ as in

Equation (36) of the text. This leads to $pp' = 2yp$, or $\int dp = \int 2y\,dy$, or $p = y^2 + A$,

that is, $\dfrac{dy}{dx} = y^2 + A$. Separating variables once again yields $\int \dfrac{1}{y^2 + A}\,dy = \int dx$, or

$A\arctan\dfrac{y}{A} = x + C$, or $\dfrac{y}{A} = \tan(Ax + B)$, or finally the solution $y(x) = A\tan(Ax + B)$.

55.   The proposed substitution $v = ax + by + c$ implies that $y = \dfrac{1}{b}(v - ax - c)$, so that

$y' = \dfrac{1}{b}(v' - a)$. Substituting into the given differential equation gives $\dfrac{1}{b}(v' - a) = F(v)$,

that is $\dfrac{dv}{dx} = bF(v) + a$, a separable equation for v as a function of x.

Problems 57-62 illustrate additional substitutions that are helpful in solving certain types of first-order differential equation.

57.   The proposed substitution $v = \ln y$ implies that $y = e^v$, and thus that $\dfrac{dy}{dx} = e^v\dfrac{dv}{dx}$. Substi-

tuting into the given equation yields $e^v\dfrac{dv}{dx} + P(x)e^v = Q(x)ve^v$. Cancellation of the fac-

tor e^v then yields the linear differential equation $\dfrac{dv}{dx} - Q(x)v = P(x)$.

59.   The substitution $y = v + k$ implies that $\dfrac{dy}{dx} = \dfrac{dv}{dx}$, leading to

$$\frac{dv}{dx} = \frac{x - (v + k) - 1}{x + (v + k) + 3} = \frac{x - v - (k + 1)}{x + v + (k + 3)}.$$

Likewise the substitution $x = u + h$ implies that $u = x - h$ and thus that $\dfrac{dv}{dx} = \dfrac{dv}{du}\dfrac{du}{dx} = \dfrac{dv}{du}$

(since $\dfrac{du}{dx} = 1$), giving

$$\frac{dv}{du} = \frac{(u + h) - (v + k) - 1}{(u + h) + (v + k) + 3} = \frac{u - v + (h - k - 1)}{u + v + (h + k + 3)}.$$

Thus h and k must be chosen to satisfy the system
$$h - k - 1 = 0$$
$$h + k + 3 = 0,$$
which means that $h = -1$ and $k = -2$. These choices for h and k lead to the homogeneous equation
$$\frac{dv}{du} = \frac{u - v}{u + v} = \frac{1 - \dfrac{v}{u}}{1 + \dfrac{v}{u}},$$
which calls for the further substitution $p = \dfrac{v}{u}$, so that $v = pu$ and thus $\dfrac{dv}{du} = p + u\dfrac{dp}{du}$.

Substituting gives $p + u\dfrac{dp}{du} = \dfrac{1 - p}{1 + p}$, or
$$u\frac{dp}{du} = \frac{1 - p}{1 + p} - \frac{p + p^2}{1 + p} = \frac{1 - 2p - p^2}{1 + p}.$$

Separating variables yields $\displaystyle\int \frac{1 + p}{1 - 2p - p^2}\,dp = \int \frac{1}{u}\,du$, or $-\dfrac{1}{2}\ln\left(1 - 2p - p^2\right) = \ln u + C$,

or $\left(p^2 + 2p - 1\right)u^2 = C$. Back-substituting $\dfrac{v}{u}$ for p leads to $\left[\left(\dfrac{v}{u}\right)^2 + 2\dfrac{v}{u} - 1\right]u^2 = C$, or

$v^2 + 2uv - u^2 = C$. Finally, back-substituting $x + 1$ for u and $y + 2$ for v gives the implicit solution
$$(y + 2)^2 + 2(x + 1)(y + 2) - (x + 1)^2 = C,$$
which reduces to $y^2 + 2xy - x^2 + 2x + 6y = C$.

61. The expression $x - y$ appearing on the right-hand side suggests that we try the substitution $v = x - y$, which implies that $y = x - v$, and thus that $\dfrac{dy}{dx} = 1 - \dfrac{dv}{dx}$. This gives the

separable equation $1 - \dfrac{dv}{dx} = \sin v$, or $\dfrac{dv}{dx} = 1 - \sin v$. Separating variables leads to

$\displaystyle\int \frac{1}{1 - \sin v}\,dv = \int dx$. The left-hand integral is carried out with the help of the trigonometric identities
$$\frac{1}{1 - \sin v} = \frac{1 + \sin v}{\cos^2 v} = \sec^2 v + \sec v \tan v;$$

the solution is given by $\displaystyle\int \sec^2 v + \sec v \tan v\,dv = \int dx$, or $x = \tan v + \sec v + C$. Finally,

back-substituting $x - y$ for v gives the implicit solution $x = \tan(x - y) + \sec(x - y) + C$. However, for no value of the constant C does this general solution include the "basic" solution $y(x) = x - \dfrac{\pi}{2}$. The reason is that for this solution, $v = x - y$ is the constant $\dfrac{\pi}{2}$, so that the expression $1 - \sin v$ (by which we divided above) is identically zero. Thus the solution $y(x) = x - \dfrac{\pi}{2}$ is singular for this solution procedure.

63.   The substitution $y = y_1 + \dfrac{1}{v}$, which implies that $\dfrac{dy}{dx} = y_1' - \dfrac{1}{v^2}\dfrac{dv}{dx}$, gives

$$y_1' - \frac{1}{v^2}\frac{dv}{dx} = A(x)\left(y_1 + \frac{1}{v}\right)^2 + B(x)\left(y_1 + \frac{1}{v}\right) + C(x),$$

which upon expanding becomes

$$\underline{y_1'} - \frac{1}{v^2}\frac{dv}{dx} = A(x)\left(y_1^2 + 2\frac{y_1}{v} + \frac{1}{v^2}\right) + B(x)y_1 + B(x)\frac{1}{v} + C(x)$$

$$= \underline{A(x)y_1^2 + B(x)y_1 + C(x)} + A(x)\left(2\frac{y_1}{v} + \frac{1}{v^2}\right) + B(x)\frac{1}{v}.$$

The underlined terms cancel because y_1 is a solution of the given equation

$\dfrac{dy}{dx} = A(x)y^2 + B(x)y + C(x)$, resulting in

$$-\frac{1}{v^2}\frac{dv}{dx} = A(x)\left(2\frac{y_1}{v} + \frac{1}{v^2}\right) + B(x)\frac{1}{v},$$

or $\dfrac{dv}{dx} = -A(x)(2vy_1 + 1) - B(x)v$, that is, $\dfrac{dv}{dx} + (B + 2Ay_1)v = -A$, a linear equation for v as a function of x.

In Problem 65 we outline the application of the method of Problem 63 to the given Riccati equation.

65.   Here $A(x) = 1$, $B(x) = -2x$, and $C(x) = 1 + x^2$. Thus the substitution

$y = y_1 + \dfrac{1}{v} = x + \dfrac{1}{v}$ yields the trivial linear equation $\dfrac{dv}{dx} = -1$, with immediate solution

$v(x) = C - x$. Hence the general solution of our Riccati equation is given by

$$y(x) = x + \frac{1}{C - x}.$$

67.   First, the line $y = Cx - \dfrac{1}{4}C^2$ has slope C and passes through the point $\left(\frac{1}{2}C, \frac{1}{4}C^2\right)$; the

same is true of the parabola $y = x^2$ at the point $\left(\frac{1}{2}C, \frac{1}{4}C^2\right)$, because

$\dfrac{dy}{dx} = 2x = 2 \cdot \frac{1}{2}C = C$. Thus the line is tangent to the parabola at this point. It follows that $y = x^2$ is in fact a solution to the differential equation, since for each x, the parabola has the same values of y and y' as the known solution $y = Cx - \dfrac{1}{4}C^2$. Finally, $y = x^2$ is a singular solution with respect to the general solution $y = Cx - \dfrac{1}{4}C^2$, since for no value of C does $Cx - \dfrac{1}{4}C^2$ equal x^2 for all x.

69. With $a = 100$ and $k = \dfrac{1}{10}$, Equation (19) in the text is $y = 50 \left[\left(\dfrac{x}{100} \right)^{9/10} - \left(\dfrac{x}{100} \right)^{11/10} \right]$.

We find the maximum northward displacement of plane by setting

$$y'(x) = 50 \left[\frac{9}{10} \left(\frac{x}{100} \right)^{-1/10} - \frac{11}{10} \left(\frac{x}{100} \right)^{1/10} \right] = 0,$$

which yields $\left(\dfrac{x}{100} \right)^{1/10} = \left(\dfrac{9}{11} \right)^{1/2}$. Because

$$y''(x) = 50 \left[\frac{-9}{100} \left(\frac{x}{100} \right)^{-11/10} - \frac{11}{100} \left(\frac{x}{100} \right)^{-9/10} \right] < 0$$

for all x, this critical point in fact represents the absolute maximum value of y. Substituting this value of x into $y(x)$ gives $y_{\max} = 50 \left[\left(\dfrac{9}{11} \right)^{9/2} - \left(\dfrac{9}{11} \right)^{11/2} \right] \approx 3.68\,\text{mi}$.

71. Equations (12)-(19) apply to this situation as with the airplane in flight.

(a) With $a = 100$ and $k = \dfrac{w}{v_0} = \dfrac{2}{4} = \dfrac{1}{2}$, the solution given by Equation (19) is

$y = 50 \left[\left(\dfrac{x}{100} \right)^{1/2} - \left(\dfrac{x}{100} \right)^{3/2} \right]$. The fact that $y(0) = 0$ means that this trajectory goes through the origin where the tree is located.

(b) With $k = \dfrac{4}{4} = 1$, the solution is $y = 50 \left[1 - \left(\dfrac{x}{100} \right)^2 \right]$, and we see that the dog hits the bank at a distance $y(0) = 50\,\text{ft}$ north of the tree.

(c) With $k = \dfrac{6}{4} = \dfrac{3}{2}$, the solution is $y = 50 \left[\left(\dfrac{x}{100} \right)^{-1/2} - \left(\dfrac{x}{100} \right)^{5/2} \right]$. This trajectory is asymptotic to the positive x-axis, so we see that the dog never reaches the west bank of the river.

CHAPTER 1 Review Problems

The main objective of this set of review problems is practice in the identification of the different types of first-order differential equations discussed in this chapter. In each of Problems 1–36 we identify the type of the given equation and indicate one or more appropriate method(s) of solution.

1.   We first rewrite the differential equation for $x > 0$ as $y' - \dfrac{3}{x} y = x^2$, showing that the

equation is *linear*. An integrating factor is given by $\rho = \exp\left(-\int \dfrac{3}{x} dx\right) = e^{-3\ln x} = x^{-3}$, and

multiplying the equation by ρ gives $x^{-3} \cdot y' - 3x^{-4} y = x^{-1}$, or $D_x\left(x^{-3} \cdot y\right) = x^{-1}$. Integrat-

ing then leads to $x^{-3} \cdot y = \ln x + C$, and thus to the general solution $y = x^3 \left(\ln x + C\right)$.

3.   Rewriting the differential equation for $x \neq 0$ as $y' = \dfrac{xy + y^2}{x^2} = \dfrac{y}{x} + \left(\dfrac{y}{x}\right)^2$ shows that the

equation is *homogeneous*. Actually the equation is identical to Problem 9 in Section 1.6;

the general solution found there is $y = \dfrac{x}{C - \ln|x|}$.

5.   We first rewrite the differential equation for $x, y \neq 0$ as $\dfrac{y'}{y} = \dfrac{2x - 3}{x^4}$, showing that the

equation is *separable*. Separating variables yields $\int \dfrac{1}{y} dy = \int \dfrac{2x - 3}{x^4} dx$, or

$\ln|y| = -\dfrac{1}{x^2} + \dfrac{1}{x^3} + C = \dfrac{1-x}{x^3} + C$, leading to the general solution $y = C\exp\left(\dfrac{1-x}{x^3}\right)$,

where C is an arbitrary nonzero constant.

7.   We first rewrite the differential equation for $x > 0$ as $y' + \dfrac{2}{x} y = \dfrac{1}{x^3}$, showing that the

equation is *linear*. An integrating factor is given by $\rho = \exp\left(\int \dfrac{2}{x} dx\right) = e^{2\ln x} = x^2$, and

multiplying the equation by ρ gives $x^2 \cdot y' + 2xy = \dfrac{1}{x}$, or $D_x\left(x^2 \cdot y\right) = \dfrac{1}{x}$. Integrating

then leads to $x^2 \cdot y = \ln x + C$, and thus to the general solution $y = \dfrac{\ln x + C}{x^2}$.

9. We first rewrite the differential equation for $x, y > 0$ as $y' + \dfrac{2}{x}y = 6xy^{1/2}$, showing that it is a *Bernoulli* equation with $n = 1/2$. The substitution $v = y^{1/2}$ implies that $y = v^2$ and thus that $y' = 2vv'$. Substituting gives $2vv' + \dfrac{2}{x}v^2 = 6xv$, or $v' + \dfrac{1}{x}v = 3x$, a linear equation for v as a function of x. An integrating factor is given by $\rho = \exp\left(\displaystyle\int \dfrac{1}{x}\,dx\right) = x$, and multiplying the differential equation by ρ gives $xv' + v = 3x^2$, or $D_x(xv) = 3x^2$. Integrating then leads to $xv = x^3 + C$, or $v = x^2 + \dfrac{C}{x}$. Finally, back-substituting $y^{1/2}$ for v gives the general solution $y^{1/2} = x^2 + \dfrac{C}{x}$, or $y = \left(x^2 + \dfrac{C}{x}\right)^2$.

11. We first rewrite the differential equation for $x, y > 0$ as $\dfrac{dy}{dx} = \dfrac{y}{x} + 3\left(\dfrac{y}{x}\right)^2$, showing that it is *homogeneous*. Substituting $v = \dfrac{y}{x}$ then gives $v + x\dfrac{dv}{dx} = v + 3v^2$, or $x\dfrac{dv}{dx} = 3v^2$. Separating variables leads to $\displaystyle\int \dfrac{1}{v^2}\,dv = \int \dfrac{3}{x}\,dx$, or $-\dfrac{1}{v} = 3\ln x + C$, or $v = \dfrac{1}{C - 3\ln x}$. Back-substituting $\dfrac{y}{x}$ for v then gives the solution $\dfrac{y}{x} = \dfrac{1}{C - 3\ln x}$, or $y = \dfrac{x}{C - 3\ln x}$.

Alternatively, writing the equation in the form $y' - \dfrac{1}{x}y = \dfrac{3}{x^2}y^2$ for $x, y > 0$ shows that it is also a *Bernoulli* equation with $n = 2$. The substitution $v = y^{-1}$ implies that $y = v^{-1}$ and thus that $y' = -v^{-2}v'$. Substituting gives $-v^{-2}v' - \dfrac{1}{x}v^{-1} = \dfrac{3}{x^2}v^{-2}$, or $v' + \dfrac{1}{x}v = -\dfrac{3}{x^2}$, a linear equation for v as a function of x. An integrating factor is given by $\rho = \exp\left(\displaystyle\int \dfrac{1}{x}\,dx\right) = x$, and multiplying the differential equation by ρ gives $xv' + v = -\dfrac{3}{x}$, or $D_x(xv) = -\dfrac{3}{x}$. Integrating then leads to $xv = -3\ln x + C$, or $v = \dfrac{-3\ln x + C}{x}$. Finally, back-substituting y^{-1} for v gives the same general solution as found above.

13.   We first rewrite the differential equation for $y > 0$ as $\dfrac{y'}{y^2} = 5x^4 - 4x$, showing that the

equation is *separable*. Separating variables yields $\displaystyle\int \frac{1}{y^2}\,dy = \int 5x^4 - 4x\,dx$, or

$-\dfrac{1}{y} = x^5 - 2x^2 + C$, leading to the general solution $y = \dfrac{1}{C + 2x^2 - x^5}$.

15.   This is a *linear* differential equation. An integrating factor is given by

$\rho = \exp\left(\displaystyle\int 3\,dx\right) = e^{3x}$, and multiplying the equation by ρ gives $e^{3x} \cdot y' + 3e^{3x}y = 3x^2$, or

$D_x\left(e^{3x} \cdot y\right) = 3x^2$. Integrating then leads to $e^{3x} \cdot y = x^3 + C$, and thus to the general solu-

tion $y = \left(x^3 + C\right)e^{-3x}$.

17.   Rewriting the differential equation in differential form gives

$$\left(e^x + ye^{xy}\right)dx + \left(e^y + xe^{xy}\right)dy = 0,$$

and because $\dfrac{\partial}{\partial y}\left(e^x + ye^{xy}\right) = xye^{xy} = \dfrac{\partial}{\partial x}\left(e^y + xe^{xy}\right)$, the given equation is *exact*. We ap-

ply the method of Example 9 in Section 1.6 to find a solution in the form $F(x, y) = C$.

First, the condition $F_x = M$ implies that

$$F(x, y) = \int e^x + ye^{xy}\,dx = e^x + e^{xy} + g(y),$$

and then the condition $F_y = N$ implies that $xe^{xy} + g'(y) = e^y + xe^{xy}$, or $g'(y) = e^y$, or

$g(y) = e^y$. Thus the solution is given by $e^x + e^{xy} + e^y = C$.

19.   We first rewrite the differential equation for $x, y \neq 0$ as $\dfrac{y'}{y^2} = 2x^{-3} - 3x^2$, showing that

the equation is *separable*. Separating variables yields $\displaystyle\int \frac{1}{y^2}\,dy = \int 2x^{-3} - 3x^2\,dx$, or

$-\dfrac{1}{y} = -x^{-2} - x^3 + C$, leading to the general solution $y = \dfrac{1}{x^{-2} + x^3 + C} = \dfrac{x^2}{x^5 + Cx^2 + 1}$.

21.   We first rewrite the differential equation for $x > 1$ as $y' + \dfrac{1}{x+1}y = \dfrac{1}{x^2 - 1}$, showing that

the equation is *linear*. An integrating factor is given by $\rho = \exp\left(\displaystyle\int \frac{1}{x+1}\,dx\right) = x + 1$, and

multiplying the equation by ρ gives $(x+1)\,y' + y = \dfrac{1}{x-1}$, or $D_x\left[(x+1)\,y\right] = \dfrac{1}{x-1}$. Integrating then leads to $(x+1)y = \ln(x-1) + C$, and thus to the general solution

$$y = \frac{1}{x+1}\bigl[\ln(x-1) + C\bigr].$$

23.   Rewriting the differential equation in differential form gives

$$(e^y + y\cos x)dx + (xe^y + \sin x)dy = 0,$$

and because $\dfrac{\partial}{\partial y}(e^y + y\cos x) = e^y + \cos x = \dfrac{\partial}{\partial x}(xe^y + \sin x)$, the given equation is exact.
We apply the method of Example 9 in Section 1.6 to find a solution in the form
$F(x,y) = C$. First, the condition $F_x = M$ implies that

$$F(x,y) = \int e^y + y\cos x \, dx = xe^y + y\sin x + g(y),$$

and then the condition $F_y = N$ implies that $xe^y + \sin x + g'(y) = xe^y + \sin x$, or
$g'(y) = 0$, that is, g is constant. Thus the solution is given by $xe^y + y\sin x = C$.

25.   We first rewrite the differential equation for $x > -1$ as $y' + \dfrac{2}{x+1}y = 3$, showing that the

equation is *linear*. An integrating factor is given by $\rho = \exp\!\left(\displaystyle\int \frac{2}{x+1}\,dx\right) = (x+1)^2$, and

multiplying the equation by ρ gives $(x+1)^2 y' + 2(x+1)y = 3(x+1)^2$, or

$D_x\!\left[(x+1)^2 \cdot y\right] = 3(x+1)^2$. Integrating then leads to $(x+1)^2 \cdot y = (x+1)^3 + C$, and thus

to the general solution $y = x+1 + \dfrac{C}{(x+1)^2}$.

27.   Writing the given equation for $x > 0$ as $\dfrac{dy}{dx} + \dfrac{1}{x}y = -\dfrac{x^2}{3}y^4$ shows that it is a *Bernoulli*

equation with $n = 4$. The substitution $v = y^{-3}$ implies that $y = v^{-1/3}$ and thus that

$y' = -\dfrac{1}{3}v^{-4/3}v'$. Substituting gives $-\dfrac{1}{3}v^{-4/3}v' + \dfrac{1}{x}v^{-1/3} = -\dfrac{x^2}{3}v^{-4/3}$, or $v' - \dfrac{3}{x}v = x^2$, a line-

ar equation for v as a function of x. An integrating factor is given by

$\rho = \exp\!\left(\displaystyle\int -\frac{3}{x}\,dx\right) = x^{-3}$, and multiplying the differential equation by ρ gives

$x^{-3} \cdot v' - 3x^{-4}v = x^{-1}$, or $D_x\!\left(x^{-3} \cdot v\right) = x^{-1}$. Integrating then leads to $x^{-3} \cdot v = \ln x + C$, or

$v = x^3(\ln x + C)$. Finally, back-substituting y^{-3} for v gives the general solution

$y = x^{-1}(\ln x + C)^{-1/3}$.

29.   We first rewrite the differential equation for $x > -\dfrac{1}{2}$ as $y' + \dfrac{1}{2x+1}y = (2x+1)^{1/2}$, show-

ing that the equation is linear. An integrating factor is given by

$\rho = \exp\left(\displaystyle\int \dfrac{1}{2x+1}\,dx\right) = (2x+1)^{1/2}$, and multiplying the equation by ρ gives

$(2x+1)^{1/2}\,y' + (2x+1)^{-1/2}\,y = 2x+1$, or $D_x\left[(2x+1)^{1/2}\cdot y\right] = 2x+1$. Integrating then

leads to $(2x+1)^{1/2}\cdot y = x^2 + x + C$, and thus to the general solution

$y = \left(x^2 + x + C\right)(2x+1)^{-1/2}$.

31.   Rewriting the differential equation as $y' - 3x^2 y = 21x^2$ shows that it is *linear*. An inte-

grating factor is given by $\rho = \exp\left(\displaystyle\int -3x^2\,dx\right) = e^{-x^3}$, and multiplying the equation by ρ

gives $e^{-x^3}\cdot y' - 3x^2 e^{-x^3} y = 21x^2 e^{-x^3}$, or $D_x\left(e^{-x^3}\cdot y\right) = 21x^2 e^{-x^3}$. Integrating then leads to

$e^{-x^3}\cdot y = -7e^{-x^3} + C$, and thus to the general solution $y = -7 + Ce^{x^3}$.

Alternatively, writing the equation for $y > -7$ as $\dfrac{dy}{y+7} = 3x^2\,dx$ shows that it is *separa-*

ble. Integrating yields the general solution $\ln(y+7) = x^3 + C$, that is, $y = Ce^{x^3} - 7$, as

found above.

(Note that the restriction $y > -7$ in the second solution causes no loss of generality. The

general solution as found by the first method shows that either $y < -7$ for all x or $y > -7$

for all x. Of course, the second solution could be carried out under the assumption

$y < -7$ as well.)

33.   Rewriting the differential equation for $x, y > 0$ in differential form gives

$$\left(3x^2 + 2y^2\right)dx + 4xy\,dy = 0,$$

and because $\dfrac{\partial}{\partial y}\left(3x^2 + 2y^2\right) = 4y = \dfrac{\partial}{\partial x}4xy$, the given equation is *exact*. We apply the

method of Example 9 in Section 1.6 to find a solution in the form $F(x,y) = C$. First, the

condition $F_x = M$ implies that

$$F(x,y) = \int 3x^2 + 2y^2\,dx = x^3 + 2xy^2 + g(y),$$

and then the condition $F_y = N$ implies that $4xy + g'(y) = 4xy$, or $g'(y) = 0$, that is, g is

constant. Thus the solution is given by $x^3 + 2xy^2 = C$.

Alternatively, rewriting the given equation for $x, y > 0$ as $\dfrac{dy}{dx} = -\dfrac{3}{4}\dfrac{x}{y} - \dfrac{1}{2}\dfrac{y}{x}$ shows that it is *homogeneous*. Substituting $v = \dfrac{y}{x}$ then gives $v + x\dfrac{dv}{dx} = -\dfrac{3}{4v} - \dfrac{1}{2}v$, or

$x\dfrac{dv}{dx} = -\dfrac{3}{4v} - \dfrac{3}{2}v = -\dfrac{3 + 6v^2}{4v}$. Separating variables leads to $\displaystyle\int \dfrac{4v}{6v^2 + 3}\, dv = -\int \dfrac{1}{x}\, dx$, or

$\ln\left(6v^2 + 3\right) = -3\ln x + C$, or $\left(2v^2 + 1\right)x^3 = C$. Back-substituting $\dfrac{y}{x}$ for v then gives the

solution $\left[2\left(\dfrac{y}{x}\right)^2 + 1\right]x^3 = C$, or finally $2y^2 x + x^3 = C$, as found above.

Still another solution arises from writing the differential equation for $x, y > 0$ as

$\dfrac{dy}{dx} + \dfrac{1}{2x}y = -\dfrac{3x}{4}y^{-1}$, which shows that it is *Bernoulli* with $n = -1$. The substitution

$v = y^2$ implies that $y = v^{1/2}$ and thus that $y' = \dfrac{1}{2}v^{-1/2}v'$. Substituting gives

$\dfrac{1}{2}v^{-1/2}v' + \dfrac{1}{2x}v^{1/2} = -\dfrac{3x}{4}v^{-1/2}$, or $v' + \dfrac{1}{x}v = -\dfrac{3x}{2}$, a linear equation for v as a function of x.

An integrating factor is given by $\rho = \exp\left(\displaystyle\int \dfrac{1}{x}\, dx\right) = x$, and multiplying the differential

equation by ρ gives $x \cdot v' + v = -\dfrac{3x^2}{2}$, or $D_x(x \cdot v) = -\dfrac{3x^2}{2}$. Integrating then leads to

$x \cdot v = -\dfrac{x^3}{2} + C$. Finally, back-substituting y^2 for v leads to the general solution

$x \cdot y^2 = -\dfrac{x^3}{2} + C$, that is, $2xy^2 + x^3 = C$, as found above.

35. Rewriting the differential equation as $\dfrac{dy}{dx} = \dfrac{2x}{x^2 + 1}(y + 1)$ shows that it is *separable*. For

$y > -1$ separating variables gives $\displaystyle\int \dfrac{1}{y + 1}\, dy = \int \dfrac{2x}{x^2 + 1}\, dx$, or $\ln(y + 1) = \ln\left(x^2 + 1\right) + C$,

leading to the general solution $y = C\left(x^2 + 1\right) - 1$.

Alternatively, writing the differential equation as $y' - \dfrac{2x}{x^2 + 1}y = \dfrac{2x}{x^2 + 1}$ shows that it is

linear. An integrating factor is given by $\rho = \exp\left(\displaystyle\int -\dfrac{2x}{x^2 + 1}\, dx\right) = \dfrac{1}{x^2 + 1}$, and multiply-

ing the equation by ρ gives $\dfrac{1}{x^2 + 1}y' - \dfrac{2x}{\left(x^2 + 1\right)^2}y = \dfrac{2x}{\left(x^2 + 1\right)^2}$, or

$D_x \left(\dfrac{1}{x^2+1} y \right) = \dfrac{2x}{\left(x^2+1\right)^2}$. Integrating then leads to $\dfrac{1}{x^2+1} y = -\dfrac{1}{x^2+1} + C$, or thus to the general solution $y = -1 + C\left(x^2+1\right)$ found above.

CHAPTER 2

MATHEMATICAL MODELS AND NUMERICAL METHODS

SECTION 2.1

POPULATION MODELS

Section 2.1 introduces the first of the two major classes of mathematical models studied in the textbook, and is a prerequisite to the discussion of equilibrium solutions and stability in Section 2.2. In Problems 1-8 we find the desired particular solution and sketch some typical solution curves, with the desired particular solution highlighted.

1. Separating variables gives $\int \frac{1}{x(1-x)} dx = \int dt$. By the method of partial fractions

    $$\int \frac{1}{x(1-x)} dx = \int \frac{1}{x} - \frac{1}{x-1} dx = \ln|x| - \ln|x-1|,$$

    and so the general solution of the differential equation is $\ln|x| - \ln|x-1| = t + C$, or

    $\frac{x}{x-1} = Ce^t$. The initial condition $x(0) = 2$ implies that $C = 2$, leading to the particular

    solution $\frac{x}{x-1} = 2e^t$, or $x = 2(x-1)e^t$, or finally $x(t) = \frac{2e^t}{2e^t - 1} = \frac{2}{2 - e^{-t}}$.

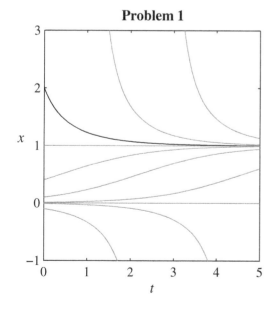

Problem 1

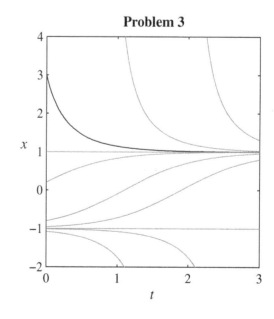

Problem 3

3.   Separating variables gives $\int \dfrac{1}{(1+x)(1-x)}\,dx = \int dt$. By the method of partial fractions

$$\int \frac{1}{(x+1)(x-1)}\,dx = -\frac{1}{2}\int \frac{1}{x-1} - \frac{1}{x+1}\,dx = -\frac{1}{2}\left(\ln|x-1| - \ln|x+1|\right),$$

and so the general solution of the differential equation is $\ln|x-1| - \ln|x+1| = -2t + C$, or

$\dfrac{x-1}{x+1} = Ce^{-2t}$. The initial condition $x(0) = 3$ implies that that $C = \dfrac{1}{2}$, leading to the

particular solution $\dfrac{x-1}{x+1} = \dfrac{1}{2}e^{-2t}$, or $2(x-1) = (x+1)e^{-2t}$, or finally

$$x(t) = \frac{2 + e^{-2t}}{2 - e^{-2t}} = \frac{2e^{2t} + 1}{2e^{2t} - 1}.$$

5.   Separating variables gives $\int \dfrac{1}{x(x-5)}\,dx = \int -3\,dt$. By the method of partial fractions,

$$\int \frac{1}{x(x-5)}\,dx = -\frac{1}{5}\int \frac{1}{x} - \frac{1}{x-5}\,dx = -\frac{1}{5}\left(\ln|x| - \ln|x-5|\right),$$

and so the general solution of the differential equation is $-\dfrac{1}{5}\left(\ln|x| - \ln|x-5|\right) = -3t + C$,

or $\dfrac{x}{x-5} = Ce^{15t}$. The initial condition $x(0) = 8$ implies that $C = \dfrac{8}{3}$, leading to the

particular solution $\dfrac{x}{x-5} = \dfrac{8}{3}e^{15t}$, or $3x = 8(x-5)e^{15t}$, or finally

$$x(t) = \frac{-40e^{15t}}{3 - 8e^{15t}} = \frac{40}{8 - 3e^{-15t}}.$$

Problem 5

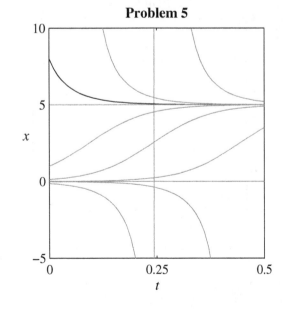

Problem 7

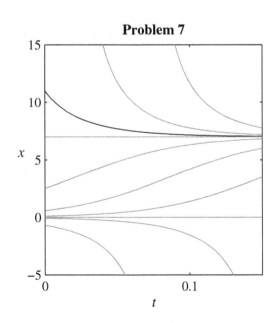

7. Separating variables gives $\int \dfrac{1}{x(x-7)}\,dx = \int -4\,dt$. By the method of partial fractions,

$$\int \frac{1}{x(x-7)}\,dx = -\frac{1}{7}\int \frac{1}{x} - \frac{1}{x-7}\,dx = -\frac{1}{7}\left(\ln|x| - \ln|x-7|\right),$$

and so the general solution of the differential equation is $\ln|x| - \ln|x-7| = 28t + C$, or

$\dfrac{x}{x-7} = Ce^{28t}$. The initial condition $x(0) = 11$ implies that $C = \dfrac{11}{4}$, leading to the

particular solution $\dfrac{x}{x-7} = \dfrac{11}{4}e^{28t}$, or $4x = 11(x-7)e^{28t}$, or finally

$$x(t) = \frac{-77e^{28t}}{4 - 11e^{28t}} = \frac{77}{11 - 4e^{-28t}}.$$

9. Substitution of $P(0) = 100$ and $P'(0) = 20$ into $P' = k\sqrt{P}$ yields $k = 2$, so the

differential equation is $P' = 2\sqrt{P}$. Separation of variables gives $\int \dfrac{1}{2\sqrt{P}}\,dP = \int dt$,

which upon integrating is $\sqrt{P} = t + C$. Then $P(0) = 100$ implies $C = 10$, so that

$P(t) = (t+10)^2$. Hence the number of rabbits after one year is $P(12) = 484$.

11. **(a)** Substituting our assumptions that $\beta = \dfrac{k_1}{\sqrt{p}}$ and $\delta = \dfrac{k_2}{\sqrt{P}}$ into the general population

equation gives $\dfrac{dP}{dt} = \dfrac{(k_1 - k_2)}{\sqrt{P}} \cdot P = k\sqrt{P}$. Separation of variables leads to

$\int \dfrac{1}{\sqrt{P}}\,dP = \int k\,dt$, which upon integrating is $2\sqrt{P} = kt + C$, or $P = \left(\dfrac{kt}{2} + C\right)^2$. The

initial condition $P(0) = P_0$ then gives $C = \sqrt{P_0}$.

(b) Our assumption implies that $C = \sqrt{P_0} = 10$, so that $P = \left(\dfrac{kt}{2} + 10\right)^2$. Measuring t in

months, we conclude from $P(6) = 169$ that $k = 1$, so that $P = \left(\dfrac{t}{2} + 10\right)^2$. Hence there

are $P(12) = 256$ fish after 12 months.

13. **(a)** Substituting our assumptions that $\beta = k_1 P$ and $\delta = k_2 P$ into the general population

equation gives $\dfrac{dP}{dt} = \left[(k_1 - k_2)P\right]P = kP^2$, where $k = k_1 - k_2 > 0$ by our assumption that

$\beta > \delta$. Solving as in Problem 12 leads to $P = \dfrac{1}{C - kt}$. The initial condition $P(0) = P_0$

implies that $C = \dfrac{1}{P_0}$, so that $P(t) = \dfrac{P_0}{1 - kP_0 t}$. As $t \to \dfrac{1}{kP_0}$ we find that $P(t) \to \infty$.

(b) Our assumption that $P_0 = 6$ gives $P(t) = \dfrac{6}{1 - 6kt}$. Then, with t measured in months,

we conclude from $P(10) = 9$ that $k = 180$, so that $P(t) = \dfrac{6}{1 - (t/30)} = \dfrac{180}{30 - t}$. From this

we can see that doomsday occurs after 30 months.

15.   Writing $\dfrac{dP}{dt} = bP\left(\dfrac{a}{b} - P\right)$ shows that the limiting population M is $\dfrac{a}{b}$. Then the facts that

$B_0 = aP_0$ and $D_0 = aP_0^2$ give $\dfrac{B_0 P_0}{D_0} = \dfrac{(aP_0)P_0}{bP_0^2} = \dfrac{a}{b} = M$. With Problems 16 and 17 in

mind, we note also that $a = \dfrac{B_0}{P_0}$ and $b = \dfrac{D_0}{P_0^2} = k$.

17.   The relations in Problem 15 give $k = \dfrac{D_0}{P_0^2} = \dfrac{12}{240^2} = \dfrac{1}{2400}$ and a limiting population of

$M = \dfrac{B_0 P_0}{D_0} = \dfrac{9 \cdot 240}{12} = 180$ rabbits. The solution is then

$$P(t) = \dfrac{180 \cdot 240}{240 + (180 - 240)e^{-t/15}} = \dfrac{43200}{120 - 60e^{-t/15}},$$

again by Equation (7). Setting $P(t) = 1.05M = 189$ rabbits yields $t \approx 44.22$ months.

19.   The relations in Problem 18 give $k = \dfrac{B_0}{P_0^2} = \dfrac{10}{100^2} = \dfrac{1}{1000}$ and $M = \dfrac{D_0 P_0}{B_0} = \dfrac{9 \cdot 100}{10} = 90$.

Problem 33 below then gives the solution

$$P(t) = \dfrac{90 \cdot 100}{100 + (90 - 100)e^{9t/100}} = \dfrac{9000}{100 - 10e^{9t/100}}.$$

Setting $P(t) = 10M = 900$ rabbits yields $t \approx 24.41$ months.

21.   Separating variables in our assumption that $\dfrac{dP}{dt} = kP(200 - P)$ gives

$\displaystyle\int \dfrac{1}{P(200 - P)}\, dP = \int k\, dt$. By the method of partial fractions

$$\int \frac{1}{P(200-P)} dP = \frac{1}{200} \int \frac{1}{P} + \frac{1}{200-P} dP = \frac{1}{200} \left(\ln P - \ln|200-P| \right),$$

and so the general solution of the differential equation is $\ln P - \ln|200 - P| = 200kt$, or

$\ln \dfrac{P}{|200-P|} = 200kt + C$, or $\dfrac{P}{200-P} = Ce^{200kt}$. The initial condition $P(0) = 100$ (taking

$t = 0$ in 1960) implies that $C = 1$. Further, $P'(0) = 1$, when substituted into the original differential equation along with $P(0) = 100$, implies that $1 = k \cdot 100(200 - 100)$, or

$k = \dfrac{1}{10000}$. Substituting these values into the general solution gives $\dfrac{P}{200-P} = e^{t/50}$, or

$P = e^{t/50}(200 - P)$, or $P(t) = \dfrac{200}{1 + e^{-t/50}}$. Finally, in the year 2020 the country's

population will be $P(60) = \dfrac{200}{1 + e^{-6/5}} \approx 153.7$ million.

23.   **(a)** The given differential equation implies that

$$x' = 0.8x - 0.004x^2 = 0.004x(200 - x),$$

which is positive for $0 < x < 200$ and negative for $x > 200$; thus the maximum amount that will dissolve is $M = 200$ g.

(b) Since the given equation conforms to Equation (6) in the text, the solution is given there by Equation (7), with $M = 200$, $P_0 = 50$, and $k = 0.004$:

$$x(t) = \frac{10000}{50 + 150e^{-0.8t}}.$$

Substituting $x = 100$, we solve for $t = 1.25 \ln 3 \approx 1.37$ sec.

25.   **(a)** Following the suggestions (and thus taking $t = 0$ in 1925), we estimate the rate of population growth in 1925 to be

$$P'(0) = \frac{P(1) - P(-1)}{2} = \frac{25.38 - 24.63}{2} = 0.375$$

million people annually. The corresponding estimate for the year 1975, corresponding to $t = 50$, is

$$P'(50) = \frac{P(51) - P(49)}{2} = \frac{48.04 - 47.04}{2} = 0.5$$

million people annually. Substituting these values, together with $P(0) = 25$ and $P(50) = 47.54$, into the logistic equation (3) leads to the system of equations

$$0.375 = 25k(M - 25)$$
$$0.5 = 47.54k(M - 47.54).$$

As in Example 3 in the text, we solve these equations to find $M = 100$ and $k = 0.0002$. Then Equation (7) gives the population function

$$P(t) = \frac{100 \cdot 25}{25 + (100 - 25)e^{-0.0002 \cdot 100 t}} = \frac{2500}{25 + 75e^{-0.02t}}.$$

(b) We find that $P = 75$ when $t = 50\ln 9 \approx 110$, that is, in 2035 A.D.

27. Our assumptions lead to the differential equation $\dfrac{dP}{dt} = kP^2 - 0.01P$ for the animal population $P(t)$. Substituting $P(0) = 200$ and $P'(0) = 2$, we find that $k = 0.0001$, so that

$$\frac{dP}{dt} = 0.0001P^2 - 0.01P = 0.0001P(P - 100).$$

Separating variables gives $\displaystyle\int \frac{1}{P(P - 100)}\,dP = \int 0.0001\,dt$. By the method of partial fractions

$$\int \frac{1}{P(P - 100)}\,dP = -\frac{1}{100}\int \frac{1}{P} - \frac{1}{P - 100}\,dP = \frac{1}{100}\big(\ln|P - 100| - \ln P\big),$$

and so the general solution of the differential equation is

$$\frac{1}{100}\ln|P - 100| - \ln P = 0.0001t + C,$$

or $\ln\dfrac{|P - 100|}{P} = \dfrac{t}{100} + C$, or $\dfrac{P - 100}{P} = Ce^{t/100}$. The initial condition $P(0) = 200$ gives $C = \dfrac{1}{2}$, and so $\dfrac{P - 100}{P} = \dfrac{1}{2}e^{t/100}$, leading to the general solution $P(t) = \dfrac{200}{2 - e^{t/100}}$.

(a) Setting $P = 1000$ gives $t = 100\ln\dfrac{9}{5} \approx 58.78$ months.

(b) Doomsday occurs as the denominator $2 - e^{t/100}$ approaches zero, that is, as t approaches $100\ln 2 \approx 69.31$ months, since the population P becomes infinite then.

29. Here we have the logistic equation

$$\frac{dP}{dt} = 0.03135P - 0.0001489P^2 = 0.0001489P(210.544 - P),$$

where $k = 0.0001489$ and $P = 210.544$. With $P_0 = 3.9$ as well, Eq. (7) in the text gives

$$P(t) = \frac{(210.544)(3.9)}{3.9 + (210.544 - 3.9)e^{-(0.0001489)(210.544)t}} = \frac{821.122}{3.9 + 206.644e^{-0.03135t}}.$$

(a) This solution gives $P(140) \approx 127.008$, fairly close to the actual 1930 U.S. census figure of 123.2 million.

(b) As t grows without bound, $P(t)$ approaches $\dfrac{821.122}{3.9} = 210.544$ million.

(c) Since the actual U.S. population in 2000 was about 281 million—already exceeding the maximum population predicted by the logistic equation—we see that that this model did *not* continue to hold throughout the 20th century.

31. Substituting $P(0) = 10^6$ and $P'(0) = 3 \times 10^5$ into the differential equation $P'(t) = \beta_0 e^{-\alpha t} P$ yields $\beta_0 = 0.3$. Hence the solution given in Problem 30 is

$P(t) = P_0 \exp\left[\dfrac{0.3}{\alpha}\left(1 - e^{-\alpha t}\right)\right]$. The fact that $P(6) = 2P_0$ now yields the equation

$$0.3\left(1 - e^{-6\alpha}\right) - \alpha \ln 2 = 0,$$

which we seek to solve for the constant α. We let $f(\alpha)$ denote the left-hand side $0.3\left(1 - e^{-6\alpha}\right) - \alpha \ln 2$ of this equation and apply Newton's iterative formula

$$\alpha_{n+1} = \alpha_n - \frac{f(\alpha_n)}{f'(\alpha_n)}$$

with initial guess $\alpha_0 = 1$ (suggested by a plot of $f(\alpha)$), leading quickly to $\alpha \approx 0.3915$. Therefore the limiting cell population as t grows without bound is

$$P_0 \exp\left(\frac{\beta_0}{\alpha}\right) = 10^6 \exp\left(\frac{0.3}{0.3915}\right) \approx 2.15 \times 10^6.$$

Thus the tumor does not grow much further after 6 months.

33. **(a)** Separating variables in the extinction-explosion equation gives

$\displaystyle\int \frac{1}{P(P - M)}\,dP = \int k\,dt$. By the method of partial fractions

$$\int \frac{1}{P(P - M)}\,dP = \frac{1}{M}\int \frac{1}{P - M} - \frac{1}{P}\,dP = \frac{1}{M}\left(\ln|P - M| - \ln P\right),$$

and so the general solution is

$$\frac{1}{M}\left(\ln|P - M| - \ln P\right) = kt + C,$$

or $\ln\dfrac{|P - M|}{P} = kMt + C$, or $\dfrac{|P - M|}{P} = Ce^{kMt}$. The initial condition $P(0) = P_0$ gives

$C = \dfrac{|P_0 - M|}{P_0}$. If the initial population P_0 is less than the threshold population M, then

$C = \dfrac{M - P_0}{P_0}$. Moreover, as in Problem 32, in this case $P < M$ for all t. Thus for $P_0 < M$

the solution of the extinction-explosion initial value problem is $\dfrac{M-P}{P}=\dfrac{M-P_0}{P_0}e^{kMt}$.

Similarly, if $P_0 > M$, then $P > M$ for all t, and so the solution is $\dfrac{P-M}{P}=\dfrac{P_0-M}{P_0}e^{kMt}$.

Solving either of these equivalent expressions for P yields
$$P_0(P-M)=P(P_0-M)e^{kMt},$$

or
$$\left[P_0+(M-P_0)e^{kMt}\right]P=MP_0,$$

or finally
$$P(t)=\frac{MP_0}{P_0+(M-P_0)e^{kMt}}.$$

(b) If $P_0 < M$, then the coefficient $M-P_0$ is positive and the denominator increases without bound, so $P(t)\to 0$ as $t\to\infty$ But if $P_0 > M$, then the denominator $P_0-(P_0-M)e^{kMt}$ approaches zero—so $P(t)\to +\infty$—as t approaches the positive value $\dfrac{1}{kM}\ln\dfrac{P_0}{P_0-M}$ from the left. Thus the population either becomes extinct or explodes.

35. Any way you look at it, you should conclude that the larger the parameter $k > 0$, the faster the logistic population $P(t)$ approaches its limiting population M:

To examine the question *geometrically*, we will assume that $M=10$ and that $k_1 = 1$ and $k_2 = 2$, leading to the logistic equations $\dfrac{dP}{dt}=P(10-P)$ and $\dfrac{dP}{dt}=2P(10-P)$. We draw slope fields and solution curves for each of these equations, using the same initial values $P(0)$ in both cases:

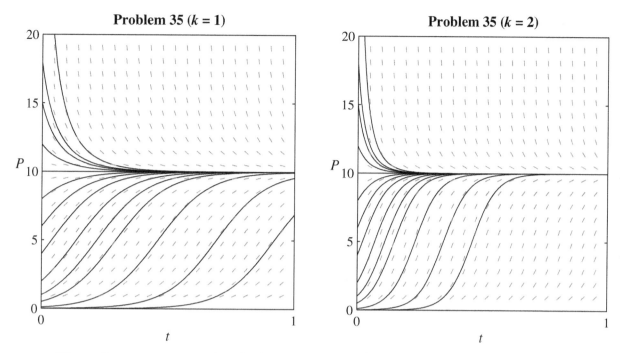

These diagrams suggest that the larger the value of k, the more rapidly the population $P(t)$ approaches the limiting population M.

To look at things *analytically*, we examine the distance between the solution (7) in the text of the logistic initial value problem and the limiting population M:

$$\left| M - \frac{MP_0}{P_0 + (M - P_0)e^{-kMt}} \right| = \left| \frac{M\left[P_0 + (M - P_0)e^{-kMt}\right] - MP_0}{P_0 + (M - P_0)e^{-kMt}} \right| = \frac{M|M - P_0|}{P_0\left(e^{kMt} - 1\right) + M}.$$

For fixed M, t, and P_0 this distance decreases as k increases; thus, the larger the value of k, the more rapidly $P(t)$ approaches M.

Finally, *numerically*, we tabulate values of $P(t)$, $t = 0, 0.1, 0.2, \ldots, 0.9, 1$, for the two solutions illustrated graphically above, using $P_0 = 0.1$ in both cases. Once again the evidence is that the larger value of k leads to the more rapid approach to M:

$k = 1$											
t	0	0.1	0.2	0.3	0.4	0.5	0.6	0.7	0.8	0.9	1.0
$P(t)$	0.1	0.267	0.695	1.687	3.555	5.999	8.030	9.172	9.679	9.879	9.955

$k = 2$											
t	0	0.1	0.2	0.3	0.4	0.5	0.6	0.7	0.8	0.9	1.0
$P(t)$	0.1	0.695	3.555	8.03	9.679	9.955	9.994	9.999	9.999	10.00	10.00

37 $k = 0.0000668717$ and $M = 338.027$, so that $P(t) = \dfrac{25761.7}{76.212 + 261.815 e^{-0.0226045t}}$, which

predicts that $P = 192.525$ in the year 2000.

39.   Separating variables gives

$\int \dfrac{1}{P} dP = \int (k + b \cos 2\pi t) dt$, or

$\ln P = kt + \dfrac{b}{2\pi} \sin 2\pi t + C$. The initial

condition $P(0) = P_0$ implies that $C = \ln P_0$,

so the desired particular solution is

$P = P_0 \exp\left(kt + \dfrac{b}{2\pi} \sin 2\pi t \right)$. Of course the

natural growth equation $P' = kP$ with the
same initial condition has solution

$P(t) = P_0 e^{kt}$. The results of both growth

patterns are indicated in the graph shown
with the typical numerical values $P_0 = 100$,

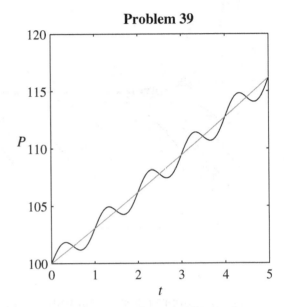

Problem 39

$k = 0.03$, and $b = 0.06$. Under the periodic growth law the population oscillates about
the curve representing natural growth. We see that the two agree at the end of each full
year.

SECTION 2.2

EQUILIBRIUM SOLUTIONS AND STABILITY

In Problems 1-12 we identify the stable and unstable critical points as well as the funnels and
spouts along the equilibrium solutions. In each problem the indicated solution satisfying
$x(0) = x_0$ is derived by separation of variables, and we show typical solution curves
corresponding to different values of x_0.

1.   The unstable critical point $x = 4$ leads to a spout along the equilibrium solution $x(t) = 4$.

Separating variables gives $\int \dfrac{1}{x-4} dx = \int dt$, or $\ln|x - 4| = t + C$, where C is an arbitrary

constant. Thus the general solution is $x = Ce^t + 4$, where C is an arbitrary nonzero
constant. The initial condition $x(0) = x_0$ then gives $x_0 = C + 4$, or $C = x_0 - 4$. Thus the
solution is given by $x(t) = (x_0 - 4)e^t + 4$.

Problem 1

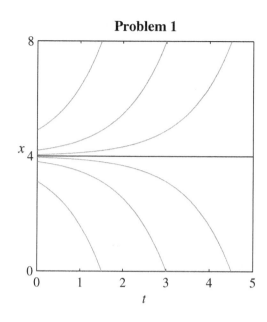

Problem 3

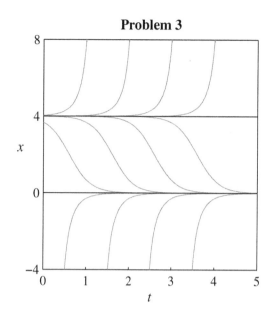

3.   The stable critical point $x = 0$ leads to a funnel along the equilibrium solution $x(t) = 0$. The unstable critical point $x = 4$ leads to a spout along the equilibrium solution $x(t) = 4$.

Separating variables gives $\int \dfrac{1}{x^2 - 4x}\,dx = \int dt$, or $\int \dfrac{1}{x - 4} - \dfrac{1}{x}\,dx = \int 4\,dt$. Integrating

gives $\ln|x - 4| - \ln|x| = 4t + C$, or $\dfrac{x - 4}{x} = Ce^{4t}$, where C is an arbitrary nonzero constant.

The initial condition $x(0) = x_0$ gives $\dfrac{x_0 - 4}{x_0} = C$, leading to $\dfrac{x - 4}{x} = \dfrac{x_0 - 4}{x_0}e^{4t}$, or finally

the solution $x(t) = \dfrac{4x_0}{x_0 + (4 - x_0)e^{4t}}$.

5.   The stable critical point $x = -2$ leads to a funnel along the equilibrium solution $x(t) = -2$. The unstable critical point $x = 2$ leads to a spout along the equilibrium

solution $x(t) = 2$. Separating variables gives $\int \dfrac{1}{x^2 - 4}\,dx = \int dt$, or

$\int \dfrac{1}{x - 2} - \dfrac{1}{x + 2}\,dx = \int 4\,dt$. Integrating gives $\ln|x - 2| - \ln|x + 2| = -4t + C$, or

$\dfrac{x - 2}{x + 2} = Ce^{-4t}$, where C is an arbitrary nonzero constant. The initial condition $x(0) = x_0$

gives $\dfrac{x_0 - 2}{x_0 + 2} = C$, leading to $\dfrac{x - 2}{x + 2} = \dfrac{x_0 - 2}{x_0 + 2}e^{-3t}$, or finally the solution

$x(t) = 2\dfrac{(x_0 + 2) + (x_0 - 2)e^{4t}}{(x_0 + 2) - (x_0 - 2)e^{4t}}$.

Problem 5

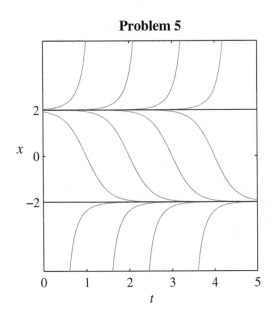

Problem 7

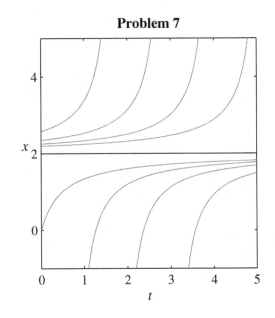

7.　The lone critical point $x = 2$ is *semi-stable*; solutions with $x_0 > 2$ approach $+\infty$ as t increases, whereas those with $x_0 < 2$ approach 2 as t increases. Separating variables gives $\int \dfrac{1}{(x-2)^2} dx = \int dt$, or $\dfrac{1}{x-2} = -t + C$, where C is an arbitrary nonzero constant.

The initial condition $x(0) = x_0$ gives $\dfrac{1}{x_0 - 2} = C$, leading to

$$\frac{1}{x-2} = -t + \frac{1}{x_0 - 2} = \frac{1 - t(x_0 - 2)}{x_0 - 2}, \text{ or finally the solution}$$

$$x(t) = 2 + \frac{x_0 - 2}{1 - t(x_0 - 2)} = \frac{x_0(2t-1) - 4t}{(x_0 - 2)t - 1}.$$

9.　Factoring gives $x^2 - 5x + 4 = (x-4)(x-1)$. The stable critical point $x = 1$ leads to a funnel along the equilibrium solution $x(t) = 1$. The unstable critical point $x = 4$ leads to a spout along the equilibrium solution $x(t) = 4$. Separating variables gives

$$\int \frac{1}{(x-4)(x-1)} dx = \int dt, \text{ or } \int \frac{1}{x-4} - \frac{1}{x-1} dx = \int 3 dt. \text{ Integrating gives}$$

$\ln|x-4| - \ln|x-1| = 3t + C$, or $\dfrac{x-4}{x-1} = Ce^{3t}$, where C is an arbitrary nonzero constant.

The initial condition $x(0) = x_0$ gives $\dfrac{x_0 - 4}{x_0 - 1} = C$, leading to $\dfrac{x-4}{x-1} = \dfrac{x_0 - 4}{x_0 - 1} e^{3t}$, or finally

the solution $x(t) = \dfrac{4(1-x_0) + (x_0 - 4)e^{3t}}{(1-x_0) + (x_0 - 4)e^{3t}}$.

Problem 9

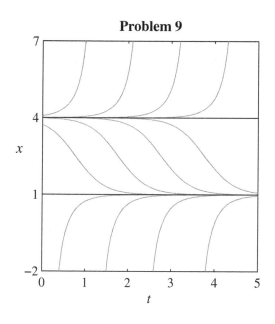

Problem 11

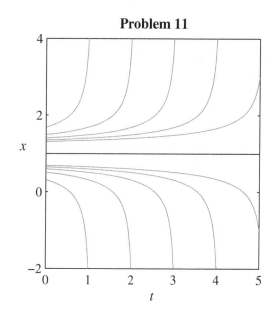

11.   The unstable critical point $x = 1$ leads to a spout along the equilibrium solution $x(t) = 4$.

Separating variables gives $\int \dfrac{1}{(x-1)^3} dx = \int dt$, and integrating gives $\dfrac{1}{2(x-1)^2} = -t + C$,

where C is an arbitrary constant. The initial condition $x(0) = x_0$ gives $\dfrac{1}{2(x_0-1)^2} = C$,

leading to $\dfrac{1}{2(x-1)^2} = -t + \dfrac{1}{2(x_0-1)^2}$, or $(x-1)^2 = \dfrac{(x_0-1)^2}{1-2t(x_0-1)^2}$, or finally the solution

$x(t) = 1 \pm \dfrac{x_0-1}{\sqrt{1-2t(x_0-1)^2}}$.

In each of Problems 13–18 we present the figure showing the slope field and typical solution curves, and then record the visually apparent classification of critical points for the given differential equation.

13.   The critical points $x = 2$ and $x = -2$ are both unstable.

Problem 13

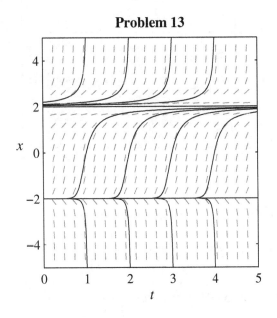

Problem 15

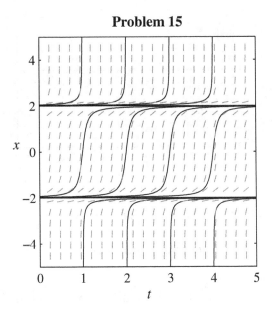

15.   The critical points $x = 2$ and $x = -2$ are both unstable.

17.   The critical points $x = 2$ and $x = 0$ are unstable, while the critical point $x = -2$ is stable.

Problem 17

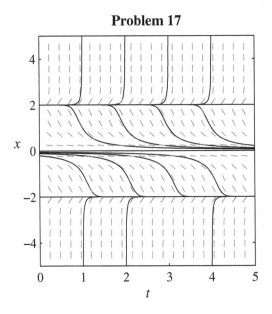

19. The critical points of the given differential equation are the roots of the quadratic equation $\frac{1}{10}x(10-x)-h=0$, that is, $x^2-10x+10h=0$. Thus a critical point c is given in terms of h by

$$c = \frac{10\pm\sqrt{100-40h}}{2} = 5\pm\sqrt{25-10h}.$$

It follows that there is no critical point if $h>\frac{5}{2}$, only the single critical point $c=0$ if $h=\frac{5}{2}$, and two distinct critical points if $h<\frac{5}{2}$, so that $10-25h>0$. Hence the bifurcation diagram in the hc-plane is the parabola $(c-5)^2 = 25-10h$ that is obtained upon squaring to eliminate the square root above.

21. **(a)** If $k=-a^2$, where $a\geq 0$, then $kx-x^3 = -a^2x-x^3 = -x(a^2+x^2)=0$ only if $x=0$, so the only critical point is $c=0$. If $a>0$, then we can solve the differential equation by writing

$$\int \frac{a^2}{x(a^2+x^2)}dx = \int \frac{1}{x}-\frac{x}{a^2+x^2}dx = -\int a^2\,dt,$$

or $\ln x - \frac{1}{2}\ln(a^2+x^2) = -a^2t+C$, or $\frac{x^2}{a^2+x^2} = Ce^{-2a^2t}$, where C is an arbitrary nonzero constant. Solving for x^2 gives $x^2 = \frac{a^2Ce^{-2a^2t}}{1-Ce^{-2a^2t}}$, from it follows that $x\to 0$ as $t\to\infty$, so the critical point $c=0$ is *stable*.

(b) If $k=a^2$, where $a>0$, then $kx-x^3 = +a^2x-x^3 = -x(x+a)(x-a)=0$ if either $x=0$ or $x=\pm a=\pm\sqrt{k}$. Thus we have the three critical points $c=0$ and $c=\pm\sqrt{k}$; this observation, together with part **(a)**, yields the pitchfork bifurcation diagram shown in Fig. 2.2.13 of the textbook. If $x(0)\neq 0$, then we can solve the differential equation by writing

$$\int \frac{2a^2}{x(x-a)(x+a)}dx = \int -\frac{2}{x}+\frac{1}{x-a}+\frac{1}{x+a}dx = -\int 2a^2\,dt,$$

or $-2\ln x + \ln(x-a) + \ln(x-a) = -2a^2t$, or $\frac{x^2-a^2}{x^2} = Ce^{-2a^2t}$, where C is an arbitrary nonzero constant. Solving for x^2 gives $x^2 = \frac{a^2}{1-Ce^{-2a^2t}}$, and so $x = \frac{\pm\sqrt{k}}{\sqrt{1-Ce^{-2a^2t}}}$. It follows that if $x(0)\neq 0$, then $x\to\sqrt{k}$ if $x>0$ and $x\to-\sqrt{k}$ if $x<0$. This implies that the critical point $c=0$ is *unstable*, while the critical points $c=\pm\sqrt{k}$ are *stable*.

23.   **(a)** If $h < kM$, then writing the differential equation as

$$x' = kx(M-x) - hx = kx\left[\left(M - \frac{h}{k}\right) - x\right],$$

still a logistic equation but with the *reduced* limiting population $M - \dfrac{h}{k}$.

(b) If $h > kM$, then the differential equation can be rewritten in the form $x' = -ax - bx^2$, with a and b both positive. The solution of this equation is $x(t) = \dfrac{ax_0}{(a + bx_0)e^{at} - bx_0}$, so it is clear that $x(t) \to 0$ as $t \to \infty$.

25.   In the first alternative form that is given, all of the coefficients within parentheses are positive if $H < x_0 < N$. Hence it is clear that $x(t) \to N$ as $t \to \infty$, which confirms (17).

In the second alternative form, all of the coefficients within parentheses are positive if $x_0 < H$. Hence the denominator is initially equal to $N - H > 0$, but decreases as t increases, and reaches the value 0 when $t = t_1 = \dfrac{1}{k(N-H)}\ln\dfrac{N - x_0}{H - x_0} > 0$. Meanwhile the numerator is initially $(N - H)x_0$, but approaches $(H - N)(H - x_0) < 0$ as $t \to t_1$. Conclusion (18) follows.

27.   Separation of variables in the differential equation $x' = -k\left[(x-a)^2 + b^2\right]$ yields

$$x(t) = a - b\tan\left(bkt + \tan^{-1}\frac{a - x_0}{b}\right).$$

It follows that $x(t) \to -\infty$ in a finite period of time.

29.   This is simply a matter of analyzing the signs of x' in the various cases $x < a$, $a < x < b$, $b < x < c$, and $c > x$. Alternatively, plot slope fields and typical solution curves for the two differential equations using typical numerical values such as $a = -1$, $b = 1$, and $c = 2$.

SECTION 2.3

ACCELERATION-VELOCITY MODELS

This section consists of three essentially independent subsections that can be studied separately: resistance proportional to velocity, resistance proportional to velocity-squared, and inverse-square gravitational acceleration.

1. The velocity v of the car (in km/hr) is related to the time t (in seconds) by the initial value problem $v' = k(250 - v)$, $v(0) = 0$, $v(10) = 100$. Separating variables gives

$$\int \frac{1}{250 - v} dv = \int k \, dt,$$ and integration yields $\ln|250 - v| = kt + C$, or $250 - v = Ce^{kt}$, or finally $v = Ce^{kt} + 250$, where C is an arbitrary nonzero constant. The initial condition $v(0) = 0$ gives $C = -250$, so that $v = 250(1 - e^{kt})$, and the condition $v(10) = 100$ implies that $k = \frac{1}{10} \ln\left(\frac{250}{150}\right) \approx 0.0511$. Finally, solving the equation $v(t) = 200$ for t gives $t = -\frac{\ln 50}{250k} \approx 31.5 \sec$.

3. The velocity v of the boat (in ft/s) is related to the time t (in seconds) by the initial value problem $v' = -kv$, $v(0) = 40$, $v(10) = 20$. By Problem 2a, $v(t) = 40^{-kt}$, and the condition $v(10) = 20$ implies that $k = \frac{1}{10} \ln 2 \approx 0.0693$. By Problem 2b, then, the boat travels a distance of $\frac{v_0}{k} = \frac{40 \cdot 10}{\ln 2} \approx 577$ ft altogether.

5. We are assuming that the velocity v of the motorboat satisfies the initial value problem $v' = -kv^2$, $v(0) = 40$, with $v(10) = 20$ as well. We seek $x(60)$. The result of Problem 4 gives $v = \frac{v_0}{1 + v_0 kt} = \frac{40}{1 + 40kt}$, and then the condition $v(10) = 20$ implies that

$20 = \frac{40}{1 + 400k}$, or $k = \frac{1}{400}$. Thus $v(t) = \frac{400}{10 + t}$, and then

$$x = \int v(t) \, dt = 400 \ln(t + 10) + C.$$

The initial condition $x(0) = 0$ implies that $C = -400 \ln 10$, so that $x(t) = 400 \ln \frac{t + 10}{10}$. It follows that $x(60) = 400 \ln 7 \approx 778$ ft.

7. The car satisfies the initial value problem $v' = 10 - 0.1v$, $v(0) = 0$. Separating variables gives $\int \frac{1}{10 - 0.1v} dv = \int dt$, or $\ln(10 - 0.1v) = -\frac{t}{10} + C$. The initial condition $v(0) = 0$ gives $C = \ln 10$, so that $\ln(10 - 0.1v) = -\frac{t}{10} + \ln 10$, or $\ln(1 - 0.01v) = -\frac{t}{10}$, or $v(t) = 100(1 - e^{-t/10})$. As $t \to \infty$, we find $v(t) \to 100$ ft/sec, the answer to **a**. Further, setting $v(t) = 90$ ft/sec (that is, 90% of limiting velocity) gives $t = 10 \ln 10 \approx 23.0259 \sec$.

Since $x = \int 100\left(1 - e^{-t/10}\right) dt = 100t + 1000 e^{-t/10} + C'$, where the initial condition $x(0) = 0$ gives $C' = -1000$, we find that $x(23.0259) \approx 1402.59\,\text{ft}$, the answer to **b**.

9.   Separating variables gives $\int \dfrac{1000}{5000 - 100v}\,dv = \int dt$, or $10\ln\left(5000 - 100v\right) = -t + C$, or $v = 50 + Ce^{-t/10}$. The initial condition $v(0) = 0$ implies that or $C = -50$, so that $v(t) = 50\left(1 - e^{-t/10}\right)$. As $t \to \infty$, $v(t) \to 50\,\text{ft/sec} \approx 34$ mph.

11.   If the paratrooper's terminal velocity was $100\,\text{mph} = \dfrac{440}{3}\,\text{ft/sec}$, then Equation (7) in the text yields $\dfrac{g}{\rho} = \dfrac{440}{3}$, or $\rho = \dfrac{3}{440} \cdot 32 = \dfrac{12}{55}$. Equation (9) then becomes

$$y(t) = -1200 + \frac{440}{3}t - \frac{55}{12} \cdot \frac{440}{3}\left(1 - e^{-12t/55}\right),$$

and solving the equation $y(t) = 0$ *via* technology gives $t \approx 12.5\,\text{sec}$. Thus the newspaper account is inaccurate.

Given the hints and integrals provided in the text, Problems 13–16 are fairly straightforward (and fairly tedious) integration problems.

17.   Equation (13) from the text gives

$$v(t) = \sqrt{\frac{9.8}{0.0011}}\,\tan\left(C_1 - t\sqrt{0.0011 \cdot 9.8}\right) = 94.3880\tan\left(C_1 - 0.1038267t\right),$$

where $C_1 = \tan^{-1}\left(49\sqrt{\dfrac{0.0011}{9.8}}\right) \approx 0.4788372$. Thus

$$v(t) = 94.3880\tan\left(0.4788372 - 0.1038267t\right).$$

Then Equation (14) gives

$$y(t) = \frac{1}{0.0011}\ln\left|\frac{\cos\left(0.4788372 - \sqrt{0.0011 \cdot 9.8}t\right)}{\cos 0.4788372}\right| \approx 909.0909\ln\left|\frac{\cos\left(0.4788372 - 0.103827t\right)}{\cos 0.4788372}\right|.$$

Setting $v(t) = 0$ leads to $t \approx 4.612\,\text{sec}$, at which time $y \approx 108.465\,\text{m}$.

19.   The initial value problem for the velocity of the motorboat is $v' = 4 - \dfrac{1}{400}v^2$, $v(0) = 0$.

Separating variables gives $\displaystyle\int \frac{1}{4 - \dfrac{1}{400}v^2}\,dv = \int dt$, or $\displaystyle\int \frac{1/40}{1 - \left(v/40\right)^2}\,dv = \int \frac{1}{10}\,dt$, or

$\tanh^{-1}\dfrac{v}{40}=\dfrac{t}{10}+C$. The initial condition $v(0)=0$ gives $C=0$, so that

$v(t)=40\tanh\dfrac{t}{10}$. Finally, $v(10)=\tanh 1\approx 30.46$ ft/sec and

$\lim\limits_{t\to\infty}v(t)=40\lim\limits_{t\to\infty}\tanh\dfrac{t}{10}=40$ ft/sec.

21. The initial value problem for the velocity of the ball is $v'=-g-\rho v^{2}$, $v(0)=v_{0}$, with the added condition that $y(0)=0$, where y is the height of the ball. Separating variables

gives $\displaystyle\int\dfrac{dv}{g+\rho v^{2}}\,dv=-\int dt$, or $\displaystyle\int\dfrac{\sqrt{\rho/g}}{1+\left(\sqrt{\rho/g}\,v\right)^{2}}\,dv=-\int\sqrt{g\rho}\,dt$, or

$\tan^{-1}\left(\sqrt{\rho/g}\,v\right)=-\sqrt{g\rho}\,t+C$. The initial condition $v(0)=v_{0}$ implies that $C=\tan^{-1}\left(\sqrt{\rho/g}\,v_{0}\right)$, and so

$$v(t)=-\sqrt{\dfrac{g}{\rho}}\tan\left[t\sqrt{g\rho}-\tan^{-1}\left(v_{0}\sqrt{\dfrac{\rho}{g}}\right)\right].$$

We solve $v(t)=0$ for $t=\dfrac{1}{\sqrt{g\rho}}\tan^{-1}\left(v_{0}\sqrt{\dfrac{\rho}{g}}\right)$ and substitute in Equation (17) for $y(t)$:

$$y_{\max}=\dfrac{1}{\rho}\ln\left|\dfrac{\cos\left(\tan^{-1}v_{0}\sqrt{\rho/g}-\tan^{-1}v_{0}\sqrt{\rho/g}\right)}{\cos\left(\tan^{-1}v_{0}\sqrt{\rho/g}\right)}\right|=\dfrac{1}{\rho}\ln\left[\sec\left(\tan^{-1}v_{0}\sqrt{\rho/g}\right)\right]$$

$$=\dfrac{1}{\rho}\ln\sqrt{1+\dfrac{\rho v_{0}^{2}}{g}}=\dfrac{1}{2\rho}\ln\left(1+\dfrac{\rho v_{0}^{2}}{g}\right).$$

23. Before the parachute opens, the paratrooper's descent is modeled by the initial value problem $v'=-32+0.00075v^{2}$, $v(0)=0$, with $y(0)=10000$. Solving gives $v(t)=-206.559\tanh(0.154919t)$, and then $v(30)=-206.521$ ft/sec. Integrating once again gives $y(t)=10000-1333.33\ln(\cosh 0.154919t)$, with $y(30)=4727.30$ ft. After the parachute opens, the initial value problem becomes $v'=-32+0.075v^{2}$, $v(0)=-206.521$, with $y(0)=4727.30$. Solving gives

$$v(t)=-20.6559\tanh(1.54919t+0.00519595),$$

followed by

$$y(t)=4727.30-13.3333\ln(\cosh 1.54919t+0.00519595).$$

We find that $y = 0$ when $t = 229.304$. Thus he opens his parachute after 30 sec at a height of 4727 feet, and the total time of descent is $30 + 229.304 = 259.304$ sec, about 4 minutes and 19.3 seconds.

25.   **(a)** The rocket's apex occurs when $v = 0$. We get the desired formula when we set $v = 0$ in Eq. (23), $v^2 = v_0^2 + 2GM\left(\dfrac{1}{r} - \dfrac{1}{R}\right)$, and solve for r.

(b) We substitute $v = 0$, $r = R + 10^5$ (note $100\,\text{km} = 10^5\,\text{m}$), and the mks values $G = 6.6726 \times 10^{-11}$, $M = 5.975 \times 10^{24}$, and $R = 6.378 \times 10^6$ in Eq. (23) and solve for $v_0 = 1389.21\,\text{m/s} \approx 1.389\,\text{km/s}$.

(c) When we substitute $v_0 = 0.9\sqrt{2GM/R}$ in the formula derived in part **a**, we find that

$$r_{max} = \frac{100}{19}R.$$

27.   **(a)** Substitution of $v_0^2 = \dfrac{2GM}{R} = \dfrac{k^2}{R}$ in Eq. (23) of the textbook gives

$$\frac{dr}{dt} = v = \sqrt{\frac{2GM}{r}} = \frac{k}{\sqrt{r}}.$$

We separate variables and proceed to integrate: $\int \sqrt{r}\, dr = \int k\, dt$ implies that

$\dfrac{2}{3} r^{3/2} = kt + \dfrac{2}{3} R^{3/2}$, since $r = R$ when $t = 0$. We solve for $r(t) = \left(\dfrac{2}{3} kt + R^{3/2}\right)^{2/3}$ and note that $r(t) \to \infty$ as $t \to \infty$.

(b) If $v_0 > \dfrac{2GM}{R}$, then Eq. (23) gives

$$\frac{dr}{dt} = v = \sqrt{\frac{2GM}{r} + \left(v_0^2 - \frac{2GM}{R}\right)} = \sqrt{\frac{k^2}{r} + \alpha} > \frac{k}{\sqrt{r}}.$$

Therefore, at every instant in its ascent, the upward velocity of the projectile in this part is greater than the velocity at the same instant of the projectile of part (a). It's as though the projectile of part (a) is the fox, and the projectile of this part is a rabbit that runs faster. Since the fox goes to infinity, so does the faster rabbit.

29.   Integration of $v\dfrac{dv}{dy} = -\dfrac{GM}{(y + R)^2}$, $y(0) = 0$, $v(0) = v_0$ gives

$$\frac{1}{2}v^2 = \frac{GM}{y + R} - \frac{GM}{R} + \frac{1}{2}v_0^2,$$

which simplifies to the desired formula for v^2. Then substitution of $G = 6.6726 \times 10^{-11}\,\text{N} \cdot (\text{m/kg})^2$, $M = 5.975 \times 10^{24}\,\text{kg}$, $R = 6.378 \times 10^6\,\text{m}$, $v = 0$, and $v_0 = 1$ yields an equation that we easily solve for $y = 51427.3\,\text{m}$, that is, about 51.427 km.

SECTION 2.4

NUMERICAL APPROXIMATION: EULER'S METHOD

In each of Problems 1–10 we also give first the explicit form of Euler's iterative formula for the given differential equation $y' = f(x, y)$. As we illustrate in Problem 1, the desired iterations are readily implemented, either manually or with a computer system or graphing calculator. Then we list the indicated values of $y(\tfrac{1}{2})$ rounded off to 3 decimal places.

1.   For the differential equation $y' = f(x, y)$ with $f(x, y) = -y$, the iterative formula of Euler's method is $y_{n+1} = y_n + h(-y_n)$. The TI-83 screen on the left shows a graphing calculator implementation of this iterative formula.

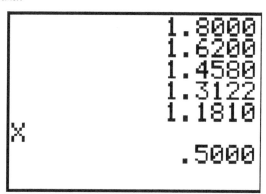

After the variables are initialized (in the first line), and the formula is entered, each press of the enter key carries out an additional step. The screen on the right shows the results of 5 steps from $x = 0$ to $x = 0.5$ with step size $h = 0.1$—winding up with $y(0.5) \approx 1.181$. Similarly, using $h = 0.25$ gives $y(0.5) \approx 1.125$. The true value is $y(\tfrac{1}{2}) \approx 1.213$.

The following *Mathematica* instructions produce precisely the line of data shown:

```
f[x_,y_] = -y;
g[x_] = 2*Exp[-x];
y0 = 2;
h = 0.25;
x = 0;
y1 = y0;

Do[k = f[x,y1];        (* the left-hand slope *)
   y1 = y1 + h*k;        (* Euler step to update y *)
   x = x + h,        (* update x *)
   {i,1,2}]

h = 0.1;
x = 0;
```

```
y2 = y0;

Do[k = f[x,y2];        (* the left-hand slope *)
    y2 = y2 + h*k;       (* Euler step to update y *)
    x = x + h,          (* update x *)
    {i,1,5}]

Print[x,"      ",y1,"      ",y2,"       ",g[0.5]]

0.5      1.125      1.18098      1.21306
```

3.   Iterative formula: $y_{n+1} = y_n + h(y_n + 1)$; approximate values 2.125 and 2.221; true value
      $y\left(\frac{1}{2}\right) \approx 2.297$.

5.   Iterative formula: $y_{n+1} = y_n + h(y_n - x_n - 1)$; approximate values 0.938 and 0.889; true
      value $y\left(\frac{1}{2}\right) \approx 0.851$.

7.   Iterative formula: $y_{n+1} = y_n + h\left(-3x_n^2 y_n\right)$; approximate values 2.859 and 2.737; true value
      $y\left(\frac{1}{2}\right) \approx 2.647$.

9.   Iterative formula: $y_{n+1} = y_n + h\dfrac{1+y_n^2}{4}$; approximate values 1.267 and 1.278; true value
      $y\left(\frac{1}{2}\right) \approx 1.287$.

The tables of approximate and actual values called for in Problems 11–16 were produced using
the following MATLAB script (appropriately altered for each problem).

```
% Section 2.4, Problems 11-16
x0 = 0;
y0 = 1;
% first run:
h = 0.01;
x = x0;
y = y0;
y1 = y0;
for n = 1:100
    y = y + h*(y-2);
    y1 = [y1,y];
    x = x + h;
    end
% second run:
h = 0.005;
x = x0;   y = y0;   y2 = y0;
```

```
for n = 1:200
   y = y + h*(y-2);
   y2 = [y2,y];
   x = x + h;
   end
% exact values
x = x0 : 0.2 : x0+1;
ye = 2 - exp(x);
% display table
ya = y2(1:40:201);
err = 100*(ye-ya)./ye;
[x; y1(1:20:101); ya; ye; err]
```

11. The iterative formula of Euler's method is $y_{n+1} = y_n + h(y_n - 2)$, and the exact solution is $y(x) = 2 - e^x$. The resulting table of approximate and actual values is

x	0.0	0.2	0.4	0.6	0.8	1.0
y ($h = 0.01$)	1.0000	0.7798	0.5111	0.1833	−0.2167	−0.7048
y ($h = 0.005$)	1.0000	0.7792	0.5097	0.1806	−0.2211	−0.7115
y actual	1.0000	0.7786	0.5082	0.1779	−0.2255	−0.7183
error	0%	−0.08%	−0.29%	−1.53%	1.97%	0.94%

13. Iterative formula: $y_{n+1} = y_n + 2h\dfrac{x_n^3}{y_n}$; exact solution: $y(x) = (8 + x^4)^{1/2}$.

x	1.0	1.2	1.4	1.6	1.8	2.0
y ($h = 0.01$)	3.0000	3.1718	3.4368	3.8084	4.2924	4.8890
y ($h = 0.005$)	3.0000	3.1729	3.4390	3.8117	4.2967	4.8940
y actual	3.0000	3.1739	3.4412	3.8149	4.3009	4.8990
error	0%	0.03%	0.06%	0.09%	0.10%	0.10%

15.   Iterative formula: $y_{n+1} = y_n + h\left(3 - \dfrac{2y_n}{x_n}\right)$; exact solution: $y(x) = x + \dfrac{4}{x^2}$.

x	2.0	2.2	2.4	2.6	2.8	3.0
$y\,(h = 0.01)$	3.0000	3.0253	3.0927	3.1897	3.3080	3.4422
$y\,(h = 0.005)$	3.0000	3.0259	3.0936	3.1907	3.3091	3.4433
y actual	3.0000	3.0264	3.0944	3.1917	3.3102	3.4444
error	0%	0.019%	0.028%	0.032%	0.033%	0.032%

The tables of approximate values called for in Problems 17–24 were produced using a MATLAB script similar to the one listed preceding the Problem 11 solution above.

17.

x	0.0	0.2	0.4	0.6	0.8	1.0
$y\,(h = 0.1)$	0.0000	0.0010	0.0140	0.0551	0.1413	0.2925
$y\,(h = 0.02)$	0.0000	0.0023	0.0198	0.0688	0.1672	0.3379
$y\,(h = 0.004)$	0.0000	0.0026	0.0210	0.0717	0.1727	0.3477
$y\,(h = 0.0008)$	0.0000	0.0027	0.0213	0.0723	0.1738	0.3497

These data indicate that $y(1) \approx 0.35$, in contrast with Example 5 in the text, where the initial condition is $y(0) = 1$.

In Problems 18–24 we give only the final approximate values of y obtained using Euler's method with step sizes $h = 0.1$, $h = 0.02$, $h = 0.004$, and $h = 0.0008$.

19.   With $x_0 = 0$ and $y_0 = 1$, the approximate values of $y(2)$ obtained are:

h	0.1	0.02	0.004	0.0008
y	6.1831	6.3653	6.4022	6.4096

21.   With $x_0 = 1$ and $y_0 = 2$, the approximate values of $y(2)$ obtained are:

h	0.1	0.02	0.004	0.0008
y	2.8508	2.8681	2.8716	2.8723

23. With $x_0 = 0$ and $y_0 = 0$, the approximate values of $y(1)$ obtained are:

h	0.1	0.02	0.004	0.0008
y	1.2262	1.2300	1.2306	1.2307

25. Here $f(t,v) = 32 - 1.6v$ and $t_0 = 0$, $v_0 = 0$. With $h = 0.01$, 100 iterations of $v_{n+1} = v_n + hf(t_n, v_n)$ yield $v(1) \approx 16.014$, and 200 iterations with $h = 0.005$ yield $v(1) \approx 15.998$. Thus we observe an approximate velocity of 16.0 ft/sec after 1 second — 80% of the limiting velocity of 20 ft/sec.

With $h = 0.01$, 200 iterations yield $v(2) \approx 19.2056$, and 400 iterations with $h = 0.005$ yield $v(2) \approx 19.1952$. Thus we observe an approximate velocity of 19.2 ft/sec after 2 seconds — 96% of the limiting velocity of 20 ft/sec.

27. Here $f(x,y) = x^2 + y^2 - 1$ and $x_0 = 0$, $y_0 = 0$. The following table gives the approximate values for the successive step sizes h and corresponding numbers n of steps. It appears likely that $y(2) = 1.00$ rounded off accurate to 2 decimal places.

h	0.1	0.01	0.001	0.0001	0.00001
n	20	200	2000	20000	200000
$y(2)$	0.7772	0.9777	1.0017	1.0042	1.0044

29. With step sizes $h = 0.15$, $h = 0.03$, and $h = 0.006$, we get the following results:

x	y with $h = 0.15$	y with $h = 0.03$	y with $h = 0.006$
-1.0	1.0000	1.0000	1.0000
-0.7	1.0472	1.0512	1.0521
-0.4	1.1213	1.1358	1.1390
-0.1	1.2826	1.3612	1.3835
$+0.2$	0.8900	1.4711	0.8210
$+0.5$	0.7460	1.2808	0.7192

While the values for $h = 0.15$ alone are not conclusive, a comparison of the values of y for all three step sizes with $x > 0$ suggests some anomaly in the transition from negative to positive values of x.

31. With step sizes $h = 0.1$ and $h = 0.01$ we get the following results:

x	y with $h = 0.1$	y with $h = 0.01$
0.0	1.0000	1.0000
0.1	1.2000	1.2200
0.2	1.4428	1.4967
⋮	⋮	⋮
0.7	4.3460	6.4643
0.8	5.8670	11.8425
0.9	8.3349	39.5010

Clearly there is some difficulty near $x = 0.9$.

SECTION 2.5

A CLOSER LOOK AT THE EULER METHOD

In each of Problems 1–10 we give first the predictor formula for u_{n+1} and then the improved Euler corrector for y_{n+1}. These predictor-corrector iterations are readily implemented, either manually or with a computer system or graphing calculator (as we illustrate in Problem 1). We give in each problem a table showing the approximate values obtained, as well as the corresponding values of the exact solution.

```
Y-H*Y→U:Y+(H/2)*
(-Y-U)→Y
              1.8100
              1.6381
              1.4824
              1.3416
              1.2142
```

```
0.1→H:0→X:2→Y
              2.0000
Y-H*Y→U:Y+(H/2)*
(-Y-U)→Y
              1.8100
              1.6381
              1.4824
```

1. $u_{n+1} = y_n + h(-y_n)$; $y_{n+1} = y_n + \dfrac{h}{2}(-y_n - u_{n+1})$

The TI-83 screen on the left above shows a graphing calculator implementation of this iteration. After the variables are initialized (in the first line), and the formulas are entered, each press of the enter key carries out an additional step. The screen on the right shows the results of 5 steps from $x = 0$ to $x = 0.5$ with step size $h = 0.1$ — winding up with $y(0.5) \approx 1.2142$ — and we see the approximate values shown in the second row of the table below.

x	0.0	0.1	0.2	0.3	0.4	0.5
y with $h = 0.1$	2.0000	1.8100	1.6381	1.4824	1.3416	1.2142
y actual	2.0000	1.8097	1.6375	1.4816	1.3406	1.2131

3. $u_{n+1} = y_n + h(y_n + 1)$; $y_{n+1} = y_n + \dfrac{h}{2}\left[(y_n + 1) + (u_{n+1} + 1)\right]$

x	0.0	0.1	0.2	0.3	0.4	0.5
y with $h = 0.1$	1.0000	1.2100	1.4421	1.6985	1.9818	2.2949
y actual	1.0000	1.2103	1.4428	1.6997	1.9837	2.2974

5. $u_{n+1} = y_n + h(y_n - x_n - 1)$; $y_{n+1} = y_n + \dfrac{h}{2}\left[(y_n - x_n - 1) + (u_{n+1} - x_n - h - 1)\right]$

x	0.0	0.1	0.2	0.3	0.4	0.5
y with $h = 0.1$	1.0000	0.9950	0.9790	0.9508	0.9091	0.8526
y actual	1.0000	0.9948	0.9786	0.9501	0.9082	0.8513

7. $u_{n+1} = y_n - 3x_n^2 y_n h$; $y_{n+1} = y_n - \dfrac{h}{2}\left[3x_n^2 y_n + 3(x_n + h)^2 u_{n+1}\right]$

x	0.0	0.1	0.2	0.3	0.4	0.5
y with $h = 0.1$	3.0000	2.9955	2.9731	2.9156	2.8082	2.6405
y actual	3.0000	2.9970	2.9761	2.9201	2.8140	2.6475

9. $u_{n+1} = y_n + h \cdot \dfrac{1 + y_n^2}{4}$; $y_{n+1} = y_n + h \cdot \dfrac{1 + y_n^2 + 1 + u_{n+1}^2}{8}$

x	0.0	0.1	0.2	0.3	0.4	0.5
y with $h = 0.1$	1.0000	1.0513	1.1053	1.1625	1.2230	1.2873
y actual	1.0000	1.0513	1.1054	1.1625	1.2231	1.2874

The results given below for Problems 11–16 were computed using the following MATLAB script.

```
% Section 2.5, Problems 11-16
x0 = 0;  y0 = 1;
% first run:
h = 0.01;
x = x0;  y = y0;  y1 = y0;
for  n = 1:100
   u = y + h*f(x,y);                        %predictor
   y = y + (h/2)*(f(x,y)+f(x+h,u));         %corrector
   y1 = [y1,y];
   x = x + h;
   end
% second run:
h = 0.005;
x = x0;  y = y0;  y2 = y0;

for  n = 1:200
   u = y + h*f(x,y);                        %predictor
   y = y + (h/2)*(f(x,y)+f(x+h,u));         %corrector
   y2 = [y2,y];
   x = x + h;
end

% exact values
x = x0 : 0.2 : x0+1;
ye = g(x);

% display table
ya = y2(1:40:201);
err = 100*(ye-ya)./ye;
x = sprintf('%10.5f',x), sprintf('\n');
y1 = sprintf('%10.5f',y1(1:20:101)), sprintf('\n');
ya = sprintf('%10.5f',ya), sprintf('\n');
ye = sprintf('%10.5f',ye), sprintf('\n');
err = sprintf('%10.5f',err), sprintf('\n');
table = [x; y1; ya; ye; err]
```

For each problem the differential equation $y' = f(x, y)$ and the known exact solution $y = g(x)$ are stored in the files **f.m** and **g.m** — for instance, the files

```
function yp = f(x,y)
yp = y-2;
```

```
function ye = g(x,y)
ye = 2-exp(x);
```

for Problem 11. (The exact solutions for Problems 11–16 here are given in the solutions for Problems 11–16 in Section 2.4.)

11.

x	0.0	0.2	0.4	0.6	0.8	1.0
$y\,(h=0.01)$	1.00000	0.77860	0.50819	0.17790	−0.22551	−0.71824
y $(h=0.005)$	1.00000	0.77860	0.50818	0.17789	−0.22553	−0.71827
y actual	1.00000	0.77860	0.50818	0.17788	−0.22554	−0.71828
error	0.000%	−0.000%	−0.000%	−0.003%	0.003%	0.002%

13.

x	1.0	1.2	1.4	1.6	1.8	2.0
$y\,(h=0.01)$	3.00000	3.17390	3.44118	3.81494	4.30091	4.89901
y $(h=0.005)$	3.00000	3.17390	3.44117	3.81492	4.30089	4.89899
y actual	3.00000	3.17389	3.44116	3.81492	4.30088	4.89898
error	0.0000%	−0.0001%	−0.0001%	-0.0001%	−0.0002%	−0.0002%

15.

x	2.0	2.2	2.4	2.6	2.8	3.0
$y\,(h=0.01)$	3.000000	3.026448	3.094447	3.191719	3.310207	3.444448
y $(h=0.005)$	3.000000	3.026447	3.094445	3.191717	3.310205	3.444445
y actual	3.000000	3.026446	3.094444	3.191716	3.310204	3.444444
error	0.00000%	−0.00002%	−0.00002%	−0.00002%	−0.00002%	−0.00002%

17.

With $h = $ 0.1:	$y(1) \approx 0.35183$
With $h = $ 0.02:	$y(1) \approx 0.35030$
With $h = $ 0.004:	$y(1) \approx 0.35023$
With $h = $ 0.0008:	$y(1) \approx 0.35023$

The table of numerical results is

x	y with $h = 0.1$	y with $h = 0.02$	y with $h = 0.004$	y with $h = 0.0008$
0.0	0.00000	0.00000	0.00000	0.00000
0.2	0.00300	0.00268	0.00267	0.00267
0.4	0.02202	0.02139	0.02136	0.02136
0.6	0.07344	0.07249	0.07245	0.07245
0.8	0.17540	0.17413	0.17408	0.17408
1.0	0.35183	0.35030	0.35023	0.35023

In Problems 18–24 we give only the final approximate values of y obtained using the improved Euler method with step sizes $h = 0.1$, $h = 0.02$, $h = 0.004$, and $h = 0.0008$.

19.

Value of h	Estimated value of $y(2)$
0.1	6.40834
0.02	6.41134
0.004	6.41147
0.0008	6.41147

21.

Value of h	Estimated value of $y(2)$
0.1	2.87204
0.02	2.87245
0.004	2.87247
0.0008	2.87247

23.

Value of h	Estimated value of $y(1)$
0.1	1.22967
0.02	1.23069
0.004	1.23073
0.0008	1.23073

25. Here $f(t,v) = 32 - 1.6v$ and $t_0 = 0$, $v_0 = 0$. With $h = 0.01$, 100 iterations of

$$k_1 = f(t, v_n), \quad k_2 = f(t + h, v_n + hk_1), \quad v_{n+1} = v_n + \frac{h}{2}(k_1 + k_2)$$

yield $v(1) \approx 15.9618$, and 200 iterations with $h = 0.005$ yield $v(1) \approx 15.9620$. Thus we observe an approximate velocity of 15.962 ft/sec after 1 second — 80% of the limiting velocity of 20 ft/sec.

With $h = 0.01$, 200 iterations yield $v(2) \approx 19.1846$, and 400 iterations with $h = 0.005$ yield $v(2) \approx 19.1847$. Thus we observe an approximate velocity of 19.185 ft/sec after 2 seconds — 96% of the limiting velocity of 20 ft/sec.

27. Here $f(x, y) = x^2 + y^2 - 1$ and $x_0 = 0$, $y_0 = 0$. The following table gives the approximate values for the successive step sizes h and corresponding numbers n of steps. It appears likely that $y(2) = 1.0045$ rounded off accurate to 4 decimal places.

h	0.1	0.01	0.001	0.0001
n	20	200	2000	20000
$y(2)$	1.01087	1.00452	1.00445	1.00445

In the solutions for Problems 29 and 30 we illustrate the following general MATLAB ode solver.

```
function  [t,y] = ode(method, yp, t0,b, y0, n)
%  [t,y] = ode(method, yp, t0,b, y0, n)
%  calls the method described by 'method' for the
%  ODE 'yp' with function header
%
%              y' = yp(t,y)
%
%  on the interval  [t0,b]  with initial (column)
%  vector  y0.  Choices for method are 'euler',
%  'impeuler', 'rk' (Runge-Kutta), 'ode23', 'ode45'.
%  Results are saved at the endPoints of n subintervals,
```

```
%   that is, in steps of length  h = (b - t0)/n.  The
%   result  t  is an (n+1)-column vector from b to t1,
%   while  y  is a matrix with  n+1  rows (one for each
%   t-value) and one column for each dependent variable.

h = (b - t0)/n;              % step size
t = t0 : h : b;
t = t';                              % col. vector of t-values
y = y0';                         % 1st row of result matrix
for  i = 2 : n+1             % for i=2 to i=n+1
   t0 = t(i-1);               % old t
   t1 = t(i);                      % new t
   y0 = y(i-1,:)';                 % old y-row-vector
   [T,Y] = feval(method, yp, t0,t1, y0);
   y = [y;Y'];                      % adjoin new y-row-vector
end
```

To use the improved Euler method, we call as **'method'** the following function.

```
function [t,y] = impeuler(yp, t0,t1, y0)
%
%   [t,y] = impeuler(yp, t0,t1, y0)
%   Takes one improved Euler step for
%
%         y' = yprime( t,y ),
%
%   from t0  to  t1  with initial value  the
%   column vector  y0.

h = t1 - t0;
k1 = feval( yp, t0, y0          );
k2 = feval( yp, t1, y0 + h*k1 );
k  = (k1 + k2)/2;
t = t1;
y = y0 + h*k;
```

29.   Here our differential equation is described by the MATLAB function

```
function  vp = vpbolt1(t,v)
vp = -0.04*v - 9.8;
```

Then the commands

```
n = 50;
[t1,v1] = ode('impeuler','vpbolt1',0,10,49,n);
n = 100;
[t2,v2] = ode('impeuler','vpbolt1',0,10,49,n);
t = (0:10)';
ve = 294*exp(-t/25)-245;
```

```
[t, v1(1:5:51), v2(1:10:101), ve]
```

generate the table

t	with $n = 50$	with $n = 100$	actual v
0	49.0000	49.0000	49.0000
1	37.4722	37.4721	37.4721
2	26.3964	26.3963	26.3962
3	15.7549	15.7547	15.7546
4	5.5307	5.5304	5.5303
5	-4.2926	-4.2930	-4.2932
6	-13.7308	-13.7313	-13.7314
7	-22.7989	-22.7994	-22.7996
8	-31.5115	-31.5120	-31.5122
9	-39.8824	-39.8830	-39.8832
10	-47.9251	-47.9257	-47.9259

We notice first that the final two columns agree to 3 decimal places (each difference being less than 0.0005). Scanning the $n = 100$ column for sign changes, we suspect that $v = 100$ (at the bolt's apex) occurs just after $t = 4.5 \sec$. Then interpolation between $t = 4.5$ and $t = 4.6$ in the table

```
[t2(40:51),v2(40:51)]
```

3.9000	6.5345
4.0000	5.5304
4.1000	4.5303
4.2000	3.5341
4.3000	2.5420
4.4000	1.5538
4.5000	0.5696
4.6000	-0.4108
4.7000	-1.3872
4.8000	-2.3597
4.9000	-3.3283
5.0000	-4.2930

indicates that $t = 4.56$ at the bolt's apex. Finally, interpolation in

`[t2(95:96),v2(95:96)]`

9.4000	-43.1387
9.5000	-43.9445

gives the impact velocity $v(9.41) \approx -43.22 \, \text{m/s}$.

SECTION 2.6

THE RUNGE-KUTTA METHOD

Each problem can be solved with a "template" of computations like those listed in Problem 1. We include a table showing the slope values k_1, k_2, k_3, k_4 and the xy-values at the ends of two successive steps of size $h = 0.25$.

1.   To make the first step of size $h = 0.25$ we start with the function defined by

```
f[x_, y_] := -y
```

and the initial values

```
x = 0;      y = 2;      h = 0.25;
```

and then perform the calculations

```
k1 = f[x, y]
k2 = f[x + h/2, y + h*k1/2]
k3 = f[x + h/2, y + h*k2/2]
k4 = f[x + h, y + h*k3]
y  = y + (h/6)*(k1 + 2*k2 + 2*k3 + k4)
x  = x + h
```

in turn. Here we are using *Mathematica* notation that translates transparently to standard mathematical notation describing the corresponding manual computations. A repetition of this same block of calculations carries out a second step of size $h = 0.25$. The following table lists the intermediate and final results obtained in these two steps.

k_1	k_2	k_3	k_4	x	Approx. y	Actual y
−2	−1/75	−1.78125	−1.55469	0.25	1.55762	1.55760
−1.55762	−1.36292	−1.38725	−1.2108	0.5	1.21309	1.21306

3.

k_1	k_2	k_3	k_4	x	Approx. y	Actual y
2	2.25	2.28125	2.57031	0.25	1.56803	1.56805
2.56803	2.88904	2.92916	3.30032	0.5	2.29740	2.29744

5.

k_1	k_2	k_3	k_4	x	Approx. y	Actual y
0	−0.125	−0.14063	−0.28516	0.25	0.96598	0.96597
−28402	−0.44452	−0.46458	−0.65016	0.5	0.85130	0.85128

7.

k_1	k_2	k_3	k_4	x	Approx. y	Actual y
0	−0.14063	−0.13980	−0.55595	0.25	2.95347	2.95349
−0.55378	−1.21679	−1.18183	−1.99351	0.5	2.6475	2.64749

9.

k_1	k_2	k_3	k_4	x	Approx. y	Actual y
0.5	0.53223	0.53437	0.57126	0.25	1.13352	1.13352
0.57122	0.61296	0.61611	0.66444	0.5	1.28743	1.28743

The results given below for Problems 11–16 were computed using the following MATLAB script.

```
% Section 2.6, Problems 11-16
x0 = 0;  y0 = 1;

% first run:
h = 0.2;
x = x0;  y = y0;  y1 = y0;
for  n = 1:5
   k1 = f(x,y);
   k2 = f(x+h/2,y+h*k1/2);
   k3 = f(x+h/2,y+h*k2/2);
   k4 = f(x+h,y+h*k3);
   y = y +(h/6)*(k1+2*k2+2*k3+k4);
   y1 = [y1,y];
```

```
    x = x + h;
    end

% second run:
h = 0.1;
x = x0;   y = y0;   y2 = y0;
for  n = 1:10
    k1 = f(x,y);
    k2 = f(x+h/2,y+h*k1/2);
    k3 = f(x+h/2,y+h*k2/2);
    k4 = f(x+h,y+h*k3);
    y = y +(h/6)*(k1+2*k2+2*k3+k4);
    y2 = [y2,y];
    x = x + h;
end

% exact values
x = x0 : 0.2 : x0+1;
ye = g(x);

% display table
y2 = y2(1:2:11);
err = 100*(ye-y2)./ye;
x = sprintf('%10.6f',x), sprintf('\n');
y1 = sprintf('%10.6f',y1), sprintf('\n');
y2 = sprintf('%10.6f',y2), sprintf('\n');
ye = sprintf('%10.6f',ye), sprintf('\n');
err = sprintf('%10.6f',err), sprintf('\n');
table = [x;y1;y2;ye;err]
```

For each problem the differential equation $y' = f(x,y)$ and the known exact solution $y = g(x)$ are stored in the files **f.m** and **g.m** — for instance, the files

```
function  yp = f(x,y)
yp = y-2;
```

and

```
function ye = g(x,y)
ye = 2-exp(x);
```

for Problem 11.

11.

x	0.0	0.2	0.4	0.6	0.8	1.0
$y(h=0.2)$	1.000000	0.778600	0.508182	0.177894	–0.225521	–0.718251
$y(h=0.1)$	1.000000	0.778597	0.508176	0.177882	–0.225540	–0.718280
y actual	1.000000	0.778597	0.508175	0.177881	–0.225541	–0.718282
error	0.00000%	–0.00002%	–0.00009%	–0.00047%	–0.00061%	–0.00029%

13.

x	1.0	1.2	1.4	1.6	1.8	2.0
$y(h=0.2)$	3.000000	3.173896	3.441170	3.814932	4.300904	4.899004
$y(h=0.1)$	3.000000	3.173894	3.441163	3.814919	4.300885	4.898981
y actual	3.000000	3.173894	3.441163	3.814918	4.300884	4.898979
error	0.00000%	–0.00001%	–0.00001%	–0.00002%	–0.00003%	–0.00003%

15.

x	2.0	2.2	2.4	2.6	2.9	3.0
$y(h=0.2)$	3.000000	3.026448	3.094447	3.191719	3.310207	3.444447
$y(h=0.1)$	3.000000	3.026446	3.094445	3.191716	3.310204	3.444445
y actual	3.000000	3.026446	3.094444	3.191716	3.310204	3.444444
error	–0.000000%	–0.000004%	–0.000005%	–0.000005%	–0.000005%	–0.000004%

17.

Value of h	Estimated value of $y(1)$
0.2	0.350258
0.1	0.350234
0.05	0.350232
0.025	0.350232

The table of numerical results is

x	y with $h=0.2$	y with $h=0.1$	y with $h=0.05$	y with $h=0.025$
0.0	0.000000	0.000000	0.000000	0.000000
0.2	0.002667	0.002667	0.002667	0.002667
0.4	0.021360	0.021359	0.021359	0.021359
0.6	0.072451	0.072448	0.072448	0.072448
0.8	0.174090	0.174081	0.174080	0.174080
1.0	0.350258	0.350234	0.350232	0.350232

In Problems 18−24 we give only the final approximate values of y obtained using the Runge-Kutta method with step sizes $h=0.2$, $h=0.1$, $h=0.05$, and $h=0.025$.

19.

Value of h	Estimated value of $y(2)$
0.2	1.679459
0.1	6.411474
0.05	6.411474
0.025	6.411474
0.025	−1.259993

21.

Value of h	Estimated value of $y(2)$
0.2	2.872467
0.1	2.872468
0.05	2.872468
0.025	2.872468

23.

Value of h	Estimated value of $y(1)$
0.2	1.230735
0.1	1.230731
0.05	1.230731
0.025	1.230731

25. Here $f(t,v) = 32 - 1.6v$ and $t_0 = 0$, $v_0 = 0$. With $h = 0.1$, 10 iterations of

$$k_1 = f(t_n, v_n) \qquad k_2 = f\left(t_n + \frac{1}{2}h, v_n + \frac{1}{2}hk_1\right)$$

$$k_3 = f\left(t_n + \frac{1}{2}h, v_n + \frac{1}{2}hk_2\right) \qquad k_4 = f(t_n + h, v_n + hk_3)$$

$$k = \frac{1}{6}(k_1 + 2k_2 + 2k_3 + k_4) \qquad v_{n+1} = v_n + hk$$

yield $v(1) \approx 15.9620$, and 20 iterations with $h = 0.05$ yield $v(1) \approx 15.9621$. Thus we observe an approximate velocity of 15.962 ft/sec after 1 second — 80% of the limiting velocity of 20 ft/sec.

With $h = 0.1$, 20 iterations yield $v(2) \approx 19.1847$, and 40 iterations with $h = 0.05$ yield $v(2) \approx 19.1848$. Thus we observe an approximate velocity of 19.185 ft/sec after 2 seconds — 96% of the limiting velocity of 20 ft/sec.

27. Here $f(x,y) = x^2 + y^2 - 1$ and $x_0 = 0$, $y_0 = 0$. The following table gives the approximate values for the successive step sizes h and corresponding numbers n of steps. It appears likely that $y(2) = 1.00445$ rounded off accurate to 5 decimal places.

h	1	0.1	0.01	0.001
n	2	20	200	2000
$y(2)$	1.05722	1.00447	1.00445	1.00445

In the solutions for Problems 29 and 30 we use the general MATLAB solver **ode** that was listed prior to the Problem 29 solution in Section 2.5. To use the Runge-Kutta method, we call as **'method'** the following function.

```
function [t,y] = rk(yp, t0,t1, y0)
```

```
%   [t, y] = rk(yp, t0, t1, y0)
%   Takes one Runge-Kutta step for
%
%          y' = yp( t,y ),
%
%   from t0  to  t1  with initial value  the
%   column vector  y0.

h = t1 - t0;
k1 = feval(yp, t0        , y0                  );
k2 = feval(yp, t0 + h/2, y0 + (h/2)*k1 );
k3 = feval(yp, t0 + h/2, y0 + (h/2)*k2 );
k4 = feval(yp, t0 + h  ,y0 +     h *k3 );
k  = (1/6)*(k1 + 2*k2 + 2*k3 + k4);
t = t1;
y = y0 + h*k;
```

29.   Here our differential equation is described by the MATLAB function

```
function  vp = vpbolt1(t,v)
vp = -0.04*v - 9.8;
```

Then the commands

```
n = 100;
[t1,v1] = ode('rk','vpbolt1',0,10,49,n);
n = 200;
[t2,v] = ode('rk','vpbolt1',0,10,49,n);
t = (0:10)';
ve = 294*exp(-t/25)-245;
[t, v1(1:n/20:1+n/2), v(1:n/10:n+1), ve]
```

generate the table

t	with $n = 100$	with $n = 200$	actual v
0	49.0000	49.0000	49.0000
1	37.4721	37.4721	37.4721
2	26.3962	26.3962	26.3962
3	15.7546	15.7546	15.7546
4	5.5303	5.5303	5.5303
5	–4.2932	–4.2932	–4.2932
6	–13.7314	–13.7314	–13.7314
7	–22.7996	–22.7996	–22.7996
8	–31.5122	–31.5122	–31.5122
9	–39.8832	–39.8832	–39.8832
10	–47.9259	–47.9259	–47.9259

We notice first that the final three columns agree to the 4 displayed decimal places. Scanning the last column for sign changes in v, we suspect that $v = 0$ (at the bolt's apex) occurs just after $t = 4.5 \sec$. Then interpolation between $t = 4.55$ and $t = 4.60$ in the table

```
[t2(91:95),v(91:95)]
```

4.5000	0.5694
4.5500	0.0788
4.6000	–0.4109
4.6500	–0.8996
4.7000	–1.3873

indicates that $t = 4.56$ at the bolt's apex. Now the commands

```
y = zeros(n+1,1);
h = 10/n;


for j = 2:n+1
   y(j) = y(j-1) + v(j-1)*h + 0.5*(-.04*v(j-1) - 9.8)*h^2;
end
ye = 7350*(1 - exp(-t/25)) - 245*t;
[t, y(1:n/10:n+1), ye]
```

generate the table

t	approx. y	actual y
0	0	0
1	43.1974	43.1976
2	75.0945	75.0949
3	96.1342	96.1348
4	106.7424	106.7432
5	107.3281	107.3290
6	98.2842	98.2852
7	79.9883	79.9895
8	52.8032	52.8046
9	17.0775	17.0790
10	−26.8540	−26.8523

We see at least 2-decimal place agreement between approximate and actual values of y. Finally, interpolation between $t = 9$ and $t = 10$ here suggests that $y = 0$ just after $t = 9.4$. Then interpolation between $t = 9.40$ and $t = 9.45$ in the table

`[t2(187:191),y(187:191)]`

9.3000	4.7448
9.3500	2.6182
9.4000	0.4713
9.4500	−1.6957
9.5000	−3.8829

indicates that the bolt is aloft for about 9.41 seconds.

CHAPTER 3

LINEAR SYSTEMS AND MATRICES

SECTION 3.1

INTRODUCTION TO LINEAR SYSTEMS

You should remember from high school mathematics the method of elimination for 2 (or 3) linear equations in 2 (or 3) unknowns. However, high school algebra courses generally emphasize only the case in which a single solution exists. Here we treat on an equal footing the other two cases — in which either no solution exists or infinitely many solutions exist.

1.   Subtraction of twice the first equation from the second equation gives $-5y = -10$, so $y = 2$, and it follows that $x = 3$.

3.   Subtraction of 3/2 times the first equation from the second equation gives $\frac{1}{2}y = \frac{3}{2}$, so $y = 3$, and it follows that $x = -4$.

5.   Subtraction of twice the first equation from the second equation gives $0 = 1$, so no solution exists.

7.   The second equation is -2 times the first equation, so we can choose $y = t$ arbitrarily. The first equation then gives $x = -10 + 4t$.

9.   Subtraction of twice the first equation from the second equation gives $-9y - 4z = -3$. Subtraction of the first equation from the third equation gives $2y + z = 1$. Solution of these latter two equations gives $y = -1$, $z = 3$. Finally substitution in the first equation gives $x = 4$.

11.   First we interchange the first and second equations. Then subtraction of twice the new first equation from the new second equation gives $y - z = 7$, and subtraction of three times the new first equation from the third equation gives $-2y + 3z = -18$. Solution of these latter two equations gives $y = 3$, $z = -4$. Finally substitution in the (new) first equation gives $x = 1$.

13.   First we subtract the second equation from the first equation to get the new first equation $x + 2y + 3z = 0$. Then subtraction of twice the new first equation from the second equation gives $3y - 2z = 0$, and subtraction of twice the new first equation from the

third equation gives $2y - z = 0$. Solution of these latter two equations gives $y = 0$, $z = 0$. Finally substitution in the (new) first equation gives $x = 0$ also.

15. Subtraction of the first equation from the second equation gives $-4y + z = -2$. Subtraction of three times the first equation from the third equation gives (after division by 2) $-4y + z = -5/2$. These latter two equations obviously are inconsistent, so the original system has no solution.

17. First we subtract the first equation from the second equation to get the new first equation $x + 3y - 6z = -4$. Then subtraction of three times the new first equation from the second equation gives $-7y + 16z = 15$, and subtraction of five times the new first equation from the third equation gives (after division by 2) $-7y + 16z = 35/2$. These latter two equations obviously are inconsistent, so the original system has no solution.

19. Subtraction of twice the first equation from the second equation gives $3y - 6z = 9$. Subtraction of the first equation from the third equation gives $y - 2z = 3$. Obviously these latter two equations are scalar multiples of each other, so we can choose $z = t$ arbitrarily. It follows first that $y = 3 + 2t$ and then that $x = 8 + 3t$.

21. Subtraction of three times the first equation from the second equation gives $3y - 6z = 9$. Subtraction of four times the first equation from the third equation gives $-3y + 9z = -6$. Obviously these latter two equations are both scalar multiples of the equation $y - 3z = 2$, so we can choose $z = t$ arbitrarily. It follows first that $y = 2 + 3t$ and then that $x = 3 - 2t$.

23. The initial conditions $y(0) = 3$ and $y'(0) = 8$ yield the equations $A = 3$ and $2B = 8$, so $A = 3$ and $B = 4$. It follows that $y(x) = 3\cos 2x + 4\sin 2x$.

25. The initial conditions $y(0) = 10$ and $y'(0) = 20$ yield the equations $A + B = 10$ and $5A - 5B = 20$ with solution $A = 7$, $B = 3$. Thus $y(x) = 7e^{5x} + 3e^{-5x}$.

27. The initial conditions $y(0) = 40$ and $y'(0) = -16$ yield the equations $A + B = 40$ and $3A - 5B = -16$ with solution $A = 23$, $B = 17$. Thus $y(x) = 23e^{3x} + 17e^{-5x}$.

29. The initial conditions $y(0) = 7$ and $y'(0) = 11$ yield the equations $A + B = 7$ and $\frac{1}{2}A + \frac{1}{3}B = 11$ with solution $A = 52$, $B = -45$. Thus $y(x) = 52e^{x/2} - 45e^{x/3}$.

31. The graph of each of these linear equations in x and y is a straight line through the origin $(0, 0)$ in the xy-plane. If these two lines are distinct then they intersect only at the origin, so the two equations have the unique solution $x = y = 0$. If the two lines coincide, then each of the infinitely many different points (x, y) on this common line provides a solution of the system.

33. **(a)** The three lines have no common point of intersection, so the system has no solution.

(b) The three lines have a single point of intersection, so the system has a unique solution.

(c) The three lines — two of them parallel — have no common point of intersection, so the system has no solution.

(d) The three distinct parallel lines have no common point of intersection, so the system has no solution.

(e) Two of the lines coincide and intersect the third line in a single point, so the system has a unique solution.

(f) The three lines coincide, and each of the infinitely many different points (x, y, z) on this common line provides a solution of the system.

SECTION 3.2

MATRICES AND GAUSSIAN ELIMINATION

Because the linear systems in Problems 1–10 are already in echelon form, we need only start at the end of the list of unknowns and work backwards.

1. Starting with $x_3 = 2$ from the third equation, the second equation gives $x_2 = 0$, and then the first equation gives $x_1 = 1$.

3. If we set $x_3 = t$ then the second equation gives $x_2 = 2 + 5t$, and next the first equation gives $x_1 = 13 + 11t$.

5. If we set $x_4 = t$ then the third equation gives $x_3 = 5 + 3t$, next the second equation gives $x_2 = 6 + t$, and finally the first equation gives $x_1 = 13 + 4t$.

7. If we set $x_3 = s$ and $x_4 = t$, then the second equation gives $x_2 = 7 + 2s - 7t$, and next the first equation gives $x_1 = 3 - 8s + 19t$.

9. Starting with $x_4 = 6$ from the fourth equation, the third equation gives $x_3 = -5$, next the second equation gives $x_2 = 3$, and finally the first equation gives $x_1 = 1$.

In each of Problems 11–22, we give just the first two or three steps in the reduction. Then we display a resulting echelon form $\mathbf{E}$ of the augmented coefficient matrix $\mathbf{A}$ of the given linear system, and finally list the resulting solution (if any). The student should understand that the echelon matrix **E** is not unique, so a different sequence of elementary row operations may produce a different echelon matrix.

11. Begin by interchanging rows 1 and 2 of **A**. Then subtract twice row 1 both from row 2 and from row 3.

$$\mathbf{E} = \begin{bmatrix} 1 & 3 & 2 & 5 \\ 0 & 1 & 0 & -2 \\ 0 & 0 & 1 & 4 \end{bmatrix}; \quad x_1 = 3, \ x_2 = -2, \ x_3 = 4$$

13. Begin by subtracting twice row 1 of **A** both from row 2 and from row 3. Then add row 2 to row 3.

$$\mathbf{E} = \begin{bmatrix} 1 & 3 & 3 & 13 \\ 0 & 1 & 2 & 3 \\ 0 & 0 & 0 & 0 \end{bmatrix}; \quad x_1 = 4 + 3t, \ x_2 = 3 - 2t, \ x_3 = t$$

15. Begin by interchanging rows 1 and 2 of **A**. Then subtract three times row 1 from row 2, and five times row 1 from row 3.

$$\mathbf{E} = \begin{bmatrix} 1 & 1 & 1 & 1 \\ 0 & 1 & 3 & 3 \\ 0 & 0 & 0 & 1 \end{bmatrix}. \qquad \text{The system has no solution.}$$

17.
$$\begin{aligned} x_1 - 4x_2 - 3x_3 - 3x_4 &= 4 \\ 2x_1 - 6x_2 - 5x_3 - 5x_4 &= 5 \\ 3x_1 - x_2 - 4x_3 - 5x_4 &= -7 \end{aligned}$$

Begin by subtracting twice row 1 from row 2 of **A**, and three times row 1 from row 3.

$$\mathbf{E} = \begin{bmatrix} 1 & -4 & -3 & -3 & 4 \\ 0 & 1 & 0 & -1 & -4 \\ 0 & 0 & 1 & 3 & 5 \end{bmatrix}; \quad x_1 = 3 - 2t, \ x_2 = -4 + t, \ x_3 = 5 - 3t, \ x_4 = t$$

19. Begin by interchanging rows 1 and 2 of **A**. Then subtract three times row 1 from row 2, and four times row 1 from row 3.

$$\mathbf{E} = \begin{bmatrix} 1 & -2 & 5 & -5 & -7 \\ 0 & 1 & -2 & 3 & 5 \\ 0 & 0 & 0 & 0 & 0 \end{bmatrix}; \quad x_1 = 3 - s - t, \ x_2 = 5 + 2s - 3t, \ x_3 = s, \ x_4 = t$$

21.   Begin by subtracting twice row 1 from row 2, three times row 1 from row 3, and four times row 1 from row 4.

$$\mathbf{E} = \begin{bmatrix} 1 & 1 & 1 & 0 & 6 \\ 0 & 1 & 5 & 1 & 20 \\ 0 & 0 & 1 & 0 & 3 \\ 0 & 0 & 0 & 1 & 4 \end{bmatrix}; \quad x_1 = 2, \ x_2 = 1, \ x_3 = 3, \ x_4 = 4$$

23.   If we subtract twice the first row from the second row, we obtain the echelon form

$$\mathbf{E} = \begin{bmatrix} 3 & 2 & 1 \\ 0 & 0 & k-2 \end{bmatrix}$$

of the augmented coefficient matrix. It follows that the given system has no solutions unless $k = 2$, in which case it has infinitely many solutions given by $x_1 = \frac{1}{3}(1-2t), \ x_2 = t$.

25.   If we subtract twice the first row from the second row, we obtain the echelon form

$$\mathbf{E} = \begin{bmatrix} 3 & 2 & 11 \\ 0 & k-4 & -1 \end{bmatrix}$$

of the augmented coefficient matrix. It follows that the given system has a unique solution if $k \neq 4$, but no solution if $k = 4$.

27.   If we first subtract twice the first row from the second row, then subtract 4 times the first row from the third row, and finally subtract the second row from the third row , we obtain the echelon form

$$\mathbf{E} = \begin{bmatrix} 1 & 2 & 1 & 3 \\ 0 & 5 & 5 & 1 \\ 0 & 0 & 0 & k-11 \end{bmatrix}$$

of the augmented coefficient matrix. It follows that the given system has no solution unless $k = 11$, in which case it has infinitely many solutions with x_3 arbitrary.

29.   In each of parts (a)-(c), we start with a typical 2×2 matrix $\mathbf{A}$ and carry out two row successive operations as indicated, observing that we wind up with the original matrix $\mathbf{A}$.

(a) $\mathbf{A} = \begin{bmatrix} s & t \\ u & v \end{bmatrix} \overset{cR2}{\rightarrow} \begin{bmatrix} s & t \\ cu & cv \end{bmatrix} \overset{(1/c)R2}{\rightarrow} \begin{bmatrix} s & t \\ u & v \end{bmatrix} = \mathbf{A}$

(b) $\quad \mathbf{A} = \begin{bmatrix} s & t \\ u & v \end{bmatrix} \xrightarrow{SWAP(R1,R2)} \begin{bmatrix} u & v \\ s & t \end{bmatrix} \xrightarrow{SWAP(R1,R2)} \begin{bmatrix} s & t \\ u & v \end{bmatrix} = \mathbf{A}$

(c) $\quad \mathbf{A} = \begin{bmatrix} s & t \\ u & v \end{bmatrix} \xrightarrow{cR1+R2} \begin{bmatrix} u & v \\ cu+s & cv+t \end{bmatrix} \xrightarrow{(-c)R1+R2} \begin{bmatrix} s & t \\ u & v \end{bmatrix} = \mathbf{A}$

Since we therefore can "reverse" any single elementary row operation, it follows that we can reverse any finite sequence of such operations — on at a time — so part (d) follows.

SECTION 3.3

REDUCED ROW-ECHELON MATRICES

Each of the matrices in Problems 1–20 can be transformed to reduced echelon form without the appearance of any fractions. The main thing is to get started right. Generally our first goal is to get a 1 in the upper left corner of **A**, then clear out the rest of the first column. In each problem we first give at least the initial steps, and then the final result **E**. The particular sequence of elementary row operations used is not unique; you might find **E** in a quite different way.

1. $\quad \begin{bmatrix} 1 & 2 \\ 3 & 7 \end{bmatrix} \xrightarrow{R2-3R1} \begin{bmatrix} 1 & 2 \\ 0 & 1 \end{bmatrix} \xrightarrow{R1-2R2} \begin{bmatrix} 1 & 0 \\ 0 & 1 \end{bmatrix}$

3. $\quad \begin{bmatrix} 3 & 7 & 15 \\ 2 & 5 & 11 \end{bmatrix} \xrightarrow{R1-R2} \begin{bmatrix} 1 & 2 & 4 \\ 2 & 5 & 11 \end{bmatrix} \xrightarrow{R2-2R1} \begin{bmatrix} 1 & 2 & 4 \\ 0 & 1 & 3 \end{bmatrix} \xrightarrow{R1-2R2} \begin{bmatrix} 1 & 0 & 2 \\ 0 & 1 & 3 \end{bmatrix}$

5. $\quad \begin{bmatrix} 1 & 2 & -11 \\ 2 & 3 & -19 \end{bmatrix} \xrightarrow{R2-2R1} \begin{bmatrix} 1 & 2 & -11 \\ 0 & -1 & 3 \end{bmatrix} \xrightarrow{(-1)R2} \begin{bmatrix} 1 & 2 & -11 \\ 0 & 1 & -3 \end{bmatrix} \xrightarrow{R1-2R2} \begin{bmatrix} 1 & 0 & -5 \\ 0 & 1 & -3 \end{bmatrix}$

7. $\quad \begin{bmatrix} 1 & 2 & 3 \\ 1 & 4 & 1 \\ 2 & 1 & 9 \end{bmatrix} \xrightarrow{R2-R1} \begin{bmatrix} 1 & 2 & 3 \\ 0 & 2 & -2 \\ 2 & 1 & 9 \end{bmatrix} \xrightarrow{R3-2R1} \begin{bmatrix} 1 & 2 & 3 \\ 0 & 2 & -2 \\ 0 & -3 & 3 \end{bmatrix}$

$\quad \xrightarrow{(1/2)R2} \begin{bmatrix} 1 & 2 & 3 \\ 0 & 1 & -1 \\ 0 & -3 & 3 \end{bmatrix} \xrightarrow{R3+3R2} \begin{bmatrix} 1 & 2 & 3 \\ 0 & 1 & -1 \\ 0 & 0 & 0 \end{bmatrix} \xrightarrow{R1-2R2} \begin{bmatrix} 1 & 0 & 5 \\ 0 & 1 & -1 \\ 0 & 0 & 0 \end{bmatrix}$

9.
$$\begin{bmatrix} 5 & 2 & 18 \\ 0 & 1 & 4 \\ 4 & 1 & 12 \end{bmatrix} \xrightarrow{R1-R3} \begin{bmatrix} 1 & 1 & 6 \\ 0 & 1 & 4 \\ 4 & 1 & 12 \end{bmatrix} \xrightarrow{R3-4R1} \begin{bmatrix} 1 & 1 & 6 \\ 0 & 1 & 4 \\ 0 & -3 & -12 \end{bmatrix}$$

$$\xrightarrow{R3+3R2} \begin{bmatrix} 1 & 1 & 6 \\ 0 & 1 & 4 \\ 0 & 0 & 0 \end{bmatrix} \xrightarrow{R1-R2} \begin{bmatrix} 1 & 0 & 2 \\ 0 & 1 & 4 \\ 0 & 0 & 0 \end{bmatrix}$$

11.
$$\begin{bmatrix} 3 & 9 & 1 \\ 2 & 6 & 7 \\ 1 & 3 & -6 \end{bmatrix} \xrightarrow{SWAP(R1,R3)} \begin{bmatrix} 1 & 3 & -6 \\ 2 & 6 & 7 \\ 3 & 9 & 1 \end{bmatrix} \xrightarrow{R2-2R1} \begin{bmatrix} 1 & 3 & -6 \\ 0 & 0 & 19 \\ 3 & 9 & 1 \end{bmatrix}$$

$$\xrightarrow{R3-3R1} \begin{bmatrix} 1 & 3 & -6 \\ 0 & 0 & 19 \\ 0 & 0 & 19 \end{bmatrix} \xrightarrow{(1/19)R2} \begin{bmatrix} 1 & 3 & -6 \\ 0 & 0 & 1 \\ 0 & 0 & 19 \end{bmatrix} \xrightarrow{R3-19R2} \cdots \begin{bmatrix} 1 & 3 & 0 \\ 0 & 0 & 1 \\ 0 & 0 & 0 \end{bmatrix}$$

13.
$$\begin{bmatrix} 2 & 7 & 4 & 0 \\ 1 & 3 & 2 & 1 \\ 2 & 6 & 5 & 4 \end{bmatrix} \xrightarrow{SWAP(R1,R2)} \begin{bmatrix} 1 & 3 & 2 & 1 \\ 2 & 7 & 4 & 0 \\ 2 & 6 & 5 & 4 \end{bmatrix} \xrightarrow{R2-2R1} \begin{bmatrix} 1 & 3 & 2 & 1 \\ 0 & 1 & 0 & -2 \\ 2 & 6 & 5 & 4 \end{bmatrix}$$

$$\xrightarrow{R2-2R1} \begin{bmatrix} 1 & 3 & 2 & 1 \\ 0 & 1 & 0 & -2 \\ 0 & 0 & 1 & 2 \end{bmatrix} \xrightarrow{R1-3R2} \cdots \to \begin{bmatrix} 1 & 0 & 0 & 3 \\ 0 & 1 & 0 & -2 \\ 0 & 0 & 1 & 2 \end{bmatrix}$$

15.
$$\begin{bmatrix} 2 & 2 & 4 & 2 \\ 1 & -1 & -4 & 3 \\ 2 & 7 & 19 & -3 \end{bmatrix} \xrightarrow{SWAP(R1,R2)} \begin{bmatrix} 1 & -1 & -4 & 3 \\ 2 & 2 & 4 & 2 \\ 2 & 7 & 19 & -3 \end{bmatrix} \xrightarrow{R2-2R1} \begin{bmatrix} 1 & -1 & -4 & 3 \\ 0 & 4 & 12 & -4 \\ 2 & 7 & 19 & -3 \end{bmatrix}$$

$$\xrightarrow{R3-2R1} \begin{bmatrix} 1 & -1 & -4 & 3 \\ 0 & 4 & 12 & -4 \\ 0 & 9 & 27 & -9 \end{bmatrix} \xrightarrow{(1/4)R2} \begin{bmatrix} 1 & -1 & -4 & 3 \\ 0 & 1 & 3 & -1 \\ 0 & 9 & 27 & -9 \end{bmatrix}$$

$$\xrightarrow{R3-9R2} \begin{bmatrix} 1 & -1 & -4 & 3 \\ 0 & 1 & 3 & -1 \\ 0 & 0 & 0 & 0 \end{bmatrix} \xrightarrow{R1+R2} \begin{bmatrix} 1 & 0 & -1 & 2 \\ 0 & 1 & 3 & -1 \\ 0 & 0 & 0 & 0 \end{bmatrix}$$

17.
$$\begin{bmatrix} 1 & 1 & 1 & -1 & -4 \\ 1 & -2 & -2 & 8 & -1 \\ 2 & 3 & -1 & 3 & 11 \end{bmatrix} \xrightarrow{R2-R1} \begin{bmatrix} 1 & 1 & 1 & -1 & -4 \\ 0 & -3 & -3 & 9 & 3 \\ 2 & 3 & -1 & 3 & 11 \end{bmatrix} \xrightarrow{R3-2R1} \begin{bmatrix} 1 & 1 & 1 & -1 & -4 \\ 0 & -3 & -3 & 9 & 3 \\ 0 & 1 & -3 & 5 & 19 \end{bmatrix}$$

$$\xrightarrow{(-1/3)R2} \begin{bmatrix} 1 & 1 & 1 & -1 & -4 \\ 0 & 1 & 1 & -3 & -1 \\ 0 & 1 & -3 & 5 & 19 \end{bmatrix} \xrightarrow{R3-R2} \begin{bmatrix} 1 & 1 & 1 & -1 & -4 \\ 0 & 1 & 1 & -3 & -1 \\ 0 & 0 & -4 & 8 & 20 \end{bmatrix}$$

$$\xrightarrow{(-1/4)R3} \begin{bmatrix} 1 & 1 & 1 & -1 & -4 \\ 0 & 1 & 1 & -3 & -1 \\ 0 & 0 & 1 & -2 & -5 \end{bmatrix} \xrightarrow{R1-R2} \cdots \rightarrow \begin{bmatrix} 1 & 0 & 0 & 2 & -3 \\ 0 & 1 & 0 & -1 & 4 \\ 0 & 0 & 1 & -2 & -5 \end{bmatrix}$$

19.
$$\begin{bmatrix} 2 & 7 & -10 & -19 & 13 \\ 1 & 3 & -4 & -8 & 6 \\ 1 & 0 & 2 & 1 & 3 \end{bmatrix} \xrightarrow{SWAP(R1,R3)} \begin{bmatrix} 1 & 0 & 2 & 1 & 3 \\ 1 & 3 & -4 & -8 & 6 \\ 2 & 7 & -10 & -19 & 13 \end{bmatrix}$$

$$\xrightarrow{R2-R1} \begin{bmatrix} 1 & 0 & 2 & 1 & 3 \\ 0 & 3 & -6 & -9 & 3 \\ 2 & 7 & -10 & -19 & 13 \end{bmatrix} \xrightarrow{R3-2R1} \begin{bmatrix} 1 & 0 & 2 & 1 & 3 \\ 0 & 3 & -6 & -9 & 3 \\ 0 & 7 & -14 & -21 & 7 \end{bmatrix}$$

$$\xrightarrow{(1/3)R2} \begin{bmatrix} 1 & 0 & 2 & 1 & 3 \\ 0 & 1 & -2 & -3 & 1 \\ 0 & 7 & -14 & -21 & 7 \end{bmatrix} \xrightarrow{R3-7R2} \begin{bmatrix} 1 & 0 & 2 & 1 & 3 \\ 0 & 1 & -2 & -3 & 1 \\ 0 & 0 & 0 & 0 & 0 \end{bmatrix}$$

In each of Problems 21–30, we give just the first two or three steps in the reduction. Then we display the resulting reduced echelon form **E** of the augmented coefficient matrix **A** of the given linear system, and finally list the resulting solution (if any).

21.   Begin by interchanging rows 1 and 2 of **A**. Then subtract twice row 1 both from row 2 and from row 3.

$$\mathbf{E} = \begin{bmatrix} 1 & 0 & 0 & 3 \\ 0 & 1 & 0 & -2 \\ 0 & 0 & 1 & 4 \end{bmatrix}; \quad x_1 = 3, \ x_2 = -2, \ x_3 = 4$$

23. Begin by subtracting twice row 1 of **A** both from row 2 and from row 3. Then add row 2 to row 3.

$$\mathbf{E} = \begin{bmatrix} 1 & 0 & -3 & 14 \\ 0 & 1 & 2 & 3 \\ 0 & 0 & 0 & 0 \end{bmatrix}; \quad x_1 = 4+3t, \ x_2 = 3-2t, \ x_3 = t$$

25. Begin by interchanging rows 1 and 2 of **A**. Then subtract three times row 1 from row 2, and five times row 1 from row 3.

$$\mathbf{E} = \begin{bmatrix} 1 & 0 & -2 & 0 \\ 0 & 1 & 3 & 0 \\ 0 & 0 & 0 & 1 \end{bmatrix}. \qquad \text{The system has no solution.}$$

27.
$$x_1 - 4x_2 - 3x_3 - 3x_4 = 4$$
$$2x_1 - 6x_2 - 5x_3 - 5x_4 = 5$$
$$3x_1 - x_2 - 4x_3 - 5x_4 = -7$$

Begin by subtracting twice row 1 from row 2 of **A**, and three times row 1 from row 3.

$$\mathbf{E} = \begin{bmatrix} 1 & 0 & 0 & 2 & 3 \\ 0 & 1 & 0 & -1 & -4 \\ 0 & 0 & 1 & 3 & 5 \end{bmatrix}; \quad x_1 = 3-2t, \ x_2 = -4+t, \ x_3 = 5-3t, \ x_4 = t$$

29. Begin by interchanging rows 1 and 2 of **A**. Then subtract three times row 1 from row 2, and four times row 1 from row 3.

$$\mathbf{E} = \begin{bmatrix} 1 & 0 & 1 & 1 & 3 \\ 0 & 1 & -2 & 3 & 5 \\ 0 & 0 & 0 & 0 & 0 \end{bmatrix}; \quad x_1 = 3-s-t, \ x_2 = 5+2s-3t, \ x_3 = s, \ x_4 = t$$

31.
$$\begin{bmatrix} 1 & 2 & 3 \\ 0 & 4 & 5 \\ 0 & 0 & 6 \end{bmatrix} \xrightarrow{(1/6)R3} \begin{bmatrix} 1 & 2 & 3 \\ 0 & 4 & 5 \\ 0 & 0 & 1 \end{bmatrix} \xrightarrow{R2-5R3} \begin{bmatrix} 1 & 2 & 3 \\ 0 & 4 & 0 \\ 0 & 0 & 1 \end{bmatrix} \xrightarrow{(1/4)R2} \begin{bmatrix} 1 & 2 & 3 \\ 0 & 1 & 0 \\ 0 & 0 & 1 \end{bmatrix}$$

$$\xrightarrow{R1-2R2} \begin{bmatrix} 1 & 0 & 3 \\ 0 & 1 & 0 \\ 0 & 0 & 1 \end{bmatrix} \xrightarrow{R1-3R3} \begin{bmatrix} 1 & 0 & 0 \\ 0 & 1 & 0 \\ 0 & 0 & 1 \end{bmatrix}$$

33.   If the upper left element of a 2×2 reduced echelon matrix is 1, then the possibilities are $\begin{bmatrix} 1 & 0 \\ 0 & 1 \end{bmatrix}$ and $\begin{bmatrix} 1 & * \\ 0 & 0 \end{bmatrix}$, depending on whether there is a nonzero element in the second row. If the upper left element is zero — so both elements of the second row are also 0, then the possibilities are $\begin{bmatrix} 0 & 1 \\ 0 & 0 \end{bmatrix}$ and $\begin{bmatrix} 0 & 0 \\ 0 & 0 \end{bmatrix}$.

35.   **(a)**   If (x_0, y_0) is a solution, then it follows that

$$a(kx_0) + b(ky_0) = k(ax_0 + by_0) = k \cdot 0 = 0,$$
$$c(kx_0) + d(ky_0) = k(cx_0 + dy_0) = k \cdot 0 = 0$$

so (kx_0, ky_0) is also a solution.

   (b)   If (x_1, y_1) and (x_2, y_2) are solutions, then it follows that

$$a(x_1 + x_2) + b(y_1 + y_2) = (ax_1 + by_1) + (ax_2 + by_2) = 0 + 0 = 0,$$
$$c(x_1 + x_2) + d(y_1 + y_2) = (cx_1 + dy_1) + (cx_2 + dy_2) = 0 + 0 = 0$$

so $(x_1 + x_2, y_1 + y_2)$ is also a solution.

37.   If $ad - bc = 0$ then, much as in Problem 32, we see that the second row of the reduced echelon form of the coefficient matrix is allzero. Hence there is a free variable, and thus the given homogeneous system has a nontrivial solution involving a parameter t.

39.   It is given that the augmented coefficient matrix of the homogeneous 3×3 system has the form

$$\begin{bmatrix} a_1 & b_1 & c_1 & 0 \\ a_2 & b_2 & c_2 & 0 \\ pa_1 + qa_2 & pb_1 + qb_2 & pc_1 + qc_2 & 0 \end{bmatrix}.$$

Upon subtracting both p times row 1 and q times row 2 from row 3, we get the matrix

$$\begin{bmatrix} a_1 & b_1 & c_1 & 0 \\ a_2 & b_2 & c_2 & 0 \\ 0 & 0 & 0 & 0 \end{bmatrix}$$

corresponding to two homogeneous linear equations in three unknowns. Hence there is at least one free variable, and thus the system has a nontrivial family of solutions.

SECTION 3.4

MATRIX OPERATIONS

The objective of this section is simple to state. It is not merely knowledge of, but complete mastery of matrix addition and multiplication (particularly the latter). Matrix multiplication must be practiced until it is carried out not only accurately but quickly and with confidence — until you can hardly look at two matrices **A** and **B** without thinking of "pouring" the ith row of **A** down the jth column of **B**.

1. $\quad 3\begin{bmatrix} 3 & -5 \\ 2 & 7 \end{bmatrix} + 4\begin{bmatrix} -1 & 0 \\ 3 & -4 \end{bmatrix} = \begin{bmatrix} 9 & -15 \\ 6 & 21 \end{bmatrix} + \begin{bmatrix} -4 & 0 \\ 12 & -16 \end{bmatrix} = \begin{bmatrix} 5 & -15 \\ 18 & 5 \end{bmatrix}$

3. $\quad -2\begin{bmatrix} 5 & 0 \\ 0 & 7 \\ 3 & -1 \end{bmatrix} + 4\begin{bmatrix} -4 & 5 \\ 3 & 2 \\ 7 & 4 \end{bmatrix} = \begin{bmatrix} -10 & 0 \\ 0 & -14 \\ -6 & 2 \end{bmatrix} + \begin{bmatrix} -16 & 20 \\ 12 & 8 \\ 28 & 16 \end{bmatrix} = \begin{bmatrix} -26 & 20 \\ 12 & -6 \\ 22 & 18 \end{bmatrix}$

5. $\quad \begin{bmatrix} 2 & -1 \\ 3 & 2 \end{bmatrix}\begin{bmatrix} -4 & 2 \\ 1 & 3 \end{bmatrix} = \begin{bmatrix} -9 & 1 \\ -10 & 12 \end{bmatrix}; \qquad \begin{bmatrix} -4 & 2 \\ 1 & 3 \end{bmatrix}\begin{bmatrix} 2 & -1 \\ 3 & 2 \end{bmatrix} = \begin{bmatrix} -2 & 8 \\ 11 & 5 \end{bmatrix}$

7. $\quad \begin{bmatrix} 1 & 2 & 3 \end{bmatrix}\begin{bmatrix} 3 \\ 4 \\ 6 \end{bmatrix} = \begin{bmatrix} 26 \end{bmatrix}; \qquad \begin{bmatrix} 3 \\ 4 \\ 6 \end{bmatrix}\begin{bmatrix} 1 & 2 & 3 \end{bmatrix} = \begin{bmatrix} 3 & 6 & 9 \\ 4 & 8 & 12 \\ 5 & 10 & 15 \end{bmatrix}$

9. $\quad \begin{bmatrix} 0 & -2 \\ 3 & 1 \\ -4 & 5 \end{bmatrix}\begin{bmatrix} 3 \\ -2 \end{bmatrix} = \begin{bmatrix} 4 \\ 7 \\ -22 \end{bmatrix}$ but the product $\begin{bmatrix} 3 \\ -2 \end{bmatrix}\begin{bmatrix} 0 & -2 \\ 3 & 1 \\ -4 & 5 \end{bmatrix}$ is not defined.

11. $\quad \mathbf{AB} = \begin{bmatrix} 3 & -5 \end{bmatrix}\begin{bmatrix} 2 & 7 & 5 & 6 \\ -1 & 4 & 2 & 3 \end{bmatrix} = \begin{bmatrix} 11 & 1 & 5 & 3 \end{bmatrix}$ but the product **BA** is not defined.

13. $\quad \mathbf{A(BC)} = \begin{bmatrix} 3 & 1 \\ -1 & 4 \end{bmatrix}\left(\begin{bmatrix} 2 & 5 \\ -3 & 1 \end{bmatrix}\begin{bmatrix} 0 & 1 \\ 2 & 3 \end{bmatrix}\right) = \begin{bmatrix} 3 & 1 \\ -1 & 4 \end{bmatrix}\begin{bmatrix} 10 & 17 \\ 2 & 0 \end{bmatrix} = \begin{bmatrix} 32 & 51 \\ -2 & -17 \end{bmatrix}$

$\quad \mathbf{(AB)C} = \left(\begin{bmatrix} 3 & 1 \\ -1 & 4 \end{bmatrix}\begin{bmatrix} 2 & 5 \\ -3 & 1 \end{bmatrix}\right)\begin{bmatrix} 0 & 1 \\ 2 & 3 \end{bmatrix} = \begin{bmatrix} 3 & 16 \\ -14 & -1 \end{bmatrix}\begin{bmatrix} 0 & 1 \\ 2 & 3 \end{bmatrix} = \begin{bmatrix} 32 & 51 \\ -2 & -17 \end{bmatrix}$

15.
$$\mathbf{A}(\mathbf{BC}) = \begin{bmatrix} 3 \\ 2 \end{bmatrix} \left(\begin{bmatrix} 1 & -1 & 2 \end{bmatrix} \begin{bmatrix} 2 & 0 \\ 0 & 3 \\ 1 & 4 \end{bmatrix} \right) = \begin{bmatrix} 3 \\ 2 \end{bmatrix} \begin{bmatrix} 4 & 5 \end{bmatrix} = \begin{bmatrix} 12 & 15 \\ 8 & 10 \end{bmatrix}$$

$$(\mathbf{AB})\mathbf{C} = \left(\begin{bmatrix} 3 \\ 2 \end{bmatrix} \begin{bmatrix} 1 & -1 & 2 \end{bmatrix} \right) \begin{bmatrix} 2 & 0 \\ 0 & 3 \\ 1 & 4 \end{bmatrix} = \begin{bmatrix} 3 & -3 & 6 \\ 2 & -2 & 4 \end{bmatrix} \begin{bmatrix} 2 & 0 \\ 0 & 3 \\ 1 & 4 \end{bmatrix} = \begin{bmatrix} 12 & 15 \\ 8 & 10 \end{bmatrix}$$

Each of the homogeneous linear systems in Problems 17–22 is already in echelon form, so it remains only to write (by back substitution) the solution, first in parametric form and then in vector form.

17. $x_3 = s,$ $x_4 = t,$ $x_1 = 5s - 4t,$ $x_2 = -2s + 7t$

$\mathbf{x} = s(5,-2,1,0) + t(-4,7,0,1)$

19. $x_4 = s,$ $x_5 = t,$ $x_1 = -3s + t,$ $x_2 = 2s - 6t,$ $x_3 = -s + 8t$

$\mathbf{x} = s(-3,2,-1,1,0) + t(1,-6,8,0,1)$

21. $x_3 = r,$ $x_4 = s,$ $x_5 = t,$ $x_1 = r - 2s - 7t,$ $x_2 = -2r + 3s - 4t$

$\mathbf{x} = r(1,-2,1,0,0) + s(-2,3,0,1,0) + t(-7,-4,0,0,1)$

23.   The matrix equation $\begin{bmatrix} 2 & 1 \\ 3 & 2 \end{bmatrix} \begin{bmatrix} a & b \\ c & d \end{bmatrix} = \begin{bmatrix} 1 & 0 \\ 0 & 1 \end{bmatrix}$ entails the four scalar equations

$$\begin{aligned} 2a + c &= 1 & 2b + d &= 0 \\ 3a + 2c &= 0 & 3b + 2d &= 1 \end{aligned}$$

that we readily solve for $a = 2,$ $b = -1,$ $c = -3,$ $d = 2.$ Hence the apparent inverse matrix of $\mathbf{A},$ such that $\mathbf{AB} = \mathbf{I},$ is $\mathbf{B} = \begin{bmatrix} 2 & -1 \\ -3 & 2 \end{bmatrix}.$ Indeed, we find that $\mathbf{BA} = \mathbf{I}$ as well.

25.   The matrix equation $\begin{bmatrix} 5 & 7 \\ 2 & 3 \end{bmatrix} \begin{bmatrix} a & b \\ c & d \end{bmatrix} = \begin{bmatrix} 1 & 0 \\ 0 & 1 \end{bmatrix}$ entails the four scalar equations

$$\begin{aligned} 5a + 7c &= 1 & 5b + 7d &= 0 \\ 2a + 3c &= 0 & 2b + 3d &= 1 \end{aligned}$$

that we readily solve for $a = 3$, $b = -7$, $c = -2$, $d = 5$. Hence the apparent inverse matrix of $\mathbf{A}$, such that $\mathbf{AB} = \mathbf{I}$, is $\mathbf{B} = \begin{bmatrix} 3 & -7 \\ -2 & 5 \end{bmatrix}$. Indeed, we find that $\mathbf{BA} = \mathbf{I}$ as well.

27.
$$\begin{bmatrix} a_1 & 0 & 0 & \cdots & 0 \\ 0 & a_2 & 0 & \cdots & 0 \\ 0 & 0 & a_3 & \cdots & 0 \\ \vdots & \vdots & \vdots & \ddots & \vdots \\ 0 & 0 & 0 & \cdots & a_n \end{bmatrix} \begin{bmatrix} b_1 & 0 & 0 & \cdots & 0 \\ 0 & b_2 & 0 & \cdots & 0 \\ 0 & 0 & b_3 & \cdots & 0 \\ \vdots & \vdots & \vdots & \ddots & \vdots \\ 0 & 0 & 0 & \cdots & b_n \end{bmatrix} = \begin{bmatrix} a_1 b_1 & 0 & 0 & \cdots & 0 \\ 0 & a_2 b_2 & 0 & \cdots & 0 \\ 0 & 0 & a_3 b_3 & \cdots & 0 \\ \vdots & \vdots & \vdots & \ddots & \vdots \\ 0 & 0 & 0 & \cdots & a_n b_n \end{bmatrix}$$

Thus the product of two diagonal matrices of the same size is obtained simply by multiplying corresponding diagonal elements. Then the commutativity of scalar multiplication immediately implies that $\mathbf{AB} = \mathbf{BA}$ for diagonal matrices.

29.
$$(a+d)\mathbf{A} - (ad-bc)\mathbf{I} = (a+d)\begin{bmatrix} a & b \\ c & d \end{bmatrix} - (ad-bc)\begin{bmatrix} 1 & 0 \\ 0 & 1 \end{bmatrix}$$
$$= \begin{bmatrix} (a^2+ad)-(ad-bc) & ab+bd \\ ac+cd & (ad+d^2)-(ad-bc) \end{bmatrix} = \begin{bmatrix} a^2+bc & ab+bd \\ ac+cd & bc+d^2 \end{bmatrix}$$
$$= \begin{bmatrix} a & b \\ c & d \end{bmatrix}\begin{bmatrix} a & b \\ c & d \end{bmatrix} = \mathbf{A}^2$$

31. **(a)** If $\mathbf{A} = \begin{bmatrix} 2 & -1 \\ -4 & 3 \end{bmatrix}$ and $\mathbf{B} = \begin{bmatrix} 1 & 5 \\ 3 & 7 \end{bmatrix}$ then

$$(\mathbf{A}+\mathbf{B})(\mathbf{A}-\mathbf{B}) = \begin{bmatrix} 3 & 4 \\ -1 & 10 \end{bmatrix}\begin{bmatrix} 1 & -6 \\ -7 & -4 \end{bmatrix} = \begin{bmatrix} -25 & -34 \\ -71 & -34 \end{bmatrix}$$

but

$$\mathbf{A}^2 - \mathbf{B}^2 = \begin{bmatrix} 8 & -5 \\ -20 & 13 \end{bmatrix} - \begin{bmatrix} 16 & 40 \\ 24 & 64 \end{bmatrix} = \begin{bmatrix} -8 & -45 \\ -44 & -51 \end{bmatrix}.$$

(b) If $\mathbf{AB} = \mathbf{BA}$ then

$$(\mathbf{A}+\mathbf{B})(\mathbf{A}-\mathbf{B}) = \mathbf{A}(\mathbf{A}-\mathbf{B})+\mathbf{B}(\mathbf{A}-\mathbf{B})$$
$$= \mathbf{A}^2 - \mathbf{AB}+\mathbf{BA}-\mathbf{B}^2 = \mathbf{A}^2 - \mathbf{B}^2.$$

33. Four different 2×2 matrices $\mathbf{A}$ with $\mathbf{A}^2 = \mathbf{I}$ are

$$\begin{bmatrix} 1 & 0 \\ 0 & 1 \end{bmatrix}, \begin{bmatrix} -1 & 0 \\ 0 & 1 \end{bmatrix}, \begin{bmatrix} 1 & 0 \\ 0 & -1 \end{bmatrix}, \text{ and } \begin{bmatrix} -1 & 0 \\ 0 & -1 \end{bmatrix}.$$

35.   If $\mathbf{A} = \begin{bmatrix} 2 & -1 \\ 2 & -1 \end{bmatrix} \neq \mathbf{0}$ then $\mathbf{A}^2 = (1)\mathbf{A} - (0)\mathbf{I} = \mathbf{A}$.

37.   If $\mathbf{A} = \begin{bmatrix} 0 & 1 \\ -1 & 0 \end{bmatrix} \neq \mathbf{0}$ then $\mathbf{A}^2 = (0)\mathbf{A} - (1)\mathbf{I} = -\mathbf{I}$.

39.   If $\mathbf{A}\mathbf{x}_1 = \mathbf{A}\mathbf{x}_2 = \mathbf{0}$, then

$$\mathbf{A}\left(c_1\mathbf{x}_1 + c_2\mathbf{x}_2\right) = c_1\left(\mathbf{A}\mathbf{x}_1\right) + c_2\left(\mathbf{A}\mathbf{x}_2\right) = c_1\left(\mathbf{0}\right) + c_2\left(\mathbf{0}\right) = \mathbf{0}.$$

41.   If $\mathbf{AB} = \mathbf{BA}$ then

$$\begin{aligned}
\left(\mathbf{A} + \mathbf{B}\right)^3 &= \left(\mathbf{A} + \mathbf{B}\right)\left(\mathbf{A} + \mathbf{B}\right)^2 = \left(\mathbf{A} + \mathbf{B}\right)\left(\mathbf{A}^2 + 2\mathbf{AB} + \mathbf{B}^2\right) \\
&= \mathbf{A}\left(\mathbf{A}^2 + 2\mathbf{AB} + \mathbf{B}^2\right) + \mathbf{B}\left(\mathbf{A}^2 + 2\mathbf{AB} + \mathbf{B}^2\right) \\
&= \left(\mathbf{A}^3 + 2\mathbf{A}^2\mathbf{B} + \mathbf{AB}^2\right) + \left(\mathbf{A}^2\mathbf{B} + 2\mathbf{AB}^2 + \mathbf{B}^3\right) \\
&= \mathbf{A}^3 + 3\mathbf{A}^2\mathbf{B} + 3\mathbf{AB}^2 + \mathbf{B}^3.
\end{aligned}$$

To compute $\left(\mathbf{A} + \mathbf{B}\right)^4$, write $\left(\mathbf{A} + \mathbf{B}\right)^4 = \left(\mathbf{A} + \mathbf{B}\right)\left(\mathbf{A} + \mathbf{B}\right)^3$ and proceed similarly, substituting the expansion of $\left(\mathbf{A} + \mathbf{B}\right)^3$ just obtained.

43.   First, matrix multiplication gives $\mathbf{A}^2 = \begin{bmatrix} 2 & -1 & -1 \\ -1 & 2 & -1 \\ -1 & -1 & 2 \end{bmatrix} = \begin{bmatrix} 6 & -3 & -3 \\ -3 & 6 & -3 \\ -3 & -3 & 6 \end{bmatrix} = 3\mathbf{A}$. Then

$$\begin{aligned}
\mathbf{A}^3 &= \mathbf{A}^2 \cdot \mathbf{A} = 3\mathbf{A} \cdot \mathbf{A} = 3\mathbf{A}^2 = 3 \cdot 3\mathbf{A} = 9\mathbf{A}, \\
\mathbf{A}^4 &= \mathbf{A}^3 \cdot \mathbf{A} = 9\mathbf{A} \cdot \mathbf{A} = 9\mathbf{A}^2 = 9 \cdot 3\mathbf{A} = 27\mathbf{A},
\end{aligned}$$

and so forth.

SECTION 3.5

INVERSES OF MATRICES

The computational objective of this section is clearcut — to find the inverse of a given invertible matrix. From a more general viewpoint, Theorem 7 on the properties of nonsingular matrices summarizes most of the basic theory of this chapter.

In Problems 1–8 we first give the inverse matrix $\mathbf{A}^{-1}$ and then calculate the solution vector $\mathbf{x}$.

1. $\mathbf{A}^{-1} = \begin{bmatrix} 3 & -2 \\ -4 & 3 \end{bmatrix}$; $\mathbf{x} = \begin{bmatrix} 3 & -2 \\ -4 & 3 \end{bmatrix}\begin{bmatrix} 5 \\ 6 \end{bmatrix} = \begin{bmatrix} 3 \\ -2 \end{bmatrix}$

3. $\mathbf{A}^{-1} = \begin{bmatrix} 6 & -7 \\ -5 & 6 \end{bmatrix}$; $\mathbf{x} = \begin{bmatrix} 6 & -7 \\ -5 & 6 \end{bmatrix}\begin{bmatrix} 2 \\ -3 \end{bmatrix} = \begin{bmatrix} 33 \\ -28 \end{bmatrix}$

5. $\mathbf{A}^{-1} = \dfrac{1}{2}\begin{bmatrix} 4 & -2 \\ -5 & 3 \end{bmatrix}$; $\mathbf{x} = \dfrac{1}{2}\begin{bmatrix} 4 & -2 \\ -5 & 3 \end{bmatrix}\begin{bmatrix} 5 \\ 6 \end{bmatrix} = \dfrac{1}{2}\begin{bmatrix} 8 \\ -7 \end{bmatrix}$

7. $\mathbf{A}^{-1} = \dfrac{1}{4}\begin{bmatrix} 7 & -9 \\ -5 & 7 \end{bmatrix}$; $\mathbf{x} = \dfrac{1}{4}\begin{bmatrix} 7 & -9 \\ -5 & 7 \end{bmatrix}\begin{bmatrix} 3 \\ 2 \end{bmatrix} = \dfrac{1}{4}\begin{bmatrix} 3 \\ -1 \end{bmatrix}$

In Problems 9–22 we give at least the first few steps in the reduction of the augmented matrix whose right half is the identity matrix of appropriate size. We wind up with its echelon form, whose left half is an identity matrix and whose right half is the desired inverse matrix.

9. $\begin{bmatrix} 5 & 6 & 1 & 0 \\ 4 & 5 & 0 & 1 \end{bmatrix} \xrightarrow{R1-R2} \begin{bmatrix} 1 & 1 & 1 & -1 \\ 4 & 5 & 0 & 1 \end{bmatrix} \xrightarrow{R2-4R1} \begin{bmatrix} 1 & 1 & 1 & -1 \\ 0 & 1 & -4 & 5 \end{bmatrix}$

$\xrightarrow{R1-R2} \begin{bmatrix} 1 & 0 & 5 & -6 \\ 0 & 1 & -4 & 5 \end{bmatrix}$;   thus $\mathbf{A}^{-1} = \begin{bmatrix} 5 & -6 \\ -4 & 5 \end{bmatrix}$

11. $\begin{bmatrix} 1 & 5 & 1 & 1 & 0 & 0 \\ 2 & 5 & 0 & 0 & 1 & 0 \\ 2 & 7 & 1 & 0 & 0 & 1 \end{bmatrix} \xrightarrow{R2-2R1} \begin{bmatrix} 1 & 5 & 1 & 1 & 0 & 0 \\ 0 & -5 & -2 & -2 & 1 & 0 \\ 2 & 7 & 1 & 0 & 0 & 1 \end{bmatrix}$

$\xrightarrow{R3-2R1} \begin{bmatrix} 1 & 5 & 1 & 1 & 0 & 0 \\ 0 & -5 & -2 & -2 & 1 & 0 \\ 0 & -3 & -1 & -2 & 0 & 1 \end{bmatrix} \xrightarrow{R1+R3} \begin{bmatrix} 1 & 2 & 0 & -1 & 0 & 1 \\ 0 & -5 & -2 & -2 & 1 & 0 \\ 0 & -3 & -1 & -2 & 0 & 1 \end{bmatrix}$

$\xrightarrow{R2-2R3} \begin{bmatrix} 1 & 2 & 0 & -1 & 0 & 1 \\ 0 & 1 & 0 & 2 & 1 & -2 \\ 0 & -3 & -1 & -2 & 0 & 1 \end{bmatrix} \xrightarrow{R3+3R2} \begin{bmatrix} 1 & 2 & 0 & -1 & 0 & 1 \\ 0 & 1 & 0 & 2 & 1 & -2 \\ 0 & 0 & -1 & 4 & 3 & -5 \end{bmatrix}$

$\underset{\longrightarrow}{\overset{(-1)R3}{}} \cdots \underset{\longrightarrow}{\overset{R1-2R2}{}} \begin{bmatrix} 1 & 0 & 0 & -5 & -2 & 5 \\ 0 & 1 & 0 & 2 & 1 & -2 \\ 0 & 0 & 1 & -4 & -3 & 5 \end{bmatrix}$;   thus $\mathbf{A}^{-1} = \begin{bmatrix} -5 & -2 & 5 \\ 2 & 1 & -2 \\ -4 & -3 & 5 \end{bmatrix}$

13.
$$\begin{bmatrix} 2 & 7 & 3 & 1 & 0 & 0 \\ 1 & 3 & 2 & 0 & 1 & 0 \\ 3 & 7 & 9 & 0 & 0 & 1 \end{bmatrix} \xrightarrow[\to]{SWAP(R1,R2)} \begin{bmatrix} 1 & 3 & 2 & 0 & 1 & 0 \\ 2 & 7 & 3 & 1 & 0 & 0 \\ 3 & 7 & 9 & 0 & 0 & 1 \end{bmatrix}$$

$$\xrightarrow[\to]{R2-2R1} \begin{bmatrix} 1 & 3 & 2 & 0 & 1 & 0 \\ 0 & 1 & -1 & 1 & -2 & 0 \\ 3 & 7 & 9 & 0 & 0 & 1 \end{bmatrix} \xrightarrow[\to]{R3-3R1} \begin{bmatrix} 1 & 3 & 2 & 0 & 1 & 0 \\ 0 & 1 & -1 & 1 & -2 & 0 \\ 0 & -2 & 3 & 0 & -3 & 1 \end{bmatrix}$$

$$\xrightarrow[\to]{R3+2R2} \cdots \to \begin{bmatrix} 1 & 0 & 0 & -13 & 42 & -5 \\ 0 & 1 & 0 & 3 & -9 & 1 \\ 0 & 0 & 1 & 2 & -7 & 1 \end{bmatrix}; \quad \text{thus } \mathbf{A}^{-1} = \begin{bmatrix} -13 & 42 & -5 \\ 3 & -9 & 1 \\ 2 & -7 & 1 \end{bmatrix}$$

15.
$$\begin{bmatrix} 1 & 1 & 5 & 1 & 0 & 0 \\ 1 & 4 & 13 & 0 & 1 & 0 \\ 3 & 2 & 12 & 0 & 0 & 1 \end{bmatrix} \xrightarrow[\to]{R2-R1} \begin{bmatrix} 1 & 1 & 5 & 1 & 0 & 0 \\ 0 & 3 & 8 & -1 & 1 & 0 \\ 3 & 2 & 12 & 0 & 0 & 1 \end{bmatrix}$$

$$\xrightarrow[\to]{R3-3R1} \begin{bmatrix} 1 & 1 & 5 & 1 & 0 & 0 \\ 0 & 3 & 8 & -1 & 1 & 0 \\ 0 & -1 & -3 & -3 & 0 & 1 \end{bmatrix} \xrightarrow[\to]{R2+2R3} \begin{bmatrix} 1 & 1 & 5 & 1 & 0 & 0 \\ 0 & 1 & 2 & -7 & 1 & 2 \\ 0 & -1 & -3 & -3 & 0 & 1 \end{bmatrix}$$

$$\xrightarrow[\to]{R3+R2} \cdots \to \begin{bmatrix} 1 & 0 & 0 & -22 & 2 & 7 \\ 0 & 1 & 0 & -27 & 3 & 8 \\ 0 & 0 & 1 & 10 & -1 & -3 \end{bmatrix}; \quad \text{thus } \mathbf{A}^{-1} = \begin{bmatrix} -22 & 2 & 7 \\ -27 & 3 & 8 \\ 10 & -1 & -3 \end{bmatrix}$$

17.
$$\begin{bmatrix} 1 & -3 & 0 & 1 & 0 & 0 \\ -1 & 2 & -1 & 0 & 1 & 0 \\ 0 & -2 & 2 & 0 & 0 & 1 \end{bmatrix} \xrightarrow[\to]{R2+R1} \begin{bmatrix} 1 & -3 & 0 & 1 & 0 & 0 \\ 0 & -1 & -1 & 1 & 1 & 0 \\ 0 & -2 & 2 & 0 & 0 & 1 \end{bmatrix}$$

$$\xrightarrow[\to]{(-1)R2} \begin{bmatrix} 1 & -3 & 0 & 1 & 0 & 0 \\ 0 & 1 & 1 & -1 & -1 & 0 \\ 0 & -2 & 2 & 0 & 0 & 1 \end{bmatrix} \xrightarrow[\to]{R3+2R2} \begin{bmatrix} 1 & -3 & 0 & 1 & 0 & 0 \\ 0 & 1 & 1 & -1 & -1 & 0 \\ 0 & 0 & 4 & -2 & -2 & 1 \end{bmatrix}$$

$$\xrightarrow[\to]{(1/4)R3} \cdots \to \begin{bmatrix} 1 & 0 & 0 & -\frac{1}{2} & -\frac{3}{2} & -\frac{3}{4} \\ 0 & 1 & 0 & -\frac{1}{2} & -\frac{1}{2} & -\frac{1}{4} \\ 0 & 0 & 1 & -\frac{1}{2} & -\frac{1}{2} & \frac{1}{4} \end{bmatrix}; \quad \text{thus } \mathbf{A}^{-1} = \frac{1}{4}\begin{bmatrix} -2 & -6 & -3 \\ -2 & -2 & -1 \\ -2 & -2 & 1 \end{bmatrix}$$

19.
$$\begin{bmatrix} 1 & 4 & 3 & 1 & 0 & 0 \\ 1 & 4 & 5 & 0 & 1 & 0 \\ 2 & 5 & 1 & 0 & 0 & 1 \end{bmatrix} \xrightarrow[\;]{R2-R1} \begin{bmatrix} 1 & 4 & 3 & 1 & 0 & 0 \\ 0 & 0 & 2 & -1 & 1 & 0 \\ 2 & 5 & 1 & 0 & 0 & 1 \end{bmatrix}$$

$$\xrightarrow[\;]{SWAP(R2,R3)} \begin{bmatrix} 1 & 4 & 3 & 1 & 0 & 0 \\ 2 & 5 & 1 & 0 & 0 & 1 \\ 0 & 0 & 2 & -1 & 1 & 0 \end{bmatrix} \xrightarrow[\;]{R2-2R1} \begin{bmatrix} 1 & 4 & 3 & 1 & 0 & 0 \\ 0 & -3 & -5 & -2 & 0 & 1 \\ 0 & 0 & 2 & -1 & 1 & 0 \end{bmatrix}$$

$$\xrightarrow[\;]{(-1/3)R2} \begin{bmatrix} 1 & 4 & 3 & 1 & 0 & 0 \\ 0 & 1 & \frac{5}{3} & \frac{2}{3} & 0 & -\frac{1}{3} \\ 0 & 0 & 2 & -1 & 1 & 0 \end{bmatrix} \xrightarrow[\;]{(1/2)R3} \cdots \to \begin{bmatrix} 1 & 0 & 0 & -\frac{7}{2} & \frac{11}{6} & \frac{4}{3} \\ 0 & 1 & 0 & \frac{3}{2} & -\frac{5}{6} & -\frac{1}{3} \\ 0 & 0 & 1 & -\frac{1}{2} & \frac{1}{2} & 0 \end{bmatrix};$$

thus $\mathbf{A}^{-1} = \dfrac{1}{6}\begin{bmatrix} -21 & 11 & 8 \\ 9 & -5 & -2 \\ -3 & 3 & 0 \end{bmatrix}$

21.
$$\begin{bmatrix} 0 & 0 & 1 & 0 & 1 & 0 & 0 & 0 \\ 1 & 0 & 0 & 0 & 0 & 1 & 0 & 0 \\ 0 & 1 & 2 & 0 & 0 & 0 & 1 & 0 \\ 3 & 0 & 0 & 1 & 0 & 0 & 0 & 1 \end{bmatrix} \xrightarrow[\;]{SWAP(R1,R2)} \begin{bmatrix} 1 & 0 & 0 & 0 & 0 & 1 & 0 & 0 \\ 0 & 0 & 1 & 0 & 1 & 0 & 0 & 0 \\ 0 & 1 & 2 & 0 & 0 & 0 & 1 & 0 \\ 3 & 0 & 0 & 1 & 0 & 0 & 0 & 1 \end{bmatrix}$$

$$\xrightarrow[\;]{SWAP(R2,R3)} \begin{bmatrix} 1 & 0 & 0 & 0 & 0 & 1 & 0 & 0 \\ 0 & 1 & 2 & 0 & 0 & 0 & 1 & 0 \\ 0 & 0 & 1 & 0 & 1 & 0 & 0 & 0 \\ 3 & 0 & 0 & 1 & 0 & 0 & 0 & 1 \end{bmatrix} \xrightarrow[\;]{R4-3R1} \begin{bmatrix} 1 & 0 & 0 & 0 & 0 & 1 & 0 & 0 \\ 0 & 1 & 2 & 0 & 0 & 0 & 1 & 0 \\ 0 & 0 & 1 & 0 & 1 & 0 & 0 & 0 \\ 0 & 0 & 0 & 1 & 0 & -3 & 0 & 1 \end{bmatrix}$$

$$\xrightarrow[\;]{R2-2R3} \begin{bmatrix} 1 & 0 & 0 & 0 & 0 & 1 & 0 & 0 \\ 0 & 1 & 0 & 0 & -2 & 0 & 1 & 0 \\ 0 & 0 & 1 & 0 & 1 & 0 & 0 & 0 \\ 0 & 0 & 0 & 1 & 0 & -3 & 0 & 1 \end{bmatrix}; \quad \text{thus } \mathbf{A}^{-1} = \begin{bmatrix} 0 & 1 & 0 & 0 \\ -2 & 0 & 1 & 0 \\ 1 & 0 & 0 & 0 \\ 0 & -3 & 0 & 1 \end{bmatrix}$$

In Problems 23–28 we first give the inverse matrix $\mathbf{A}^{-1}$ and then calculate the solution matrix $\mathbf{X}$.

23. $\mathbf{A}^{-1} = \begin{bmatrix} 4 & -3 \\ -5 & 4 \end{bmatrix}$; $\quad \mathbf{X} = \begin{bmatrix} 4 & -3 \\ -5 & 4 \end{bmatrix}\begin{bmatrix} 1 & 3 & -5 \\ -1 & -2 & 5 \end{bmatrix} = \begin{bmatrix} 7 & 18 & -35 \\ -9 & -23 & 45 \end{bmatrix}$

25. $\mathbf{A}^{-1} = \begin{bmatrix} 11 & -9 & 4 \\ -2 & 2 & -1 \\ -2 & 1 & 0 \end{bmatrix}; \quad \mathbf{X} = \begin{bmatrix} 11 & -9 & 4 \\ -2 & 2 & -1 \\ -2 & 1 & 0 \end{bmatrix} \begin{bmatrix} 1 & 0 & 3 \\ 0 & 2 & 2 \\ -1 & 1 & 0 \end{bmatrix} = \begin{bmatrix} 7 & -14 & 15 \\ -1 & 3 & -2 \\ -2 & 2 & -4 \end{bmatrix}$

27. $\mathbf{A}^{-1} = \begin{bmatrix} 7 & -20 & 17 \\ 0 & -1 & 1 \\ -2 & 6 & -5 \end{bmatrix}; \quad \mathbf{X} = \begin{bmatrix} 7 & -20 & 17 \\ 0 & -1 & 1 \\ -2 & 6 & -5 \end{bmatrix} \begin{bmatrix} 0 & 0 & 1 & 1 \\ 0 & 1 & 0 & 1 \\ 1 & 0 & 1 & 0 \end{bmatrix} = \begin{bmatrix} 17 & -20 & 24 & -13 \\ 1 & -1 & 1 & -1 \\ -5 & 6 & -7 & 4 \end{bmatrix}$

29.   **(a)**   The fact that $\mathbf{A}^{-1}$ is the inverse of $\mathbf{A}$ means that $\mathbf{A}\mathbf{A}^{-1} = \mathbf{A}^{-1}\mathbf{A} = \mathbf{I}$. That is, that when $\mathbf{A}^{-1}$ is multiplied either on the right or on the left by $\mathbf{A}$, the result is the identity matrix $\mathbf{I}$. By the same token, this means that $\mathbf{A}$ is the inverse of $\mathbf{A}^{-1}$.

   (b) $\mathbf{A}^n (\mathbf{A}^{-1})^n = \mathbf{A}^{n-1} \cdot \mathbf{A}\mathbf{A}^{-1} \cdot (\mathbf{A}^{-1})^{n-1} = \mathbf{A}^{n-1} \cdot \mathbf{I} \cdot (\mathbf{A}^{-1})^{n-1} = \cdots = \mathbf{I}$. Similarly, $(\mathbf{A}^{-1})^n \mathbf{A}^n = \mathbf{I}$, so it follows that $(\mathbf{A}^{-1})^n$ is the inverse of $\mathbf{A}^n$.

31.   Let $p = -r > 0$, $q = -s > 0$, and $\mathbf{B} = \mathbf{A}^{-1}$. Then

$$\mathbf{A}^r \mathbf{A}^s = \mathbf{A}^{-p} \mathbf{A}^{-q} = (\mathbf{A}^{-1})^p (\mathbf{A}^{-1})^q$$
$$= \mathbf{B}^p \mathbf{B}^q = \mathbf{B}^{p+q} \quad \text{(because } p, q > 0\text{)}$$
$$= (\mathbf{A}^{-1})^{p+q} = \mathbf{A}^{-p-q} = \mathbf{A}^{r+s}$$

as desired, and $(\mathbf{A}^r)^s = (\mathbf{A}^{-p})^{-q} = (\mathbf{B}^p)^{-q} = \mathbf{B}^{-pq} = \mathbf{A}^{pq} = \mathbf{A}^{rs}$ similarly.

33.   In particular, $\mathbf{A}\mathbf{e}_j = \mathbf{e}_j$ where $\mathbf{e}_j$ denotes the jth column vector of the identity matrix $\mathbf{I}$. Hence it follows from Fact 2 that $\mathbf{A}\mathbf{I} = \mathbf{I}$, and therefore $\mathbf{A} = \mathbf{I}^{-1} = \mathbf{I}$.

35.   If the jth column of $\mathbf{A}$ is all zeros and $\mathbf{B}$ is any $n \times n$ matrix, then the jth column of $\mathbf{B}\mathbf{A}$ is all zeros, so $\mathbf{B}\mathbf{A} \neq \mathbf{I}$. Hence $\mathbf{A}$ has no inverse matrix. Similarly, if the ith row of $\mathbf{A}$ is all zeros, then so is the ith row of $\mathbf{A}\mathbf{B}$.

37.   Direct multiplication shows that $\mathbf{A}\mathbf{A}^{-1} = \mathbf{A}^{-1}\mathbf{A} = \mathbf{I}$.

39. $\mathbf{E}\mathbf{A} = \begin{bmatrix} 1 & 0 & 0 \\ 0 & 1 & 0 \\ 2 & 0 & 1 \end{bmatrix} \begin{bmatrix} a_{11} & a_{12} & a_{13} \\ a_{21} & a_{22} & a_{23} \\ a_{31} & a_{32} & a_{33} \end{bmatrix} = \begin{bmatrix} a_{11} & a_{12} & a_{13} \\ a_{21} & a_{22} & a_{23} \\ a_{31} + 2a_{11} & a_{32} + a_{12} & a_{33} + a_{13} \end{bmatrix}$

41.   This follows immediately from the fact that the ijth element of $\mathbf{A}\mathbf{B}$ is the product of the ith row of $\mathbf{A}$ and the jth column of $\mathbf{B}$.

43.   Let $\mathbf{E}_1, \mathbf{E}_2, \cdots, \mathbf{E}_k$ be the elementary matrices corresponding to the elementary row operations that reduce $\mathbf{A}$ to $\mathbf{B}$. Then Theorem 5 gives $\mathbf{B} = \mathbf{E}_k \mathbf{E}_{k-1} \cdots \mathbf{E}_2 \mathbf{E}_1 \mathbf{A} = \mathbf{GA}$ where $\mathbf{G} = \mathbf{E}_k \mathbf{E}_{k-1} \cdots \mathbf{E}_2 \mathbf{E}_1$.

45.   One can simply photocopy the portion of the proof of Theorem 7 that follows Equation (20). Starting only with the assumption that $\mathbf{A}$ and $\mathbf{B}$ are square matrices with $\mathbf{AB} = \mathbf{I}$, it is proved there that $\mathbf{A}$ and $\mathbf{B}$ are then invertible.

SECTION 3.6

DETERMINANTS

1.
$$\begin{vmatrix} 0 & 0 & 3 \\ 4 & 0 & 0 \\ 0 & 5 & 0 \end{vmatrix} = +(3)\begin{vmatrix} 4 & 0 \\ 0 & 5 \end{vmatrix} = 3 \cdot 4 \cdot 5 = 60$$

3.
$$\begin{vmatrix} 1 & 0 & 0 & 0 \\ 2 & 0 & 5 & 0 \\ 3 & 6 & 9 & 8 \\ 4 & 0 & 10 & 7 \end{vmatrix} = +(1)\begin{vmatrix} 0 & 5 & 0 \\ 6 & 9 & 8 \\ 0 & 10 & 7 \end{vmatrix} = -(5)\begin{vmatrix} 6 & 8 \\ 0 & 7 \end{vmatrix} = -5(42-0) = -210$$

5.
$$\begin{vmatrix} 0 & 0 & 1 & 0 & 0 \\ 2 & 0 & 0 & 0 & 0 \\ 0 & 0 & 0 & 3 & 0 \\ 0 & 0 & 0 & 0 & 4 \\ 0 & 5 & 0 & 0 & 0 \end{vmatrix} = +1\begin{vmatrix} 2 & 0 & 0 & 0 \\ 0 & 0 & 3 & 0 \\ 0 & 0 & 0 & 4 \\ 0 & 5 & 0 & 0 \end{vmatrix} = +2\begin{vmatrix} 0 & 3 & 0 \\ 0 & 0 & 4 \\ 5 & 0 & 0 \end{vmatrix} = 2(+5)\begin{vmatrix} 3 & 0 \\ 0 & 4 \end{vmatrix} = 2 \cdot 5 \cdot 3 \cdot 4 = 120$$

7.
$$\begin{vmatrix} 1 & 1 & 1 \\ 2 & 2 & 2 \\ 3 & 3 & 3 \end{vmatrix} \overset{R2-2R1}{=} \begin{vmatrix} 1 & 1 & 1 \\ 0 & 0 & 0 \\ 3 & 3 & 3 \end{vmatrix} = 0$$

9.
$$\begin{vmatrix} 3 & -2 & 5 \\ 0 & 5 & 17 \\ 6 & -4 & 12 \end{vmatrix} \overset{R3-2R1}{=} \begin{vmatrix} 3 & -2 & 5 \\ 0 & 5 & 17 \\ 0 & 0 & 2 \end{vmatrix} = +2\begin{vmatrix} 3 & 5 \\ 0 & 2 \end{vmatrix} = 5(6-0) = 30$$

11. $\begin{vmatrix} 1 & 2 & 3 & 4 \\ 0 & 5 & 6 & 7 \\ 0 & 0 & 8 & 9 \\ 2 & 4 & 6 & 9 \end{vmatrix} \overset{R4-2R1}{=} \begin{vmatrix} 1 & 2 & 3 & 4 \\ 0 & 5 & 6 & 7 \\ 0 & 0 & 8 & 9 \\ 0 & 0 & 0 & 1 \end{vmatrix} = +1 \begin{vmatrix} 5 & 6 & 7 \\ 0 & 8 & 9 \\ 0 & 0 & 1 \end{vmatrix} = +5 \begin{vmatrix} 8 & 9 \\ 0 & 1 \end{vmatrix} = 5 \cdot 8 = 40$

13. $\begin{vmatrix} -4 & 4 & -1 \\ -1 & -2 & 2 \\ 1 & 4 & 3 \end{vmatrix} \overset{R2+R3}{=} \begin{vmatrix} -4 & 4 & -1 \\ 0 & 2 & 5 \\ 1 & 4 & 3 \end{vmatrix} \overset{R1+4R3}{=} \begin{vmatrix} 0 & 20 & 11 \\ 0 & 2 & 5 \\ 1 & 4 & 3 \end{vmatrix} = +1 \begin{vmatrix} 20 & 11 \\ 2 & 5 \end{vmatrix} = 100 - 22 = 78$

15. $\begin{vmatrix} -2 & 5 & 4 \\ 5 & 3 & 1 \\ 1 & 4 & 5 \end{vmatrix} \overset{R1+2R3}{=} \begin{vmatrix} 0 & 13 & 14 \\ 5 & 3 & 1 \\ 1 & 4 & 5 \end{vmatrix} \overset{R2-5R3}{=} \begin{vmatrix} 0 & 13 & 14 \\ 0 & -17 & -24 \\ 1 & 4 & 5 \end{vmatrix} = +1 \begin{vmatrix} 13 & 14 \\ -17 & -24 \end{vmatrix} = -74$

17. $\begin{vmatrix} 2 & 3 & 3 & 1 \\ 0 & 4 & 3 & -3 \\ 2 & -1 & -1 & -3 \\ 0 & -4 & -3 & 2 \end{vmatrix} \overset{R3-R1}{=} \begin{vmatrix} 2 & 3 & 3 & 1 \\ 0 & 4 & 3 & -3 \\ 0 & -4 & -4 & -4 \\ 0 & -4 & -3 & 2 \end{vmatrix} = 2 \begin{vmatrix} 4 & 3 & -3 \\ -4 & -4 & -4 \\ -4 & -3 & 2 \end{vmatrix} \overset{\substack{R2+R1 \\ R3+R1}}{=} 2 \begin{vmatrix} 4 & 3 & -3 \\ 0 & -1 & -7 \\ 0 & 0 & -1 \end{vmatrix} = 8$

19. $\begin{vmatrix} 1 & 0 & 0 & 3 \\ 0 & 1 & -2 & 0 \\ -2 & 3 & -2 & 3 \\ 0 & -3 & 3 & 3 \end{vmatrix} \overset{R3+2R1}{=} \begin{vmatrix} 1 & 0 & 0 & 3 \\ 0 & 1 & -2 & 0 \\ 0 & 3 & -2 & 9 \\ 0 & -3 & 3 & 3 \end{vmatrix} = 1 \begin{vmatrix} 1 & -2 & 0 \\ 3 & -2 & 9 \\ -3 & 3 & 3 \end{vmatrix} \overset{C2+2C1}{=} \begin{vmatrix} 1 & 0 & 0 \\ 3 & 4 & 9 \\ -3 & -3 & 3 \end{vmatrix} = 39$

21. $\Delta = \begin{vmatrix} 3 & 4 \\ 5 & 7 \end{vmatrix} = 1; \quad x = \frac{1}{\Delta} \begin{vmatrix} 2 & 4 \\ 1 & 7 \end{vmatrix} = 10, \quad y = \frac{1}{\Delta} \begin{vmatrix} 3 & 2 \\ 5 & 1 \end{vmatrix} = -7$

23. $\Delta = \begin{vmatrix} 17 & 7 \\ 12 & 5 \end{vmatrix} = 1; \quad x = \frac{1}{\Delta} \begin{vmatrix} 6 & 7 \\ 4 & 5 \end{vmatrix} = 2, \quad y = \frac{1}{\Delta} \begin{vmatrix} 17 & 6 \\ 12 & 4 \end{vmatrix} = -4$

25. $\Delta = \begin{vmatrix} 5 & 6 \\ 3 & 4 \end{vmatrix} = 2; \quad x = \frac{1}{\Delta} \begin{vmatrix} 12 & 6 \\ 6 & 4 \end{vmatrix} = 6, \quad y = \frac{1}{\Delta} \begin{vmatrix} 5 & 12 \\ 3 & 6 \end{vmatrix} = -3$

27. $\Delta = \begin{vmatrix} 5 & 2 & -2 \\ 1 & 5 & -3 \\ 5 & -3 & 5 \end{vmatrix} = 96; \quad x_1 = \frac{1}{\Delta} \begin{vmatrix} 1 & 2 & -2 \\ -2 & 5 & -3 \\ 2 & -3 & 5 \end{vmatrix} = \frac{1}{3},$

$$x_2 = \frac{1}{\Delta}\begin{vmatrix} 5 & 1 & -2 \\ 1 & -2 & -3 \\ 5 & 2 & 5 \end{vmatrix} = -\frac{2}{3}, \qquad x_3 = \frac{1}{\Delta}\begin{vmatrix} 5 & 2 & 1 \\ 1 & 5 & -2 \\ 5 & -3 & 2 \end{vmatrix} = -\frac{1}{3}$$

29. $\Delta = \begin{vmatrix} 3 & -1 & -5 \\ 4 & -4 & -3 \\ 1 & 0 & -5 \end{vmatrix} = 23; \qquad x_1 = \frac{1}{\Delta}\begin{vmatrix} 3 & -1 & -5 \\ -4 & -4 & -3 \\ 2 & 0 & -5 \end{vmatrix} = 2,$

$$x_2 = \frac{1}{\Delta}\begin{vmatrix} 3 & 3 & -5 \\ 4 & -4 & -3 \\ 1 & 2 & -5 \end{vmatrix} = 3, \qquad x_3 = \frac{1}{\Delta}\begin{vmatrix} 3 & -1 & 3 \\ 4 & -4 & -4 \\ 1 & 0 & 2 \end{vmatrix} = 0$$

31. $\Delta = \begin{vmatrix} 2 & 0 & -5 \\ 4 & -5 & 3 \\ -2 & 1 & 1 \end{vmatrix} = 14; \qquad x_1 = \frac{1}{\Delta}\begin{vmatrix} -3 & 0 & -5 \\ 3 & -5 & 3 \\ 1 & 1 & 1 \end{vmatrix} = -\frac{8}{7},$

$$x_2 = \frac{1}{\Delta}\begin{vmatrix} 2 & -3 & -5 \\ 4 & 3 & 3 \\ -2 & 1 & 1 \end{vmatrix} = -\frac{10}{7}, \qquad x_3 = \frac{1}{\Delta}\begin{vmatrix} 2 & 0 & -3 \\ 4 & -5 & 3 \\ -2 & 1 & 1 \end{vmatrix} = \frac{1}{7}$$

33. $\det \mathbf{A} = -4, \qquad \mathbf{A}^{-1} = \frac{1}{4}\begin{bmatrix} 4 & 4 & 4 \\ 16 & 15 & 13 \\ 28 & 25 & 23 \end{bmatrix}$

35. $\det \mathbf{A} = 35, \qquad \mathbf{A}^{-1} = \frac{1}{35}\begin{bmatrix} -15 & 25 & -26 \\ 10 & -5 & 8 \\ 15 & -25 & 19 \end{bmatrix}$

37. $\det \mathbf{A} = 29, \qquad \mathbf{A}^{-1} = \frac{1}{29}\begin{bmatrix} 11 & -14 & -15 \\ -17 & 19 & 10 \\ 18 & -15 & -14 \end{bmatrix}$

39. $\det \mathbf{A} = 37, \qquad \mathbf{A}^{-1} = \frac{1}{37}\begin{bmatrix} -21 & -1 & -13 \\ 4 & 9 & 6 \\ -6 & 5 & -9 \end{bmatrix}$

41.   If $\mathbf{A} = \begin{bmatrix} \mathbf{a}_1 \\ \mathbf{a}_2 \end{bmatrix}$ and $\mathbf{B} = \begin{bmatrix} \mathbf{b}_1 & \mathbf{b}_2 \end{bmatrix}$ in terms of the two row vectors of $\mathbf{A}$ and the two column

vectors of $\mathbf{B}$, then $\mathbf{AB} = \begin{bmatrix} \mathbf{a}_1\mathbf{b}_1 & \mathbf{a}_1\mathbf{b}_2 \\ \mathbf{a}_2\mathbf{b}_1 & \mathbf{a}_2\mathbf{b}_2 \end{bmatrix}$, so

$$\left(\mathbf{AB}\right)^T = \begin{bmatrix} \mathbf{a}_1\mathbf{b}_1 & \mathbf{a}_2\mathbf{b}_1 \\ \mathbf{a}_1\mathbf{b}_2 & \mathbf{a}_2\mathbf{b}_2 \end{bmatrix} = \begin{bmatrix} \mathbf{b}_1^T \\ \mathbf{b}_2^T \end{bmatrix}\begin{bmatrix} \mathbf{a}_1^T & \mathbf{a}_1^T \end{bmatrix} = \mathbf{B}^T\mathbf{A}^T,$$

because the rows of $\mathbf{A}$ are the columns of $\mathbf{A}^T$ and the columns of $\mathbf{B}$ are the rows of $\mathbf{B}^T$.

43.   We expand the left-hand determinant along its first column:

$$\begin{vmatrix} ka_{11} & a_{12} & a_{13} \\ ka_{21} & a_{22} & a_{23} \\ ka_{31} & a_{32} & a_{33} \end{vmatrix}$$
$$= ka_{11}\left(a_{12}a_{23} - a_{22}a_{13}\right) - ka_{21}\left(a_{12}a_{33} - a_{32}a_{13}\right) + ka_{31}\left(a_{12}a_{23} - a_{22}a_{13}\right)$$
$$= k\left[a_{11}\left(a_{12}a_{23} - a_{22}a_{13}\right) - a_{21}\left(a_{12}a_{33} - a_{32}a_{13}\right) + a_{31}\left(a_{12}a_{23} - a_{22}a_{13}\right)\right]$$
$$= k\begin{vmatrix} a_{11} & a_{12} & a_{13} \\ a_{21} & a_{22} & a_{23} \\ a_{31} & a_{32} & a_{33} \end{vmatrix}$$

45.   We expand the left-hand determinant along its third column:

$$\begin{vmatrix} a_1 & b_1 & c_1 + d_1 \\ a_2 & b_2 & c_2 + d_2 \\ a_3 & b_3 & c_3 + d_3 \end{vmatrix}$$
$$= \left(c_1 + d_1\right)\left(a_2b_3 - a_3b_2\right) - \left(c_2 + d_2\right)\left(a_1b_3 - a_3b_1\right) + \left(c_3 + d_3\right)\left(a_1b_2 - a_2b_1\right)$$
$$= c_1\left(a_2b_3 - a_3b_2\right) - c_2\left(a_1b_3 - a_3b_1\right) + c_3\left(a_1b_2 - a_2b_1\right)$$
$$\quad + d_1\left(a_2b_3 - a_3b_2\right) - d_2\left(a_1b_3 - a_3b_1\right) + d_3\left(a_1b_2 - a_2b_1\right)$$
$$= \begin{vmatrix} a_1 & b_1 & c_1 \\ a_2 & b_2 & c_2 \\ a_3 & b_3 & c_3 \end{vmatrix} + \begin{vmatrix} a_1 & b_1 & d_1 \\ a_2 & b_2 & d_2 \\ a_3 & b_3 & d_3 \end{vmatrix}$$

47.   We illustrate these properties with 2×2 matrices $\mathbf{A} = \begin{bmatrix} a_{ij} \end{bmatrix}$ and $\mathbf{B} = \begin{bmatrix} b_{ij} \end{bmatrix}$.

(a) $\left(\mathbf{A}^T \right)^T = \begin{bmatrix} a_{11} & a_{21} \\ a_{12} & a_{22} \end{bmatrix}^T = \begin{bmatrix} a_{11} & a_{12} \\ a_{22} & a_{22} \end{bmatrix} = \mathbf{A}$

(b) $\left(c\mathbf{A} \right)^T = \begin{bmatrix} ca_{11} & ca_{12} \\ ca_{21} & ca_{22} \end{bmatrix}^T = \begin{bmatrix} ca_{11} & ca_{21} \\ ca_{12} & ca_{22} \end{bmatrix} = c \begin{bmatrix} a_{11} & a_{21} \\ a_{12} & a_{22} \end{bmatrix} = c\mathbf{A}^T$

(c) $\left(\mathbf{A} + \mathbf{B} \right)^T = \begin{bmatrix} a_{11} + b_{11} & a_{12} + b_{12} \\ a_{21} + b_{21} & a_{22} + b_{22} \end{bmatrix}^T = \begin{bmatrix} a_{11} + b_{11} & a_{21} + b_{21} \\ a_{12} + b_{12} & a_{22} + b_{22} \end{bmatrix}$

$= \begin{bmatrix} a_{11} & a_{21} \\ a_{12} & a_{22} \end{bmatrix} + \begin{bmatrix} b_{11} & b_{21} \\ b_{12} & b_{22} \end{bmatrix} = \mathbf{A}^T + \mathbf{B}^T$

49.   If we write $\mathbf{A} = \begin{vmatrix} a_1 & b_1 & c_1 \\ a_2 & b_2 & c_2 \\ a_3 & b_3 & c_3 \end{vmatrix}$ and $\mathbf{A}^T = \begin{vmatrix} a_1 & a_2 & a_3 \\ b_1 & b_2 & b_3 \\ c_1 & c_2 & c_3 \end{vmatrix}$, then expansion of $|\mathbf{A}|$ along its

first row and of $\left| \mathbf{A}^T \right|$ along its first column both give the result

$a_1 \left(b_2 c_3 - b_3 c_2 \right) + b_1 \left(a_2 c_3 - a_3 c_2 \right) + c_1 \left(a_2 b_3 - a_3 b_2 \right).$

51.   If $\mathbf{A}^n = \mathbf{0}$ then $|\mathbf{A}|^n = \mathbf{0}$, so it follows immediately that $|\mathbf{A}| = \mathbf{0}$.

53.   If $\mathbf{A} = \mathbf{P}^{-1} \mathbf{B} \mathbf{P}$ then $|\mathbf{A}| = \left| \mathbf{P}^{-1} \mathbf{B} \mathbf{P} \right| = \left| \mathbf{P}^{-1} \right| |\mathbf{B}| |\mathbf{P}| = \left| \mathbf{P} \right|^{-1} |\mathbf{B}| |\mathbf{P}| = |\mathbf{B}|.$

55.   If either $\mathbf{AB} = \mathbf{I}$ or $\mathbf{BA} = \mathbf{I}$ is given, then it follows from Problem 54 that $\mathbf{A}$ and $\mathbf{B}$ are both invertible because their product (one way or the other) is invertible. Hence $\mathbf{A}^{-1}$ exists. So if (for instance) it is $\mathbf{AB} = \mathbf{I}$ that is given, then multiplication by $\mathbf{A}^{-1}$ on the right yields $\mathbf{B} = \mathbf{A}^{-1}$.

57.   If $\mathbf{A} = \begin{bmatrix} a & d & f \\ 0 & b & e \\ 0 & 0 & b \end{bmatrix}$ then $\mathbf{A}^{-1} = \dfrac{1}{abc} \begin{bmatrix} bc & -cd & de - bf \\ 0 & ac & -ae \\ 0 & 0 & ab \end{bmatrix}.$

59.   These are almost immediate computations.

61. Subtraction of the first row from both the second and the third row gives

$$\begin{vmatrix} 1 & a & a^2 \\ 1 & b & b^2 \\ 1 & c & c^2 \end{vmatrix} = \begin{vmatrix} 1 & a & a^2 \\ 0 & b-a & b^2-a^2 \\ 0 & c-a & c^2-a^2 \end{vmatrix} = (b-a)(c^2-a^2)-(c-a)(b^2-a^2)$$

$$= (b-a)(c-a)(c+a)-(c-a)(b-a)(b+a)$$
$$= (b-a)(c-a)[(c+a)-(b+a)] = (b-a)(c-a)(c-b).$$

62. Expansion of the 4×4 determinant defining $P(y)$ along its 4th row yields

$$P(y) = y^3 \begin{vmatrix} 1 & x_1 & x_1^2 \\ 1 & x_2 & x_2^2 \\ 1 & x_3 & x_3^2 \end{vmatrix} + \cdots = y^3 V(x_1,x_2,x_3) + \text{lower-degree terms in } y.$$

Because it is clear from the determinant definition of $P(y)$ that
$P(x_1) = P(x_2) = P(x_3) = 0$, the three roots of the cubic polynomial $P(y)$ are x_1, x_2, x_3.
The factor theorem therefore says that $P(y) = k(y-x_1)(y-x_2)(y-x_3)$ for some constant k, and the calculation above implies that

$$k = V(x_1,x_2,x_3) = (x_3-x_1)(x_3-x_2)(x_2-x_1).$$

Finally we see that

$$V(x_1,x_2,x_3,x_4) = P(x_4) = V(x_1,x_2,x_3)\cdot(x_4-x_1)(x_4-x_2)(x_4-x_1)$$
$$= (x_4-x_1)(x_4-x_2)(x_4-x_1)(x_3-x_1)(x_3-x_2)(x_2-x_1),$$

which is the desired formula for $V(x_1,x_2,x_3,x_4)$.

63. The same argument as in Problem 62 yields

$$P(y) = V(x_1,x_2,\cdots,x_{n-1})\cdot(y-x_1)(y-x_2)\cdots\cdots(y-x_{n-1}).$$

Therefore

$$V(x_1, x_2, \cdots, x_n) = (x_n - x_1)(x_n - x_2) \cdots \cdots (x_n - x_{n-1}) V(x_1, x_2, \cdots, x_{n-1})$$

$$= (x_n - x_1)(x_n - x_2) \cdots \cdots (x_n - x_{n-1}) \prod_{i > j}^{n-1} (x_i - x_j)$$

$$= \prod_{i > j}^{n} (x_i - x_j).$$

SECTION 3.7

LINEAR EQUATIONS AND CURVE FITTING

In Problems 1–10 we first set up the linear system in the coefficients $a, b, \ldots$ that we get by substituting each given point (x_i, y_i) into the desired interpolating polynomial equation $y = a + bx + \cdots$. Then we give the polynomial that results from solution of this linear system.

1. $y(x) = a + bx$

$$\begin{bmatrix} 1 & 1 \\ 1 & 3 \end{bmatrix} \begin{bmatrix} a \\ b \end{bmatrix} = \begin{bmatrix} 1 \\ 7 \end{bmatrix} \quad \Rightarrow \quad a = -2, \ b = 3 \quad \text{so} \quad y(x) = -2 + 3x$$

3. $y(x) = a + bx + cx^2$

$$\begin{bmatrix} 1 & 0 & 0 \\ 1 & 1 & 1 \\ 1 & 2 & 4 \end{bmatrix} \begin{bmatrix} a \\ b \\ c \end{bmatrix} = \begin{bmatrix} 3 \\ 1 \\ -5 \end{bmatrix} \quad \Rightarrow \quad a = 3, \ b = 0, \ c = -2 \quad \text{so} \quad y(x) = 3 - 2x^2$$

5. $y(x) = a + bx + cx^2$

$$\begin{bmatrix} 1 & 1 & 1 \\ 1 & 2 & 4 \\ 1 & 3 & 9 \end{bmatrix} \begin{bmatrix} a \\ b \\ c \end{bmatrix} = \begin{bmatrix} 3 \\ 3 \\ 5 \end{bmatrix} \quad \Rightarrow \quad a = 5, \ b = -3, \ c = 1 \quad \text{so} \quad y(x) = 5 - 3x + x^2$$

7. $y(x) = a + bx + cx^2 + dx^3$

$$\begin{bmatrix} 1 & -1 & 1 & -1 \\ 1 & 0 & 0 & 0 \\ 1 & 1 & 1 & 1 \\ 1 & 2 & 4 & 8 \end{bmatrix} \begin{bmatrix} a \\ b \\ c \\ d \end{bmatrix} = \begin{bmatrix} 1 \\ 0 \\ 1 \\ -4 \end{bmatrix}$$

$$\Rightarrow \quad a = 0, \; b = \frac{4}{3}, \; c = 1, \; d = -\frac{4}{3} \quad \text{so} \quad y(x) = \frac{1}{3}\left(4x + 3x^2 - 4x^3\right)$$

9. $\quad y(x) = a + bx + cx^2 + dx^3$

$$\begin{bmatrix} 1 & -2 & 4 & -8 \\ 1 & -1 & 1 & -1 \\ 1 & 1 & 1 & 1 \\ 1 & 2 & 4 & 8 \end{bmatrix} \begin{bmatrix} a \\ b \\ c \\ d \end{bmatrix} = \begin{bmatrix} -2 \\ 2 \\ 10 \\ 26 \end{bmatrix}$$

$$\Rightarrow \quad a = 4, \; b = 3, \; c = 2, \; d = 1 \quad \text{so} \quad y(x) = 4 + 3x + 2x^2 + x^3$$

In Problems 11–14 we first set up the linear system in the coefficients A, B, C that we get by substituting each given point (x_i, y_i) into the circle equation $Ax + By + C = -x^2 - y^2$ (see Eq. (9) in the text). Then we give the circle that results from solution of this linear system.

11. $\quad Ax + By + C = -x^2 - y^2$

$$\begin{bmatrix} -1 & -1 & 1 \\ 6 & 6 & 1 \\ 7 & 5 & 1 \end{bmatrix} \begin{bmatrix} A \\ B \\ C \end{bmatrix} = \begin{bmatrix} -2 \\ -72 \\ -74 \end{bmatrix} \quad \Rightarrow \quad A = -6, \; B = -4, \; C = -12$$

$$x^2 + y^2 - 6x - 4y - 12 = 0$$

$$(x-3)^2 + (y-2)^2 = 25 \qquad \text{center } (3, 2) \text{ and radius } 5$$

13. $\quad Ax + By + C = -x^2 - y^2$

$$\begin{bmatrix} 1 & 0 & 1 \\ 0 & -5 & 1 \\ -5 & -4 & 1 \end{bmatrix} \begin{bmatrix} A \\ B \\ C \end{bmatrix} = \begin{bmatrix} -1 \\ -25 \\ -41 \end{bmatrix} \quad \Rightarrow \quad A = 4, \; B = 4, \; C = -5$$

$$x^2 + y^2 + 4x + 4y - 5 = 0$$

$$(x+2)^2 + (y+2)^2 = 13 \qquad \text{center } (-3, -2) \text{ and radius } \sqrt{13}$$

In Problems 15–18 we first set up the linear system in the coefficients A, B, C that we get by substituting each given point (x_i, y_i) into the central conic equation $Ax^2 + Bxy + Cy^2 = 1$ (see Eq. (10) in the text). Then we give the equation that results from solution of this linear system.

15. $Ax^2 + Bxy + Cy^2 = 1$

$$\begin{bmatrix} 0 & 0 & 25 \\ 25 & 0 & 0 \\ 25 & 25 & 25 \end{bmatrix} \begin{bmatrix} A \\ B \\ C \end{bmatrix} = \begin{bmatrix} 1 \\ 1 \\ 1 \end{bmatrix} \quad \Rightarrow \quad A = \frac{1}{25}, \; B = -\frac{1}{25}, \; C = \frac{1}{25}$$

$$x^2 - xy + y^2 = 25$$

17. $Ax^2 + Bxy + Cy^2 = 1$

$$\begin{bmatrix} 0 & 0 & 1 \\ 1 & 0 & 0 \\ 100 & 100 & 100 \end{bmatrix} \begin{bmatrix} A \\ B \\ C \end{bmatrix} = \begin{bmatrix} 1 \\ 1 \\ 1 \end{bmatrix} \quad \Rightarrow \quad A = 1, \; B = -\frac{199}{100}, \; C = 1$$

$$100x^2 - 199xy + 100y^2 = 100$$

19. We substitute each of the two given points into the equation $y = A + \dfrac{B}{x}$.

$$\begin{bmatrix} 1 & 1 \\ 1 & \frac{1}{2} \end{bmatrix} \begin{bmatrix} A \\ B \end{bmatrix} = \begin{bmatrix} 5 \\ 4 \end{bmatrix} \quad \Rightarrow \quad A = 3, \; B = 2 \text{ so } y = 3 + \frac{2}{x}$$

In Problems 21 and 22 we fit the sphere equation $(x-h)^2 + (y-k)^2 + (z-l)^2 = r^2$ in the expanded form $Ax + By + Cz + D = -x^2 - y^2 - z^2$ that is analogous to Eq. (9) in the text (for a circle).

21. $Ax + By + Cz + D = -x^2 - y^2 - z^2$

$$\begin{bmatrix} 4 & 6 & 15 & 1 \\ 13 & 5 & 7 & 1 \\ 5 & 14 & 6 & 1 \\ 5 & 5 & -9 & 1 \end{bmatrix} \begin{bmatrix} A \\ B \\ C \\ D \end{bmatrix} = \begin{bmatrix} -277 \\ -243 \\ -257 \\ -131 \end{bmatrix} \quad \Rightarrow \quad A = -2, \; B = -4, \; C = -6, \; D = -155$$

$$x^2 + y^2 + z^2 - 2x - 4y - 6z - 155 = 0$$
$$(x-1)^2 + (y-2)^2 + (z-3)^2 = 169 \qquad \text{Center } (1, 2, 3) \text{ and radius } 13$$

In Problems 23–26 we first take $t = 0$ in 1970 to fit a quadratic polynomial $P(t) = a + bt + ct^2$. Then we write the quadratic polynomial $Q(T) = P(T - 1970)$ that expresses the predicted population in terms of the actual calendar year T.

23. $P(t) = a + bt + ct^2$

$$\begin{bmatrix} 1 & 0 & 0 \\ 1 & 10 & 100 \\ 1 & 20 & 400 \end{bmatrix} \begin{bmatrix} a \\ b \\ c \end{bmatrix} = \begin{bmatrix} 49.061 \\ 49.137 \\ 50.809 \end{bmatrix}$$

$P(t) = 49.061 - 0.0722t + 0.00798t^2$

$Q(T) = 31160.9 - 31.5134T + 0.00798T^2$

25. $P(t) = a + bt + ct^2$

$$\begin{bmatrix} 1 & 0 & 0 \\ 1 & 10 & 100 \\ 1 & 20 & 400 \end{bmatrix} \begin{bmatrix} a \\ b \\ c \end{bmatrix} = \begin{bmatrix} 62.813 \\ 75.367 \\ 85.446 \end{bmatrix}$$

$P(t) = 62.813 + 1.37915t - 0.012375t^2$

$Q(T) = -50680.3 + 50.1367T - 0.012375T^2$

In Problems 27–30 we first take $t = 0$ in 1960 to fit a cubic polynomial $P(t) = a + bt + ct^2 + dt^3$. Then we write the cubic polynomial $Q(T) = P(T - 1960)$ that expresses the predicted population in terms of the actual calendar year T.

27. $P(t) = a + bt + ct^2 + dt^3$

$$\begin{bmatrix} 1 & 0 & 0 & 0 \\ 1 & 10 & 100 & 1000 \\ 1 & 20 & 400 & 8000 \\ 1 & 30 & 900 & 27000 \end{bmatrix} \begin{bmatrix} a \\ b \\ c \\ d \end{bmatrix} = \begin{bmatrix} 44.678 \\ 49.061 \\ 49.137 \\ 50.809 \end{bmatrix}$$

$P(t) = 44.678 + 0.850417t - 0.05105t^2 + 0.000983833t^3$

$Q(T) = -7.60554 \times 10^6 + 11539.4T - 5.83599T^2 + 0.000983833T^3$

29. $P(t) = a + bt + ct^2 + dt^3$

$$\begin{bmatrix} 1 & 0 & 0 & 0 \\ 1 & 10 & 100 & 1000 \\ 1 & 20 & 400 & 8000 \\ 1 & 30 & 900 & 27000 \end{bmatrix} \begin{bmatrix} a \\ b \\ c \\ d \end{bmatrix} = \begin{bmatrix} 54.973 \\ 62.813 \\ 75.367 \\ 85.446 \end{bmatrix}$$

$P(t) = 54.973 + 0.308667t + 0.059515t^2 - 0.00119817t^3$

$Q(T) = 9.24972 \times 10^6 - 14041.6T + 7.10474T^2 - 0.00119817T^3$

In Problems 31–34 we take $t = 0$ in 1950 to fit a quartic polynomial $P(t) = a + bt + ct^2 + dt^3 + et^4$. Then we write the quartic polynomial $Q(T) = P(T - 1950)$ that expresses the predicted population in terms of the actual calendar year T.

31. $P(t) = a + bt + ct^2 + dt^3 + et^4$.

$$\begin{bmatrix} 1 & 0 & 0 & 0 & 0 \\ 1 & 10 & 100 & 1000 & 10000 \\ 1 & 20 & 400 & 8000 & 160000 \\ 1 & 30 & 900 & 27000 & 810000 \\ 1 & 40 & 1600 & 64000 & 2560000 \end{bmatrix} \begin{bmatrix} a \\ b \\ c \\ d \\ e \end{bmatrix} = \begin{bmatrix} 39.478 \\ 44.678 \\ 49.061 \\ 49.137 \\ 50.809 \end{bmatrix}$$

$P(t) = 39.478 + 0.209692t + 0.0564163t^2 - 0.00292992t^3 + 0.0000391375t^4$

$Q(T) = 5.87828 \times 10^8 - 1.19444 \times 10^6 T + 910.118T^2 - 0.308202T^3 + 0.0000391375T^4$

33. $P(t) = a + bt + ct^2 + dt^3 + et^4$.

$$\begin{bmatrix} 1 & 0 & 0 & 0 & 0 \\ 1 & 10 & 100 & 1000 & 10000 \\ 1 & 20 & 400 & 8000 & 160000 \\ 1 & 30 & 900 & 27000 & 810000 \\ 1 & 40 & 1600 & 64000 & 2560000 \end{bmatrix} \begin{bmatrix} a \\ b \\ c \\ d \\ e \end{bmatrix} = \begin{bmatrix} 47.197 \\ 54.973 \\ 62.813 \\ 75.367 \\ 85.446 \end{bmatrix}$$

$P(t) = 47.197 + 1.22537t - 0.0771921t^2 + 0.00373475t^3 - 0.0000493292t^4$

$Q(T) = -7.41239 \times 10^8 + 1.50598 \times 10^6 T - 1147.37T^2 + 0.388502T^3 - 0.0000493292T^4$

35. Expansion of the determinant along the first row gives an equation of the form $ay + bx^2 + cx + d = 0$ that can be solved for $y = Ax^2 + Bx + C$. If the coordinates of any one of the three given points $(x_1, y_1), (x_2, y_2), (x_3, y_3)$ are substituted in the first row, then the determinant has two identical rows and therefore vanishes.

37. Expansion of the determinant along the first row gives an equation of the form $a(x^2 + y^2) + bx + cy + d = 0$, and we get the desired form of the equation of a circle upon division by a. If the coordinates of any one of the three given points $(x_1, y_1), (x_2, y_2)$, and (x_3, y_3) are substituted in the first row, then the determinant has two identical rows and therefore vanishes.

39. Expansion of the determinant along the first row gives an equation of the form $ax^2 + bxy + cy^2 + d = 0$, which can be written in the central conic form $Ax^2 + Bxy + Cy^2 = 1$ upon division by $-d$. If the coordinates of any one of the three given points $(x_1, y_1), (x_2, y_2)$, and (x_3, y_3) are substituted in the first row, then the determinant has two identical rows and therefore vanishes.

CHAPTER 4

VECTOR SPACES

The treatment of vector spaces in this chapter is very concrete. Prior to the final section of the chapter, almost all of the vector spaces appearing in examples and problems are subspaces of Cartesian coordinate spaces of n-tuples of real numbers. The main motivation throughout is the fact that the solution space of a homogeneous linear system $\mathbf{Ax} = \mathbf{0}$ is precisely such a "concrete" vector space.

SECTION 4.1

THE VECTOR SPACE R³

Here the fundamental concepts of vectors, linear independence, and vector spaces are introduced in the context of the familiar 2-dimensional coordinate plane $\mathbf{R}^2$ and 3-space $\mathbf{R}^3$. The concept of a subspace of a vector space is illustrated, the proper nontrivial subspaces of $\mathbf{R}^3$ being simply lines and planes through the origin.

1.
$$|\mathbf{a} - \mathbf{b}| = |(2,5,-4) - (1,-2,-3)| = |(1,7,-1)| = \sqrt{51}$$
$$2\mathbf{a} + \mathbf{b} = 2(2,5,-4) + (1,-2,-3) = (4,10,-8) + (1,-2,-3) = (5,8,-11)$$
$$3\mathbf{a} - 4\mathbf{b} = 3(2,5,-4) - 4(1,-2,-3) = (6,15,-12) - (4,-8,-12) = (2,23,0)$$

3.
$$|\mathbf{a} - \mathbf{b}| = |(2\mathbf{i} - 3\mathbf{j} + 5\mathbf{k}) - (5\mathbf{i} + 3\mathbf{j} - 7\mathbf{k})| = |-3\mathbf{i} - 6\mathbf{j} + 12\mathbf{k}| = \sqrt{189} = 3\sqrt{21}$$
$$2\mathbf{a} + \mathbf{b} = 2(2\mathbf{i} - 3\mathbf{j} + 5\mathbf{k}) + (5\mathbf{i} + 3\mathbf{j} - 7\mathbf{k})$$
$$= (4\mathbf{i} - 6\mathbf{j} + 10\mathbf{k}) + (5\mathbf{i} + 3\mathbf{j} - 7\mathbf{k}) = 9\mathbf{i} - 3\mathbf{j} + 3\mathbf{k}$$
$$3\mathbf{a} - 4\mathbf{b} = 3(2\mathbf{i} - 3\mathbf{j} + 5\mathbf{k}) - 4(5\mathbf{i} + 3\mathbf{j} - 7\mathbf{k})$$
$$= (6\mathbf{i} - 9\mathbf{j} + 15\mathbf{k}) - (20\mathbf{i} + 12\mathbf{j} - 28\mathbf{k}) = -14\mathbf{i} - 21\mathbf{j} + 43\mathbf{k}$$

5. $\mathbf{v} = \frac{3}{2}\mathbf{u}$, so the vectors $\mathbf{u}$ and $\mathbf{v}$ are linearly dependent.

7. $a\mathbf{u} + b\mathbf{v} = a(2,2) + b(2,-2) = (2a + 2b, 2a - 2b) = \mathbf{0}$ implies $a = b = 0$, so the vectors $\mathbf{u}$ and $\mathbf{v}$ are linearly independent.

In each of Problems 9–14, we set up and solve (as in Example 2 of this section) the system

$$a\mathbf{u} + b\mathbf{v} = \begin{bmatrix} u_1 & v_1 \\ u_2 & v_2 \end{bmatrix}\begin{bmatrix} a \\ b \end{bmatrix} = \begin{bmatrix} w_1 \\ w_2 \end{bmatrix} = \mathbf{w}$$

to find the coefficient values a and b such that $\mathbf{w} = a\mathbf{u} + b\mathbf{v}$,

9. $\begin{bmatrix} 1 & -1 \\ -2 & 3 \end{bmatrix}\begin{bmatrix} a \\ b \end{bmatrix} = \begin{bmatrix} 1 \\ 0 \end{bmatrix}$ $\Rightarrow$ $a = 3,\ b = 2$ so $\mathbf{w} = 3\mathbf{u} + 2\mathbf{v}$

11. $\begin{bmatrix} 5 & 2 \\ 7 & 3 \end{bmatrix}\begin{bmatrix} a \\ b \end{bmatrix} = \begin{bmatrix} 1 \\ 1 \end{bmatrix}$ $\Rightarrow$ $a = 1,\ b = -2$ so $\mathbf{w} = \mathbf{u} - 2\mathbf{v}$

13. $\begin{bmatrix} 7 & 3 \\ 5 & 4 \end{bmatrix}\begin{bmatrix} a \\ b \end{bmatrix} = \begin{bmatrix} 5 \\ -2 \end{bmatrix}$ $\Rightarrow$ $a = 2,\ b = -2$ so $\mathbf{w} = 2\mathbf{u} - 3\mathbf{v}$

In Problems 15–18, we calculate the determinant $|\mathbf{u}\ \ \mathbf{v}\ \ \mathbf{w}|$ so as to determine (using Theorem 4) whether the three vectors $\mathbf{u}$, $\mathbf{v}$, and $\mathbf{w}$ are linearly dependent (det = 0) or linearly independent (det ≠ 0).

15. $\begin{vmatrix} 3 & 5 & 8 \\ -1 & 4 & 3 \\ 2 & -6 & -4 \end{vmatrix} = 0$ so the three vectors are linearly dependent.

17. $\begin{vmatrix} 1 & 3 & 1 \\ -1 & 0 & -2 \\ 2 & 1 & 2 \end{vmatrix} = -5 \neq 0$ so the three vectors are linearly independent.

In Problems 19–24, we attempt to solve the homogeneous system $\mathbf{Ax} = \mathbf{0}$ by reducing the coefficient matrix $\mathbf{A} = \begin{bmatrix} \mathbf{u} & \mathbf{v} & \mathbf{w} \end{bmatrix}$ to echelon form $\mathbf{E}$. If we find that the system has only the trivial solution $a = b = c = 0$, this means that the vectors $\mathbf{u}$, $\mathbf{v}$, and $\mathbf{w}$ are linearly independent. Otherwise, a nontrivial solution $\mathbf{x} = \begin{bmatrix} a & b & c \end{bmatrix}^T \neq \mathbf{0}$ provides us with a nontrivial linear combination $a\mathbf{u} + b\mathbf{v} + c\mathbf{w} \neq \mathbf{0}$ that shows the three vectors are linearly dependent.

19. $\mathbf{A} = \begin{bmatrix} 2 & -3 & 0 \\ 0 & 1 & -2 \\ 1 & -1 & -1 \end{bmatrix} \rightarrow \begin{bmatrix} 1 & 0 & -3 \\ 0 & 1 & -2 \\ 0 & 0 & 0 \end{bmatrix} = \mathbf{E}$

The nontrivial solution $a = 3$, $b = 2$, $c = 1$ gives $3\mathbf{u} + 2\mathbf{v} + \mathbf{w} = \mathbf{0}$, so the three vectors are linearly dependent.

21. $\mathbf{A} = \begin{bmatrix} 1 & -2 & 3 \\ 1 & -1 & 7 \\ -2 & 6 & 2 \end{bmatrix} \rightarrow \begin{bmatrix} 1 & 0 & 11 \\ 0 & 1 & 4 \\ 0 & 0 & 0 \end{bmatrix} = \mathbf{E}$

The nontrivial solution $a = 11$, $b = 4$, $c = -1$ gives $11\mathbf{u} + 4\mathbf{v} - \mathbf{w} = \mathbf{0}$, so the three vectors are linearly dependent.

23. $\mathbf{A} = \begin{bmatrix} 2 & 5 & 2 \\ 0 & 4 & -1 \\ 3 & -2 & 1 \end{bmatrix} \rightarrow \begin{bmatrix} 1 & 0 & 0 \\ 0 & 1 & 0 \\ 0 & 0 & 1 \end{bmatrix} = \mathbf{E}$

The system $\mathbf{Ax} = \mathbf{0}$ has only the trivial solution $a = b = c = 0$, so the vectors $\mathbf{u}$, $\mathbf{v}$, and $\mathbf{w}$ are linearly independent.

In Problems 25–28, we solve the nonhomogeneous system $\mathbf{Ax} = \mathbf{t}$ by reducing the augmented coefficient matrix $\mathbf{A} = \begin{bmatrix} \mathbf{u} & \mathbf{v} & \mathbf{w} & \mathbf{t} \end{bmatrix}$ to echelon form $\mathbf{E}$. The solution vector $\mathbf{x} = \begin{bmatrix} a & b & c \end{bmatrix}^T$ appears as the final column of $\mathbf{E}$, and provides us with the desired linear combination $\mathbf{t} = a\mathbf{u} + b\mathbf{v} + c\mathbf{w}$.

25. $\mathbf{A} = \begin{bmatrix} 1 & 3 & 1 & 2 \\ -2 & 0 & -1 & -7 \\ 2 & 1 & 2 & 9 \end{bmatrix} \rightarrow \begin{bmatrix} 1 & 0 & 0 & 2 \\ 0 & 1 & 0 & -1 \\ 0 & 0 & 1 & 3 \end{bmatrix} = \mathbf{E}$

Thus $a = 2$, $b = -1$, $c = 3$ so $\mathbf{t} = 2\mathbf{u} - \mathbf{v} + 3\mathbf{w}$.

27. $\mathbf{A} = \begin{bmatrix} 1 & -1 & 4 & 0 \\ 4 & -2 & 4 & 0 \\ 3 & 2 & 1 & 19 \end{bmatrix} \rightarrow \begin{bmatrix} 1 & 0 & 0 & 2 \\ 0 & 1 & 0 & 6 \\ 0 & 0 & 1 & 1 \end{bmatrix} = \mathbf{E}$

Thus $a = 2$, $b = 6$, $c = 1$ so $\mathbf{t} = 2\mathbf{u} + 6\mathbf{v} + \mathbf{w}$.

29. Given vectors $(0, y, z)$ and $(0, v, w)$ in V, we see that their sum $(0, y+v, z+w)$ and the scalar multiple $c(0, y, z) = (0, cy, cz)$ both have first component 0, and therefore are elements of V.

31. If (x, y, z) and (u, v, w) are in V, then
$$2(x + u) = (2x) + (2u) = (3y) + (3v) = 3(y + v),$$

so their sum $(x+u, y+v, z+w)$ is in V. Similarly,

$$2(cx) = c(2x) = c(3y) = 3(cy),$$

so the scalar multiple (cx, cy, cz) is in V.

33. $(0,1,0)$ is in V but the sum $(0,1,0)+(0,1,0) = (0,2,0)$ is not in V; thus V is not closed under addition. Alternatively, $2(0,1,0) = (0,2,0)$ is not in V, so V is not closed under multiplication by scalars.

35.   Evidently V is closed under addition of vectors. However, $(0,0,1)$ is in V but $(-1)(0,0,1) = (0,0,-1)$ is not, so V is not closed under multiplication by scalars.

37.   Pick a fixed element $\mathbf{u}$ in the (nonempty) vector space V. Then, with $c = 0$, the scalar multiple $c\mathbf{u} = 0\mathbf{u} = \mathbf{0}$ must be in V. Thus V necessarily contains the zero vector $\mathbf{0}$.

39.   It suffices to show that every vector $\mathbf{v}$ in V is a scalar multiple of the given nonzero vector $\mathbf{u}$ in V. If $\mathbf{u}$ and $\mathbf{v}$ were linearly independent, then — as illustrated in Example 2 of this section — every vector in $\mathbf{R}^2$ could be expressed as a linear combination of $\mathbf{u}$ and $\mathbf{v}$. In this case it would follow that V is all of $\mathbf{R}^2$ (since, by Problem 38, V is closed under taking linear combinations). But we are given that V is a proper subspace of $\mathbf{R}^2$, so we must conclude that $\mathbf{u}$ and $\mathbf{v}$ are linearly dependent vectors. Since $\mathbf{u} \neq \mathbf{0}$, it follows that the arbitrary vector $\mathbf{v}$ in V is a scalar multiple of $\mathbf{u}$, and thus V is precisely the set of all scalar multiples of $\mathbf{u}$. In geometric language, the subspace V is then the straight line through the origin determined by the nonzero vector $\mathbf{u}$.

41.   If the vectors $\mathbf{u}$ and $\mathbf{v}$ are in the intersection V of the subspaces V_1 and V_2, then their sum $\mathbf{u} + \mathbf{v}$ is in V_1 because both vectors are in V_1, and $\mathbf{u} + \mathbf{v}$ is in V_2 because both are in V_2. Therefore $\mathbf{u} + \mathbf{v}$ is in V, and thus V is closed under addition of vectors. Similarly, the intersection V is closed under multiplication by scalars, and is therefore itself a subspace.

SECTION 4.2

THE VECTOR SPACE R^n AND SUBSPACES

The main objective in this section is for the student to understand what types of subsets of the vector space $\mathbf{R}^n$ of n-tuples of real numbers are subspaces — playing the role in $\mathbf{R}^n$ of lines and planes through the origin in $\mathbf{R}^3$. Our first reason for studying subspaces is the fact that the *solution space* of any homogeneous linear system $\mathbf{Ax} = \mathbf{0}$ is a subspace of $\mathbf{R}^n$.

1.   If $\mathbf{x} = (x_1, x_2, 0)$ and $\mathbf{y} = (y_1, y_2, 0)$ are vectors in W, then their sum

$$\mathbf{x} + \mathbf{y} = (x_1, x_2, 0) + (y_1, y_2, 0) = (x_1 + y_1, x_2 + y_2, 0)$$

and the scalar multiple $c\mathbf{x} = (cx_1, cx_2, 0)$ both have third coordinate zero, and therefore are also elements of W. Hence W is a subspace of $\mathbf{R}^3$.

3.   The typical vector in W is of the form $\mathbf{x} = (x_1, 1, x_3)$ with second coordinate 1. But the particular scalar multiple $2\mathbf{x} = (2x_1, 2, 2x_3)$ of such a vector has second coordinate $2 \neq 1$, and thus is not in W. Hence W is not closed under multiplication by scalars, and therefore is not a subspace of $\mathbf{R}^3$. (Since $2\mathbf{x} = \mathbf{x} + \mathbf{x}$, W is not closed under vector addition either.)

5.   Suppose $\mathbf{x} = (x_1, x_2, x_3, x_4)$ and $\mathbf{y} = (y_1, y_2, y_3, y_4)$ are vectors in W, so

$$x_1 + 2x_2 + 3x_3 + 4x_4 = 0 \quad \text{and} \quad y_1 + 2y_2 + 3y_3 + 4y_4 = 0.$$

Then their sum $\mathbf{s} = \mathbf{x} + \mathbf{y} = (x_1 + y_1, x_2 + y_2, x_3 + y_3, x_4 + y_4) = (s_1, s_2, s_3, s_4)$ satisfies the same condition

$$\begin{aligned} s_1 + 2s_2 + 3s_3 + 4s_4 &= (x_1 + y_1) + 2(x_2 + y_2) + 3(x_3 + y_3) + 4(x_4 + y_4) \\ &= (x_1 + 2x_2 + 3x_3 + 4x_4) + (y_1 + 2y_2 + 3y_3 + 4y_4) = 0 + 0 = 0, \end{aligned}$$

and thus is an element of W. Similarly, the scalar multiple $\mathbf{m} = c\mathbf{x} = (cx_1, cx_2, cx_3, cx_4) = (m_1, m_2, m_3, m_4)$ satisfies the condition

$$m_1 + 2m_2 + 3m_3 + 4m_4 = cx_1 + 2cx_2 + 3cx_3 + 4cx_4 = c(x_1 + 2x_2 + 3x_3 + 4x_4) = 0,$$

and hence is also an element of W. Therefore W is a subspace of $\mathbf{R}^4$.

7.   The vectors $\mathbf{x} = (1,1)$ and $\mathbf{y} = (1,-1)$ are in W, but their sum $\mathbf{x} + \mathbf{y} = (2,0)$ is not, because $|2| \neq |0|$. Hence W is not a subspace of $\mathbf{R}^2$.

9.   The vector $\mathbf{x} = (1,0)$ is in W, but its scalar multiple $2\mathbf{x} = (2,0)$ is not, because $(2)^2 + (0)^2 = 4 \neq 1$. Hence W is not a subspace of $\mathbf{R}^2$.

11.   Suppose $\mathbf{x} = (x_1, x_2, x_3, x_4)$ and $\mathbf{y} = (y_1, y_2, y_3, y_4)$ are vectors in W, so

$$x_1 + x_2 = x_3 + x_4 \quad \text{and} \quad y_1 + y_2 = y_3 + y_4.$$

Then their sum $\mathbf{s} = \mathbf{x} + \mathbf{y} = (x_1 + y_1, x_2 + y_2, x_3 + y_3, x_4 + y_4) = (s_1, s_2, s_3, s_4)$ satisfies the same condition

$$s_1 + s_2 = (x_1 + y_1) + (x_2 + y_2) = (x_1 + x_2) + (y_1 + y_2)$$
$$= (x_3 + x_4) + (y_3 + y_4) = (x_3 + y_3) + (x_4 + y_4) = s_3 + s_4$$

and thus is an element of W. Similarly, the scalar multiple $\mathbf{m} = c\mathbf{x} = (cx_1, cx_2, cx_3, cx_4) = (m_1, m_2, m_3, m_4)$ satisfies the condition

$$m_1 + m_2 = cx_1 + cx_2 = c(x_1 + x_2) = c(x_3 + x_4) = cx_3 + cx_4 = m_3 + m_4,$$

and hence is also an element of W. Therefore W is a subspace of $\mathbf{R}^4$.

13.   The vectors $\mathbf{x} = (1, 0, 1, 0)$ and $\mathbf{y} = (0, 1, 0, 1)$ are in W (because the product of the 4 components is 0 in each case) but their sum $\mathbf{s} = \mathbf{x} + \mathbf{y} = (1, 1, 1, 1)$ is not, because $s_1 s_2 s_3 s_4 = 1 \neq 0$. Hence W is not a subspace of $\mathbf{R}^4$.

In Problems 15–22, we first reduce the coefficient matrix $\mathbf{A}$ to echelon form $\mathbf{E}$ in order to solve the given homogeneous system $\mathbf{Ax} = \mathbf{0}$.

15. $\mathbf{A} = \begin{bmatrix} 1 & -4 & 1 & -4 \\ 1 & 2 & 1 & 8 \\ 1 & 1 & 1 & 6 \end{bmatrix} \rightarrow \begin{bmatrix} 1 & 0 & 1 & 4 \\ 0 & 1 & 0 & 2 \\ 0 & 0 & 0 & 0 \end{bmatrix} = \mathbf{E}$

Thus $x_3 = s$ and $x_4 = t$ are free variables. We solve for $x_1 = -s - 4t$ and $x_2 = -2t$, so

$$\mathbf{x} = (x_1, x_2, x_3, x_4) = (-s - 4t, -2t, s, t)$$
$$= (-s, 0, s, 0) + (-4t, -2t, 0, t) = s\mathbf{u} + t\mathbf{v}$$

where $\mathbf{u} = (-1, 0, 1, 0)$ and $\mathbf{v} = (-4, -2, 0, 1)$.

17. $\mathbf{A} = \begin{bmatrix} 1 & 3 & 8 & -1 \\ 1 & -3 & -10 & 5 \\ 1 & 4 & 11 & -2 \end{bmatrix} \rightarrow \begin{bmatrix} 1 & 0 & -1 & 2 \\ 0 & 1 & 3 & -1 \\ 0 & 0 & 0 & 0 \end{bmatrix} = \mathbf{E}$

Thus $x_3 = s$ and $x_4 = t$ are free variables. We solve for $x_1 = s - 2t$ and $x_2 = -3s + t$, so

$$\mathbf{x} = (x_1, x_2, x_3, x_4) = (-s - 5t, -s - 3t, s, t)$$
$$= (s, -3s, s, 0) + (-2t, t, 0, t) = s\mathbf{u} + t\mathbf{v}$$

where $\mathbf{u} = (1, -3, 1, 0)$ and $\mathbf{v} = (-2, 1, 0, 1)$.

19. $A = \begin{bmatrix} 1 & -3 & -5 & -6 \\ 2 & 1 & 4 & -4 \\ 1 & 3 & 7 & 1 \end{bmatrix} \rightarrow \begin{bmatrix} 1 & 0 & 1 & 0 \\ 0 & 1 & 2 & 0 \\ 0 & 0 & 0 & 1 \end{bmatrix} = E$

Thus $x_3 = t$ is a free variable and $x_4 = 0.$. We solve for $x_1 = -t$ and $x_2 = -2t$, so

$$\mathbf{x} = (x_1, x_2, x_3, x_4) = (-t, -2t, t, 0) = t\,\mathbf{u}$$

where $\mathbf{u} = (-1, -2, 1, 0)$.

21. $A = \begin{bmatrix} 1 & 7 & 2 & -3 \\ 2 & 7 & 1 & -4 \\ 3 & 5 & -1 & -5 \end{bmatrix} \rightarrow \begin{bmatrix} 1 & 0 & 0 & 3 \\ 0 & 1 & 0 & -2 \\ 0 & 0 & 1 & 4 \end{bmatrix} = E$

Thus $x_4 = t$ is a free variable. We solve for $x_1 = -3t$, $x_2 = 2t$, and $x_4 = -4t$. so

$$\mathbf{x} = (x_1, x_2, x_3, x_4) = (-3t, 2t, -4t, t) = t\,\mathbf{u}$$

where $\mathbf{u} = (-3, 2, -4, 1)$.

23.   Let $\mathbf{u}$ be a vector in W. Then $0\mathbf{u}$ is also in W. But $0\mathbf{u} = (0+0)\mathbf{u} = 0\mathbf{u} + 0\mathbf{u}$, so upon subtracting $0\mathbf{u}$ from each side, we see that $0\mathbf{u} = 0$, the zero vector.

25.   If W is a subspace, then it contains the scalar multiples $a\mathbf{u}$ and $b\mathbf{v}$, and hence contains their sum $a\mathbf{u} + b\mathbf{v}$. Conversely, if the subset W is closed under taking linear combinations of pairs of vectors, then it contains $(1)\mathbf{u} + (1)\mathbf{v} = \mathbf{u} + \mathbf{v}$ and $(c)\mathbf{u} + (0)\mathbf{v} = c\mathbf{u}$, and hence is a subspace.

27.   Let $a_1\mathbf{u} + b_1\mathbf{v}$ and $a_2\mathbf{u} + b_2\mathbf{v}$ be two vectors in $W = \{a\mathbf{u} + b\mathbf{v}\}$. Then the sum

$$(a_1\mathbf{u} + b_1\mathbf{v}) + (a_2\mathbf{u} + b_2\mathbf{v}) = (a_1 + a_2)\mathbf{u} + (b_1 + b_2)\mathbf{v}$$

and the scalar multiple $c(a_1\mathbf{u} + b_1\mathbf{v}) = (ca_1)\mathbf{u} + (cb_1)\mathbf{v}$ are again scalar multiples of $\mathbf{u}$ and $\mathbf{v}$, and hence are themselves elements of W. Hence W is a subspace.

29.   If $\mathbf{Ax}_0 = \mathbf{b}$ and $\mathbf{y} = \mathbf{x} - \mathbf{x}_0$, then

$$\mathbf{Ay} = \mathbf{A}(\mathbf{x} - \mathbf{x}_0) = \mathbf{Ax} - \mathbf{Ax}_0 = \mathbf{Ax} - \mathbf{b}.$$

Hence it is clear that $\mathbf{Ay} = \mathbf{0}$ if and only if $\mathbf{Ax} = \mathbf{b}$

31. Let $\mathbf{w}_1$ and $\mathbf{w}_2$ be two vectors in the sum $U + V$. Then $\mathbf{w}_i = \mathbf{u}_i + \mathbf{v}_i$ where $\mathbf{u}_i$ is in U and $\mathbf{v}_i$ is in V $(i = 1, 2)$. Then the linear combination

$$a\mathbf{w}_1 + b\mathbf{w}_2 = a(\mathbf{u}_1 + \mathbf{v}_1) + b(\mathbf{u}_2 + \mathbf{v}_2) = (a\mathbf{u}_1 + b\mathbf{u}_2) + (a\mathbf{v}_1 + b\mathbf{v}_2)$$

is the sum of the vectors $a\mathbf{u}_1 + b\mathbf{u}_2$ in U and $a\mathbf{v}_1 + b\mathbf{v}_2$ in U, and therefore is an element of $U + V$. Thus $U + V$ is a subspace. If U and V are noncollinear lines through the origin in $\mathbf{R}^3$, then $U + V$ is a plane through the origin.

.

SECTION 4.3

LINEAR COMBINATIONS AND INDEPENDENCE OF VECTORS

In this section we use two types of computational problems as aids in understanding linear independence and dependence. The first of these problems is that of expressing a vector $\mathbf{w}$ as a linear combination of k given vectors $\mathbf{v}_1, \mathbf{v}_2, \cdots, \mathbf{v}_k$ (if possible). The second is that of determining whether k given vectors $\mathbf{v}_1, \mathbf{v}_2, \cdots, \mathbf{v}_k$ are linearly independent. For vectors in $\mathbf{R}^n$, each of these problems reduces to solving a linear system of n equations in k unknowns. Thus an abstract question of linear independence or dependence becomes a concrete question of whether or not a given linear system has a nontrivial solution.

1. $\mathbf{v}_2 = \frac{3}{2}\mathbf{v}_1$, so the two vectors $\mathbf{v}_1$ and $\mathbf{v}_2$ are linearly dependent.

3. The three vectors $\mathbf{v}_1$, $\mathbf{v}_2$, and $\mathbf{v}_3$ are linearly dependent, as are any 3 vectors in $\mathbf{R}^2$. The reason is that the vector equation $c_1\mathbf{v}_1 + c_2\mathbf{v}_2 + c_3\mathbf{v}_3 = \mathbf{0}$ reduces to a homogeneous linear system of 2 equations in the 3 unknowns c_1, c_2, and c_3, and any such system has a nontrivial solution.

5. The equation $c_1\mathbf{v}_1 + c_2\mathbf{v}_2 + c_3\mathbf{v}_3 = \mathbf{0}$ yields

$$c_1(1,0,0) + c_2(0,-2,0) + c_3(0,0,3) = (c_1, -2c_2, 3c_3) = (0,0,0),$$

and therefore implies immediately that $c_1 = c_2 = c_3 = 0$. Hence the given vectors $\mathbf{v}_1$, $\mathbf{v}_2$, and $\mathbf{v}_3$ are linearly independent.

7. The equation $c_1\mathbf{v}_1 + c_2\mathbf{v}_2 + c_3\mathbf{v}_3 = \mathbf{0}$ yields

$$c_1(2,1,0,0) + c_2(3,0,1,0) + c_3(4,0,0,1) = (2c_1 + 3c_2, c_1, c_2, c_3) = (0,0,0,0).$$

Obviously it follows immediately that $c_1 = c_2 = c_3 = 0$. Hence the given vectors $\mathbf{v}_1$, $\mathbf{v}_2$, and $\mathbf{v}_3$ are linearly independent.

In Problems 9–16 we first set up the linear system to be solved for the linear combination coefficients $\{c_i\}$, and then show the reduction of its augmented coefficient matrix $\mathbf{A}$ to reduced echelon form $\mathbf{E}$.

9. $c_1\mathbf{v}_1 + c_2\mathbf{v}_2 = \mathbf{w}$

$$
\mathbf{A} = \begin{bmatrix} 5 & 3 & 1 \\ 3 & 2 & 0 \\ 4 & 5 & -7 \end{bmatrix} \rightarrow \begin{bmatrix} 1 & 0 & 2 \\ 0 & 1 & -3 \\ 0 & 0 & 0 \end{bmatrix} = \mathbf{E}
$$

We see that the system of 3 equations in 2 unknowns has the unique solution $c_1 = 2, c_2 = -3$, so $\mathbf{w} = 2\mathbf{v}_1 - 3\mathbf{v}_2$.

11. $c_1\mathbf{v}_1 + c_2\mathbf{v}_2 = \mathbf{w}$

$$
\mathbf{A} = \begin{bmatrix} 7 & 3 & 1 \\ -6 & -3 & 0 \\ 4 & 2 & 0 \\ 5 & 3 & -1 \end{bmatrix} \rightarrow \begin{bmatrix} 1 & 0 & 1 \\ 0 & 1 & -2 \\ 0 & 0 & 0 \\ 0 & 0 & 0 \end{bmatrix} = \mathbf{E}
$$

We see that the system of 4 equations in 2 unknowns has the unique solution $c_1 = 1, c_2 = -2$, so $\mathbf{w} = \mathbf{v}_1 - 2\mathbf{v}_2$.

13. $c_1\mathbf{v}_1 + c_2\mathbf{v}_2 = \mathbf{w}$

$$
\mathbf{A} = \begin{bmatrix} 1 & 5 & 5 \\ 5 & -3 & 2 \\ -3 & 4 & -2 \end{bmatrix} \rightarrow \begin{bmatrix} 1 & 0 & 0 \\ 0 & 1 & 0 \\ 0 & 0 & 1 \end{bmatrix} = \mathbf{E}
$$

The last row of $\mathbf{E}$ corresponds to the scalar equation $0c_1 + 0c_2 = 1$, so the system of 3 equations in 2 unknowns is inconsistent. This means that $\mathbf{w}$ cannot be expressed as a linear combination of $\mathbf{v}_1$ and $\mathbf{v}_2$.

15. $c_1\mathbf{v}_1 + c_2\mathbf{v}_2 + c_3\mathbf{v}_3 = \mathbf{w}$

$$
\mathbf{A} = \begin{bmatrix} 2 & 3 & 1 & 4 \\ -1 & 0 & 2 & 5 \\ 4 & 1 & -1 & 6 \end{bmatrix} \rightarrow \begin{bmatrix} 1 & 0 & 0 & 3 \\ 0 & 1 & 0 & -2 \\ 0 & 0 & 1 & 4 \end{bmatrix} = \mathbf{E}
$$

We see that the system of 3 equations in 3 unknowns has the unique solution $c_1 = 3, c_2 = -2, c_3 = 4$, so $\mathbf{w} = 3\mathbf{v}_1 - 2\mathbf{v}_2 + 4\mathbf{v}_3$.

In Problems 17–22, $\mathbf{A} = \begin{bmatrix} \mathbf{v}_1 & \mathbf{v}_2 & \mathbf{v}_3 \end{bmatrix}$ is the coefficient matrix of the homogeneous linear system corresponding to the vector equation $c_1\mathbf{v}_1 + c_2\mathbf{v}_2 + c_3\mathbf{v}_3 = \mathbf{0}$. Inspection of the indicated reduced echelon form $\mathbf{E}$ of $\mathbf{A}$ then reveals whether or not a nontrivial solution exists.

17. $\quad \mathbf{A} = \begin{bmatrix} 1 & 2 & 3 \\ 0 & -3 & 5 \\ 1 & 4 & 2 \end{bmatrix} \rightarrow \begin{bmatrix} 1 & 0 & 0 \\ 0 & 1 & 0 \\ 0 & 0 & 1 \end{bmatrix} = \mathbf{E}$

We see that the system of 3 equations in 3 unknowns has the unique solution $c_1 = c_2 = c_3 = 0$, so the vectors $\mathbf{v}_1, \mathbf{v}_2, \mathbf{v}_3$ are linearly independent.

19. $\quad \mathbf{A} = \begin{bmatrix} 2 & 5 & 2 \\ 0 & 4 & -1 \\ 3 & -2 & 1 \\ 0 & 1 & -1 \end{bmatrix} \rightarrow \begin{bmatrix} 1 & 0 & 0 \\ 0 & 1 & 0 \\ 0 & 0 & 1 \\ 0 & 0 & 0 \end{bmatrix} = \mathbf{E}$

We see that the system of 4 equations in 3 unknowns has the unique solution $c_1 = c_2 = c_3 = 0$, so the vectors $\mathbf{v}_1, \mathbf{v}_2, \mathbf{v}_3$ are linearly independent.

21. $\quad \mathbf{A} = \begin{bmatrix} 3 & 1 & 1 \\ 0 & -1 & 2 \\ 1 & 0 & 1 \\ 2 & 1 & 0 \end{bmatrix} \rightarrow \begin{bmatrix} 1 & 0 & 1 \\ 0 & 1 & -2 \\ 0 & 0 & 0 \\ 0 & 0 & 0 \end{bmatrix} = \mathbf{E}$

We see that the system of 4 equations in 3 unknowns has a 1-dimensional solution space. If we choose $c_3 = -1$ then $c_1 = 1$ and $c_2 = -2$. Therefore $\mathbf{v}_1 - 2\mathbf{v}_2 - \mathbf{v}_3 = \mathbf{0}$.

23. Because $\mathbf{v}_1$ and $\mathbf{v}_2$ are linearly independent, the vector equation

$$c_1\mathbf{u}_1 + c_2\mathbf{u}_2 = c_1(\mathbf{v}_1 + \mathbf{v}_2) + c_2(\mathbf{v}_1 - \mathbf{v}_2) = \mathbf{0}$$

yields the homogeneous linear system

$$c_1 + c_2 = 0$$
$$c_1 - c_2 = 0.$$

It follows readily that $c_1 = c_2 = 0$, and therefore that the vectors $\mathbf{u}_1$ and $\mathbf{u}_2$ are linearly independent.

25. Because the vectors $\mathbf{v}_1, \mathbf{v}_2, \mathbf{v}_3$ are linearly independent, the vector equation
$$c_1\mathbf{u}_1 + c_2\mathbf{u}_2 + c_3\mathbf{u}_3 = c_1(\mathbf{v}_1) + c_2(\mathbf{v}_1 + 2\mathbf{v}_2) + c_3(\mathbf{v}_1 + 2\mathbf{v}_2 + 3\mathbf{v}_3) = \mathbf{0}$$

yields the homogeneous linear system

$$c_1 + c_2 + c_3 = 0$$
$$2c_2 + 2c_3 = 0$$
$$3c_3 = 0.$$

It follows by back-substitution that $c_1 = c_2 = c_3 = 0$, and therefore that the vectors $\mathbf{u}_1, \mathbf{u}_2, \mathbf{u}_3$ are linearly independent.

27. If the elements of S are $\mathbf{v}_1, \mathbf{v}_2, \cdots, \mathbf{v}_k$ with $\mathbf{v}_1 = \mathbf{0}$, then we can take $c_1 = 1$ and $c_2 = \cdots = c_k = 0$. This choice gives coefficients $c_1, c_2, \cdots, c_k$ not all zero such that $c_1\mathbf{v}_1 + c_2\mathbf{v}_2 + \cdots + c_k\mathbf{v}_k = \mathbf{0}$. This means that the vectors $\mathbf{v}_1, \mathbf{v}_2, \cdots, \mathbf{v}_k$ are linearly dependent.

29. If some subset of S were linearly dependent, then Problem 28 would imply immediately that S itself is linearly dependent (contrary to hypothesis).

31. If S is contained in span(T), then every vector in S is a linear combination of vectors in T. Hence every vector in span(S) is a linear combination of linear combinations of vectors in T. Therefore every vector in span(S) is a linear combination of vectors in T, and therefore is itself in span(T). Thus span(S) is a subset of span(T).

33. The determinant of the $k \times k$ identity matrix is nonzero, so it follows immediately from Theorem 3 in this section that the vectors $\mathbf{v}_1, \mathbf{v}_2, \cdots, \mathbf{v}_k$ are linearly independent.

35. Because the vectors $\mathbf{v}_1, \mathbf{v}_2, \cdots, \mathbf{v}_k$ are linearly independent, Theorem 3 implies that some $k \times k$ submatrix $\mathbf{A}_0$ of $\mathbf{A}$ has nonzero determinant. Let $\mathbf{A}_0$ consist of the rows $i_1, i_2, \cdots, i_k$ of the matrix $\mathbf{A}$, and let $\mathbf{C}_0$ denote the $k \times k$ submatrix consisting of the same rows of the product matrix $\mathbf{C} = \mathbf{AB}$. Then $\mathbf{C}_0 = \mathbf{A}_0\mathbf{B}$, so $|\mathbf{C}_0| = |\mathbf{A}_0||\mathbf{B}| \neq 0$ because (by hypothesis) the $k \times k$ matrix $\mathbf{B}$ is also nonsingular. Therefore Theorem 3 implies that the column vectors of $\mathbf{AB}$ are linearly independent.

SECTION 4.4

BASES AND DIMENSION FOR VECTOR SPACES

A basis $\{\mathbf{v}_1, \mathbf{v}_2, \cdots, \mathbf{v}_k\}$ for a subspace W of $\mathbf{R}^n$ enables up to visualize W as a k-dimensional plane (or "hyperplane") through the origin in $\mathbf{R}^n$. In case W is the solution space of a homogeneous linear system, a basis for W is a maximal linearly independent set of solutions of the system, and every other solution is a linear combination of these particular solutions.

1. The vectors $\mathbf{v}_1$ and $\mathbf{v}_2$ are linearly independent (because neither is a scalar multiple of the other) and therefore form a basis for $\mathbf{R}^2$.

3. Any four vectors in $\mathbf{R}^3$ are linearly dependent, so the given vectors do not form a basis for $\mathbf{R}^3$.

5. The three given vectors $\mathbf{v}_1, \mathbf{v}_2, \mathbf{v}_3$ all lie in the 2-dimensional subspace $x_1 = 0$ of $\mathbf{R}^3$. Therefore they are linearly dependent, and hence do not form a basis for $\mathbf{R}^3$.

7. $\operatorname{Det}\left(\begin{bmatrix} \mathbf{v}_1 & \mathbf{v}_2 & \mathbf{v}_3 \end{bmatrix}\right) = 1 \neq 0$, so the three vectors are linearly independent, and hence do form a basis for $\mathbf{R}^3$.

9. The single equation $x - 2y + 5z = 0$ is already a system in reduced echelon form, with free variables y and z. With $y = s$, $z = t$, $x = 2s - 5t$ we get the solution vector

$$(x, y, z) = (2s - 5t, s, t) = s(2, 1, 0) + t(-5, 0, 1).$$

Hence the plane $x - 2y + 5z = 0$ is a 2-dimensional subspace of $\mathbf{R}^3$ with basis consisting of the vectors $\mathbf{v}_1 = (2, 1, 0)$ and $\mathbf{v}_2 = (-5, 0, 1)$.

11. The line of intersection of the planes in Problems 9 and 11 is the solution space of the system
$$x - 2y + 5z = 0$$
$$y - z = 0.$$

This system is in echelon form with free variable $z = t$. With $y = t$ and $x = -3t$ we have the solution vector $(-3t, t, t) = t(-3, 1, 1)$. Thus the line is a 1-dimensional subspace of $\mathbf{R}^3$ with basis consisting of the vector $\mathbf{v} = (-3, 1, 1)$.

13. The typical vector in $\mathbf{R}^4$ of the form (a, b, c, d) with $a = 3c$ and $b = 4d$ can be written as
$$\mathbf{v} = (3c, 4d, c, d) = c(3, 0, 1, 0) + d(0, 4, 0, 1).$$

Hence the subspace consisting of all such vectors is 2-dimensional with basis consisting of the vectors $\mathbf{v}_1 = (3,0,1,0)$ and $\mathbf{v}_2 = (0,4,0,1)$.

In Problems 15–26, we show first the reduction of the coefficient matrix $\mathbf{A}$ to echelon form $\mathbf{E}$. Then we write the typical solution vector as a linear combination of basis vectors for the subspace of the given system.

15. $\mathbf{A} = \begin{bmatrix} 1 & -2 & 3 \\ 2 & -3 & 1 \end{bmatrix} \rightarrow \begin{bmatrix} 1 & 0 & -11 \\ 0 & 1 & -7 \end{bmatrix} = \mathbf{E}$

With free variable $x_3 = t$ and $x_1 = 11t$, $x_2 = 7t$ we get the solution vector
$\mathbf{x} = (11t, 7t, t) = t(11,7,1)$. Thus the solution space of the given system is 1-dimensional with basis consisting of the vector $\mathbf{v}_1 = (11,7,1)$.

17. $\mathbf{A} = \begin{bmatrix} 1 & -3 & 2 & -4 \\ 2 & -5 & 7 & -3 \end{bmatrix} \rightarrow \begin{bmatrix} 1 & 0 & 11 & 11 \\ 0 & 1 & 3 & 5 \end{bmatrix} = \mathbf{E}$

With free variables $x_3 = s$, $x_4 = t$ and with $x_1 = -11s - 11t$, $x_2 = -3s - 5t$ we get the solution vector

$$\mathbf{x} = (-11s - 11t, -3s - 5t, s, t) = s(-11, -3, 1, 0) + t(-11, -5, 0, 1).$$

Thus the solution space of the given system is 2-dimensional with basis consisting of the vectors $\mathbf{v}_1 = (-11, -3, 1, 0)$ and $\mathbf{v}_2 = (-11, -5, 0, 1)$.

19. $\mathbf{A} = \begin{bmatrix} 1 & -3 & -8 & -5 \\ 2 & 1 & -4 & 11 \\ 1 & 3 & 3 & 13 \end{bmatrix} \rightarrow \begin{bmatrix} 1 & 0 & -3 & 4 \\ 0 & 1 & 2 & 3 \\ 0 & 0 & 0 & 0 \end{bmatrix} = \mathbf{E}$

With free variables $x_3 = s$, $x_4 = t$ and with $x_1 = 3s - 4t$, $x_2 = -2s - 3t$ we get the solution vector

$$\mathbf{x} = (3s - 4t, -2s - 3t, s, t) = s(3, -2, 1, 0) + t(-4, -3, 0, 1).$$

Thus the solution space of the given system is 2-dimensional with basis consisting of the vectors $\mathbf{v}_1 = (3, -2, 1, 0)$ and $\mathbf{v}_2 = (-4, -3, 0, 1)$.

21. $\mathbf{A} = \begin{bmatrix} 1 & -4 & -3 & -7 \\ 2 & -1 & 1 & 7 \\ 1 & 2 & 3 & 11 \end{bmatrix} \rightarrow \begin{bmatrix} 1 & 0 & 1 & 5 \\ 0 & 1 & 1 & 3 \\ 0 & 0 & 0 & 0 \end{bmatrix} = \mathbf{E}$

With free variables $x_3 = s$, $x_4 = t$ and with $x_1 = -s - 5t$, $x_2 = -s - 3t$ we get the solution vector

$$\mathbf{x} = (-s - 5t, -s - 3t, s, t) = s(-1, -1, 1, 0) + t(-5, -3, 0, 1).$$

Thus the solution space of the given system is 2-dimensional with basis consisting of the vectors $\mathbf{v}_1 = (-1, -1, 1, 0)$ and $\mathbf{v}_2 = (-5, -3, 0, 1)$.

23. $A = \begin{bmatrix} 1 & 5 & 13 & 14 \\ 2 & 5 & 11 & 12 \\ 2 & 7 & 17 & 19 \end{bmatrix} \rightarrow \begin{bmatrix} 1 & 0 & -2 & 0 \\ 0 & 1 & 3 & 0 \\ 0 & 0 & 0 & 1 \end{bmatrix} = E$

With free variable $x_3 = s$ and with $x_1 = 2s$, $x_2 = -3s$, $x_4 = 0$ we get the solution vector $\mathbf{x} = (2s, -3s, s, 0) = s(2, -3, 1, 0)$. Thus the solution space of the given system is 1-dimensional with basis consisting of the vector $\mathbf{v}_1 = (2, -3, 1, 0)$.

25. $A = \begin{bmatrix} 1 & 2 & 7 & -9 & 31 \\ 2 & 4 & 7 & -11 & 34 \\ 3 & 6 & 5 & -11 & 29 \end{bmatrix} \rightarrow \begin{bmatrix} 1 & 2 & 0 & -2 & 3 \\ 0 & 0 & 1 & -1 & 4 \\ 0 & 0 & 0 & 0 & 0 \end{bmatrix} = E$

With free variables $x_2 = r$, $x_4 = s$, $x_5 = t$ and with $x_1 = -2r + 2s - 3t$, $x_3 = s - 4t$ we get the solution vector

$$\mathbf{x} = (-2r + 2s - 3t, r, s - 4t, s, t) = r(-2, 1, 0, 0, 0) + s(2, 0, 1, 1, 0) + t(-3, 0, -4, 0, 1).$$

Thus the solution space of the given system is 3-dimensional with basis consisting of the vectors $\mathbf{v}_1 = (-2, 1, 0, 0, 0)$, $\mathbf{v}_2 = (2, 0, 1, 1, 0)$, and $\mathbf{v}_3 = (-3, 0, -4, 0, 1)$.

27. If the vectors $\mathbf{v}_1, \mathbf{v}_2, \cdots, \mathbf{v}_n$ are linearly independent, and $\mathbf{w}$ is another vector in V, then the vectors $\mathbf{w}, \mathbf{v}_1, \mathbf{v}_2, \cdots, \mathbf{v}_n$ are linearly dependent (because no $n+1$ vectors in the n-dimensional vector space V are linearly independent). Hence there exist scalars $c, c_1, c_2, \cdots, c_n$ not all zero such that

$$c\mathbf{w} + c_1\mathbf{v}_1 + c_2\mathbf{v}_2 + \cdots + c_n\mathbf{v}_n = \mathbf{0}.$$

If $c = 0$ then the coefficients $c_1, c_2, \cdots, c_n$ would not all be zero, and hence this equation would say (contrary to hypothesis) that the vectors $\mathbf{v}_1, \mathbf{v}_2, \cdots, \mathbf{v}_n$ are linearly dependent. Therefore $c \neq 0$, so we can solve for $\mathbf{w}$ as a linear combination of the vectors $\mathbf{v}_1, \mathbf{v}_2, \cdots, \mathbf{v}_n$. Thus the linearly independent vectors $\mathbf{v}_1, \mathbf{v}_2, \cdots, \mathbf{v}_n$ span V, and therefore form a basis for V.

29. Suppose $c\mathbf{v} + c_1\mathbf{v}_1 + c_2\mathbf{v}_2 + \cdots + c_k\mathbf{v}_k = \mathbf{0}$. Then $c = 0$ because, otherwise, we could solve for $\mathbf{v}$ as a linear combination of the vectors $\mathbf{v}_1, \mathbf{v}_2, \cdots, \mathbf{v}_k$. But this is impossible, because $\mathbf{v}$ is not in the subspace W spanned by $\mathbf{v}_1, \mathbf{v}_2, \cdots, \mathbf{v}_k$. It follows that $c_1\mathbf{v}_1 + c_2\mathbf{v}_2 + \cdots + c_k\mathbf{v}_k = \mathbf{0}$, which implies that $c_1 = c_2 = \cdots = c_k = 0$ also, because the vectors $\mathbf{v}_1, \mathbf{v}_2, \cdots, \mathbf{v}_k$ are linearly independent. Hence we have shown that the $k+1$ vectors $\mathbf{v}, \mathbf{v}_1, \mathbf{v}_2, \cdots, \mathbf{v}_k$ are linearly independent.

31. If $\mathbf{v}_{k+1}$ is a linear combination of the vectors $\mathbf{v}_1, \mathbf{v}_2, \cdots, \mathbf{v}_k$, then obviously every linear combination of the vectors $\mathbf{v}_1, \mathbf{v}_2, \cdots, \mathbf{v}_k, \mathbf{v}_{k+1}$ is also a linear combination of $\mathbf{v}_1, \mathbf{v}_2, \cdots, \mathbf{v}_k$. But the former set of $k+1$ vectors spans V, so the latter set of k vectors also spans V.

33. If S is a maximal linearly independent set in V, the we see immediately that every other vector in V is a linear combination of the vectors in S. Thus S also spans V, and is therefore a basis for V.

35. Let $S = \{\mathbf{v}_1, \mathbf{v}_2, \cdots, \mathbf{v}_n\}$ be a uniquely spanning set for V. Then the fact, that

$$\mathbf{0} = 0\mathbf{v}_1 + 0\mathbf{v}_2 + \cdots + 0\mathbf{v}_n$$

is the unique expression of the zero vector $\mathbf{0}$ as a linear combination of the vectors in S, means that S is a linearly independent set of vectors. Hence S is a basis for V.

SECTION 4.5

ROW AND COLUMN SPACES

Conventional wisdom (at a certain level) has it that a homogeneous linear system $\mathbf{Ax} = \mathbf{0}$ of m equations in $n > m$ unknowns ought to have $n - m$ independent solutions. In Section 4.5 of the text we use row and column spaces to show that this "conventional wisdom" is valid under the condition that the m equations are irredundant — meaning that the rank of the coefficient matrix $\mathbf{A}$ is m (so its m row vectors are linearly independent).

In each of Problems 1–12 we give the *reduced* echelon form $\mathbf{E}$ of the matrix $\mathbf{A}$, a basis for the row space of $\mathbf{A}$, and a basis for the column space of $\mathbf{A}$.

1. $\mathbf{E} = \begin{bmatrix} 1 & 0 & 11 \\ 0 & 1 & -4 \\ 0 & 0 & 0 \end{bmatrix}$

   Row basis: The first and second row vectors of $\mathbf{E}$.
   Column basis: The first and second column vectors of $\mathbf{A}$.

3. $E = \begin{bmatrix} 1 & 0 & 1 & 5 \\ 0 & 1 & 1 & 3 \\ 0 & 0 & 0 & 0 \end{bmatrix}$

Row basis: The first and second row vectors of **E**.
Column basis: The first and second column vectors of **A**.

5. $E = \begin{bmatrix} 1 & 0 & -2 & 0 \\ 0 & 1 & 3 & 0 \\ 0 & 0 & 0 & 1 \end{bmatrix}$

Row basis: The three row vectors of **E**.
Column basis: The first, second, and fourth column vectors of **A**.

7. $E = \begin{bmatrix} 1 & 0 & -3 & 4 \\ 0 & 1 & 2 & 3 \\ 0 & 0 & 0 & 0 \\ 0 & 0 & 0 & 0 \end{bmatrix}$

Row basis: The first two row vectors of **E**.
Column basis: The first two column vectors of **A**.

9. $E = \begin{bmatrix} 1 & 0 & 0 & 3 \\ 0 & 1 & 0 & -2 \\ 0 & 0 & 1 & 4 \\ 0 & 0 & 0 & 0 \end{bmatrix}$

Row basis: The first three row vectors of **E**.
Column basis: The first three column vectors of **A**.

11. $E = \begin{bmatrix} 1 & 0 & 2 & 1 & 0 \\ 0 & 1 & 1 & 2 & 0 \\ 0 & 0 & 0 & 0 & 1 \\ 0 & 0 & 0 & 0 & 0 \end{bmatrix}$

Row basis: The first three row vectors of **E**.
Column basis: The first, second, and fifth column vectors of **A**.

In each of Problems 13–16 we give the *reduced* echelon form **E** of the matrix having the given vectors $\mathbf{v}_1, \mathbf{v}_2, \ldots$ as its column vectors.

13. $\quad \mathbf{E} = \begin{bmatrix} 1 & 0 & 1 \\ 0 & 1 & 2 \\ 0 & 0 & 0 \\ 0 & 0 & 0 \end{bmatrix}$

Linearly independent: $\mathbf{v}_1$ and $\mathbf{v}_2$

15. $\quad \mathbf{E} = \begin{bmatrix} 1 & 0 & 2 & 0 \\ 0 & 1 & -1 & 0 \\ 0 & 0 & 0 & 1 \\ 0 & 0 & 0 & 0 \end{bmatrix}$

Linearly independent: $\mathbf{v}_1, \mathbf{v}_2$, and $\mathbf{v}_4$

In each of Problems 17–20 the matrix **E** is the *reduced* echelon matrix of the matrix $\mathbf{A} = \begin{bmatrix} \mathbf{v}_1 & \cdots & \mathbf{v}_k & \mathbf{e}_1 & \cdots & \mathbf{e}_n \end{bmatrix}$.

17. $\quad \mathbf{E} = \begin{bmatrix} 1 & 0 & -3 & 0 & 2 \\ 0 & 1 & 2 & 0 & -1 \\ 0 & 0 & 0 & 1 & -1 \end{bmatrix}$

Basis vectors: $\mathbf{v}_1, \mathbf{v}_2, \mathbf{e}_2$

19. $\quad \mathbf{E} = \begin{bmatrix} 1 & 0 & 3 & 0 & 0 & -2 \\ 0 & 1 & -1 & 0 & 0 & 1 \\ 0 & 0 & 0 & 1 & 0 & -1 \\ 0 & 0 & 0 & 0 & 1 & -1 \end{bmatrix}$

Basis vectors: $\mathbf{v}_1, \mathbf{v}_2, \mathbf{e}_2, \mathbf{e}_3$

In each of Problems 21–24 the matrix $\mathbf{E}$ is the *reduced* echelon form of the transpose $\mathbf{A}^T$ of the coefficient matrix $\mathbf{A}$.

21. $\mathbf{E} = \begin{bmatrix} 1 & 0 & 2 \\ 0 & 1 & 1 \\ 0 & 0 & 0 \end{bmatrix}$

The first and second equations are irredundant.

23. $\mathbf{E} = \begin{bmatrix} 1 & 0 & 2 & 0 \\ 0 & 1 & 1 & 0 \\ 0 & 0 & 0 & 1 \\ 0 & 0 & 0 & 0 \end{bmatrix}$

The first, second, and fourth equations are irredundant.

25.   The row vectors of $\mathbf{A}$ are the column vectors of its transpose matrix $\mathbf{A}^T$, so

$$\text{rank}(\mathbf{A}) = \text{row rank of } \mathbf{A} = \text{column rank of } \mathbf{A}^T = \text{rank}(\mathbf{A}^T).$$

27.   The rank of the 3×5 matrix $\mathbf{A}$ is 3, so its column vectors $\mathbf{a}_1, \mathbf{a}_1, \ldots, \mathbf{a}_5$ span $\mathbf{R}^3$. Therefore any given vector $\mathbf{b}$ in $\mathbf{R}^3$ can be expressed as a linear combination $\mathbf{b} = x_1\mathbf{a}_1 + x_2\mathbf{a}_2 + \cdots + x_5\mathbf{a}_5$ of the column vectors of $\mathbf{A}$. The column vector $\mathbf{x}$ whose elements are the coefficients in this linear combination is then a solution of the equation $\mathbf{A}\mathbf{x} = \mathbf{b}$.

29.   The rank of the $m \times n$ matrix $\mathbf{A}$ is at most $m < n$, and therefore is less than the number n of its column vectors. Hence the column vectors $\mathbf{a}_1, \mathbf{a}_1, \ldots, \mathbf{a}_n$ of $\mathbf{A}$ are linearly dependent, so there exists a linear combination $y_1\mathbf{a}_1 + y_2\mathbf{a}_2 + \cdots + y_n\mathbf{a}_n = \mathbf{0}$ with not all the coefficients being zero. If $\mathbf{x} = \begin{bmatrix} x_1 & x_2 & \ldots & x_n \end{bmatrix}^T$ is one solution of the equation $\mathbf{A}\mathbf{x} = \mathbf{b}$, then $\mathbf{A}(\mathbf{x} + \mathbf{y}) = \mathbf{A}\mathbf{x} + \mathbf{A}\mathbf{y} = \mathbf{b} + \mathbf{0} = \mathbf{b}$, so $\mathbf{x} + \mathbf{y}$ is a second different solution. Thus solutions of the equation are not unique.

31.   The rank of the $m \times n$ matrix $\mathbf{A}$ is m if and only if $\mathbf{A}$ has m linearly independent column vectors — in which case these m linearly independent column vectors constitute a basis for $\mathbf{R}^m$. Hence the rank of $\mathbf{A}$ is m if and only if its column vectors $\mathbf{a}_1, \mathbf{a}_1, \ldots, \mathbf{a}_n$ span $\mathbf{R}^m$ — in which case every vector $\mathbf{b}$ in $\mathbf{R}^m$ can be expressed as a linear combination $\mathbf{b} = x_1\mathbf{a}_1 + x_2\mathbf{a}_2 + \cdots + x_n\mathbf{a}_n$, so the equation $\mathbf{A}\mathbf{x} = \mathbf{b}$ has the solution $\mathbf{x} = \begin{bmatrix} x_1 & x_2 & \ldots & x_n \end{bmatrix}^T$.

33.   Suppose that some linear combination of the k pivot column vectors $\mathbf{p}_1, \mathbf{p}_1, \ldots, \mathbf{p}_k$ in (8) equals the zero vector. Denote by $c_1, c_1, \ldots, c_k$ the coefficients in this linear combination. Then the first k scalar components of the equation $c_1\mathbf{p}_1 + c_2\mathbf{p}_2 + c_3\mathbf{p}_3 + \cdots + c_k\mathbf{p}_k = \mathbf{0}$ yield the $k \times k$ upper-triangular system

$$
\begin{aligned}
c_1 d_1 + c_2 p_{21} + c_3 p_{31} + \cdots + c_k p_{k1} &= 0 \\
c_2 d_2 + c_3 p_{32} + \cdots + c_k p_{k2} &= 0 \\
c_3 d_3 + \cdots + c_k p_{k3} &= 0 \\
&\vdots \\
c_k d_k &= 0
\end{aligned}
$$

where p_{ij} (for $i > j$) denotes the jth element of the vector $\mathbf{p}_i$ and $p_{ii} = d_i$. Because the leading entries $d_1, d_1, \ldots, d_k$ are all nonzero, it follows by back-substitution that $c_1 = c_2 = \cdots = c_k = 0$. Therefore the column vectors are linearly independent.

35.   Look at the r row vectors of the matrix **A** that are determined by its largest nonsingular $r \times r$ submatrix. Then Theorem 3 in Section 4.3 says that these r row vectors are linearly independent, whereas any $r + 1$ row vectors of **A** are linearly dependent.

SECTION 4.6

ORTHOGONAL VECTORS IN R^n

The generalization in this section, of the dot product to vectors in **R**^{n}, enables us to flesh out the algebra of vectors in **R**^{n} with the Euclidean geometry of angles and distance. We can now refer to the vector space **R**^{n} (provided with the dot product) as n-dimensional *Euclidean space*.

1. $\mathbf{v}_1 \cdot \mathbf{v}_2 = (2)(3) + (1)(-6) + (2)(1) + (1)(-2) = 6 - 6 + 2 - 2 = 0$
   $\mathbf{v}_1 \cdot \mathbf{v}_3 = (2)(3) + (1)(-1) + (2)(-5) + (1)(5) = 6 - 1 - 10 + 5 = 0$
   $\mathbf{v}_2 \cdot \mathbf{v}_3 = (3)(3) + (-6)(-1) + (1)(-5) + (-2)(5) = 9 + 6 - 5 - 10 = 0$

   Yes, the three vectors are mutually orthogonal.

3. $\mathbf{v}_1 \cdot \mathbf{v}_2 = 15 - 10 - 4 - 1 = 0$, $\mathbf{v}_1 \cdot \mathbf{v}_3 = 15 + 0 - 32 + 17 = 0$, $\mathbf{v}_2 \cdot \mathbf{v}_3 = 9 + 0 + 8 - 17 = 0$
   Yes, the three vectors are mutually orthogonal.

In each of Problems 5–8 we write $\mathbf{u} = \overrightarrow{CB}$, $\mathbf{v} = \overrightarrow{CA}$, and $\mathbf{w} = \overrightarrow{AB}$. Then we calculate $a = |\mathbf{u}|$, $b = |\mathbf{v}|$, and $c = |\mathbf{w}|$ so as to verify that $a^2 + b^2 = c^2$.

5. $\mathbf{u} = (1,1,2,-1),\ \mathbf{v} = (1,-1,1,2),\ \mathbf{w} = (0,2,1,-3);\quad a^2 = 7,\ b^2 = 7,\ c^2 = 14$

7. $\mathbf{u} = (2,1,-2,1,3),\ \mathbf{v} = (3,2,2,2,-2),\ \mathbf{w} = (-1,-1,-4,-1,5);\quad a^2 = 19,\ b^2 = 25,\ c^2 = 44$

The computations in Problems 5–8 show that in each triangle $\triangle ABC$ the angle at C is a right angle. The angles at the vertices A and B are then determined by the relations

$$\cos \angle A = \frac{\overrightarrow{AB} \cdot \overrightarrow{AC}}{|\overrightarrow{AB}||\overrightarrow{AC}|} = -\frac{\mathbf{v} \cdot \mathbf{w}}{|\mathbf{v}||\mathbf{w}|} \quad \text{and} \quad \cos \angle B = \frac{\overrightarrow{BA} \cdot \overrightarrow{BC}}{|\overrightarrow{BA}||\overrightarrow{BC}|} = +\frac{\mathbf{u} \cdot \mathbf{w}}{|\mathbf{u}||\mathbf{w}|}.$$

The fact that $\angle A + \angle B = 90°$ then serves as a check on our numerical computations.

9. $\angle A = \cos^{-1}\left(-\dfrac{-7}{\sqrt{7}\sqrt{14}}\right) = \cos^{-1}\left(\dfrac{1}{\sqrt{2}}\right) = 45°,\ \angle B = \cos^{-1}\left(+\dfrac{7}{\sqrt{7}\sqrt{14}}\right) = \cos^{-1}\left(\dfrac{1}{\sqrt{2}}\right) = 45°$

11. $\angle A = \cos^{-1}\left(-\dfrac{-25}{\sqrt{25}\sqrt{44}}\right) = \cos^{-1}\left(\sqrt{\dfrac{25}{44}}\right) = 41.08°,$

$\angle B = \cos^{-1}\left(\dfrac{19}{\sqrt{19}\sqrt{44}}\right) = \cos^{-1}\left(\sqrt{\dfrac{19}{44}}\right) = 48.92°$

In each of Problems 13–22, we denote by $\mathbf{A}$ the matrix having the given vectors as its row vectors, and by $\mathbf{E}$ the reduced echelon form of $\mathbf{A}$. From $\mathbf{E}$ we find the general solution of the homogeneous system $\mathbf{Ax} = \mathbf{0}$ in terms of parameters $s, t, \ldots$. We then get basis vectors $\mathbf{u}_1, \mathbf{u}_2, \ldots$ for the orthogonal complement $V^{\perp}$ by setting each parameter in turn equal to 1 (and the others then equal to 0).

13. $\mathbf{A} = \mathbf{E} = \begin{bmatrix} 1 & -2 & 3 \end{bmatrix};\ x_2 = s,\ x_3 = t,\ x_1 = 2s - 3t$
        $\mathbf{u}_1 = (2,1,0),\ \mathbf{u}_2 = (-3,0,1)$

15. $\mathbf{A} = \mathbf{E} = \begin{bmatrix} 1 & -2 & -3 & 5 \end{bmatrix};\ x_2 = r,\ x_3 = s,\ x_4 = t,\ x_1 = 2r + 3s - 5t$
        $\mathbf{u}_1 = (2,1,0,0),\ \mathbf{u}_2 = (3,0,1,0),\ \mathbf{u}_3 = (-5,0,0,1)$

17. $E = \begin{bmatrix} 1 & 0 & -7 & 19 \\ 0 & 1 & 3 & -5 \end{bmatrix}$

$x_3 = s,\ x_4 = t,\ x_2 = -3s + 5t,\ x_1 = 7s - 19t$

$\mathbf{u}_1 = (7, -3, 1, 0),\ \ \mathbf{u}_2 = (-19, 5, 0, 1)$

19. $E = \begin{bmatrix} 1 & 0 & 13 & -4 & 11 \\ 0 & 1 & -4 & 3 & -4 \end{bmatrix}$

$x_3 = r,\ x_4 = s,\ x_5 = t,\ x_2 = 4r - 3s + 4t,\ x_1 = -13r + 4s - 11t$

$\mathbf{u}_1 = (-13, 4, 1, 0, 0),\ \ \mathbf{u}_2 = (4, -3, 0, 1, 0),\ \ \mathbf{u}_3 = (-11, 4, 0, 0, 1)$

21. $E = \begin{bmatrix} 1 & 0 & 1 & 0 & 0 \\ 0 & 1 & 1 & 0 & 1 \\ 0 & 0 & 0 & 1 & 1 \end{bmatrix}$

$x_3 = s,\ x_5 = t,\ x_4 = -t,\ x_2 = -s - t,\ x_1 = -s$

$\mathbf{u}_1 = (-1, -1, 1, 0, 0),\ \ \mathbf{u}_2 = (0, -1, 0, -1, 1)$

23.   (a) $\begin{aligned}[t] \left|\mathbf{u} + \mathbf{v}\right|^2 + \left|\mathbf{u} - \mathbf{v}\right|^2 &= (\mathbf{u} \cdot \mathbf{u} + 2\mathbf{u} \cdot \mathbf{v} + \mathbf{v} \cdot \mathbf{v}) + (\mathbf{u} \cdot \mathbf{u} - 2\mathbf{u} \cdot \mathbf{v} + \mathbf{v} \cdot \mathbf{v}) \\ &= 2\left|\mathbf{u}\right|^2 + \left|\mathbf{v}\right|^2 = 2\mathbf{u} \cdot \mathbf{u} + 2\mathbf{u} \cdot \mathbf{u} \end{aligned}$

   (b) $\left|\mathbf{u} + \mathbf{v}\right|^2 - \left|\mathbf{u} - \mathbf{v}\right|^2 = (\mathbf{u} \cdot \mathbf{u} + 2\mathbf{u} \cdot \mathbf{v} + \mathbf{v} \cdot \mathbf{v}) - (\mathbf{u} \cdot \mathbf{u} - 2\mathbf{u} \cdot \mathbf{v} + \mathbf{v} \cdot \mathbf{v}) = 4\mathbf{u} \cdot \mathbf{v}$

25.   Suppose, for instance, that $A = \mathbf{e}_1 = (1, 0, 0, 0, 0),\ B = \mathbf{e}_3 = (0, 0, 1, 0, 0)$, and $C = \mathbf{e}_5 = (0, 0, 0, 0, 1)$ in $\mathbf{R}^5$. Then $\overrightarrow{AB} = \mathbf{e}_3 - \mathbf{e}_1 = (-1, 0, 1, 0, 0)$ and $\overrightarrow{AC} = \mathbf{e}_5 - \mathbf{e}_1 = (0, 0, 1, 0, 0, -1)$. Then $\overrightarrow{AB} \cdot \overrightarrow{AC} = 1$ while $\left|\overrightarrow{AB}\right| = \left|\overrightarrow{AC}\right| = \sqrt{2}$. It follows that $\cos \angle A = 1/(\sqrt{2})(\sqrt{2}) = \tfrac{1}{2}$, so $\angle A = 60°$. Similarly, $\angle B = \angle C = 60°$, so we see that $\triangle ABC$ is an equilateral triangle.

27.   If the $\mathbf{u}$ lines both in the subspace V and in its orthogonal complement $V^\perp$, then the vector $\mathbf{u}$ is orthogonal to itself. Hence $\mathbf{u} \cdot u = \left|\mathbf{u}\right|^2 = 0$, so it follows that $\mathbf{u} = \mathbf{0}$.

29.   If **u** is orthogonal to each vector in the set S of vectors, then it follows easily (using the dot product) that **u** is orthogonal to every linear combination of vectors in S. Therefore **u** is orthogonal to $V = \text{Span}(S)$.

31.   We want to show that any linear combination of vectors $\mathbf{u}_1, \mathbf{u}_2, \ldots, \mathbf{u}_p$ of vectors in S is orthogonal to every linear combination of vectors $\mathbf{v}_1, \mathbf{v}_2, \ldots, \mathbf{v}_q$ in T. But if each $\mathbf{u}_i$ is orthogonal to each $\mathbf{v}_j$, so $\mathbf{u}_i \cdot \mathbf{v}_j = 0$, then it follows that

$$\left(a_1\mathbf{u}_1 + a_2\mathbf{u}_2 + \cdots + a_p\mathbf{u}_p\right) \cdot \left(b_1\mathbf{v}_1 + b_2\mathbf{v}_2 + \cdots + b_q\mathbf{v}_q\right) = \sum_{i=1}^{p}\sum_{j=1}^{q} a_i b_j \mathbf{u}_i \cdot \mathbf{v}_j = 0,$$

so we see that the two linear combinations are orthogonal, as desired.

33.   This is the same as Problem 32, except with

$$\mathbf{u} = a_1\mathbf{u}_1 + a_2\mathbf{u}_2 + \cdots + a_k\mathbf{u}_k \qquad \text{and} \qquad \mathbf{v} = b_1\mathbf{v}_1 + b_2\mathbf{v}_2 + \cdots + b_m\mathbf{v}_m.$$

35.   This is one of the fundamental theorems of linear algebra. The nonhomogeneous system

$$\mathbf{Ax} = \mathbf{b}$$

is consistent if and only if the vector **b** is in the subspace $\text{Col}(\mathbf{A}) = \text{Row}(\mathbf{A}^T)$. But **b** is in $\text{Row}(\mathbf{A}^T)$ if and only if **b** is orthogonal to the orthogonal complement of $\text{Row}(\mathbf{A}^T)$. But $\text{Row}(\mathbf{A}^T)^\perp = \text{Null}(\mathbf{A}^T)$, which is the solution space of the homogeneous system

$$\mathbf{A}^T\mathbf{y} = \mathbf{0}.$$

Thus we have proved (as desired) that the nonhomogeneous system $\mathbf{Ax} = \mathbf{b}$ has a solution if and only if the constant vector **b** is orthogonal to every solution **y** of the nonhomogeneous system $\mathbf{A}^T\mathbf{y} = \mathbf{0}$.

SECTION 4.7

GENERAL VECTOR SPACES

In each of Problems 1–12, a certain subset of a vector space is described. This subset is a subspace of the vector space if and only if it is closed under the formation of linear combinations of its elements. Recall also that every subspace of a vector space must contain the zero vector.

1.   It is a subspace of M_{33}, because any linear combination of diagonal 3×3 matrices — with only zeros off the principal diagonal — obviously is again a diagonal matrix.

3.   The set of all nonsingular 3×3 matrices does not contain the zero matrix, so it is not a subspace.

5.   The set of all functions $f:\mathbf{R}\to\mathbf{R}$ with $f(0)=0$ is a vector space, because if $f(0)=g(0)=0$ then $(af+bg)(0) = af(0)+bg(0) = a\cdot0+b\cdot0 = 0.$

7.   The set of all functions $f:\mathbf{R}\to\mathbf{R}$ with $f(0)=0$ and $f(1)=1$ is not a vector space. For instance, if $g=2f$ then $g(1)=2f(1)=2\cdot1=2\neq1$, so g is not such a function. Also, this set does not contain the zero function.

For Problems 9–12, let us call a polynomial of the form $a_0+a_1x+a_2x^2+a_3x^3$ a "degree at most 3" polynomial.

9.   The set of all degree at most 3 polynomials with nonzero leading coefficient $a_3\neq0$ is not a vector space, because it does not contain the zero polynomial (with all coefficients zero).

11.   The set of all degree at most 3 polynomials with coefficient sum zero is a vector space, because any linear combination of such polynomials obviously is such a polynomial.

13.   The functions $\sin x$ and $\cos x$ are linearly independent, because neither is a scalar multiple of the other. (This follows, for instance, from the facts that $\sin(0)=0$, $\cos(0)=1$ and $\sin(\pi/2)=1$, $\cos(\pi/2)=0$, noting that any scalar multiple of a function with a zero value must have the value 0 at the same point.)

15.   If
$$c_1(1+x)+c_2(1-x)+c_3(1-x^2) = (c_1+c_2+c_3)+(c_1-c_2)x-c_3x^2 = 0,$$
then
$$c_1+c_2+c_3 = c_1-c_2 = c_3 = 0.$$

It follows easily that $c_1=c_2=c_3=0$, so we conclude that the functions $(1+x)$, $(1-x)$, and $(1-x^2)$ are linearly independent.

17. $\cos2x = \cos^2x-\sin^2x$ according to a well-known trigonometric identity. Thus these three trigonometric functions are linearly dependent.

19.   Multiplication by $(x-2)(x-3)$ yields
$$x-5 = A(x-3)+B(x-2) = (A+B)x-(3A+2B).$$

Hence $A+B=1$ and $3A+2B=5$, and it follows readily that $A=3$ and $B=-2$.

21.   Multiplication by $x(x^2+4)$ yields

$$8 = A(x^2+4)+Bx^2+Cx = 4A+Cx+(A+B)x^2.$$

Hence $4A=8$, $C=0$ and $A+B=0$. It follows readily that $A=2$, $B=-2$, and $C=0$.

23.   If $y'''(x)=0$ then

$$y''(x) = \int y'''(x)\,dx = \int (0)\,dx = A,$$
$$y'(x) = \int y''(x)\,dx = \int A\,dx = Ax+B, \text{ and}$$
$$y(x) = \int y'(x)\,dx = \int (Ax+B)\,dx = \tfrac{1}{2}Ax^2+Bx+C,$$

where A, B, and C are arbitrary constants of integration. It follows that the function $y(x)$ is a solution of the differential equation $y'''(x)=0$ if and only if it is a quadratic (at most 2nd degree) polynomial. Thus the solution space is 3-dimensional with basis $\{1, x, x^2\}$.

25.   If $y(x)$ is any solution of the second-order differential equation $y''-5y'=0$ and $v(x)=y'(x)$, then $v(x)$ is a solution of the first-order differential equation $v'(x)=5v(x)$ with the familiar exponential solution $v(x) = Ce^{5x}$. Therefore

$$y(x) = \int y'(x)\,dx = \int v(x)\,dx = \int Ce^{5x}\,dx = \tfrac{1}{5}Ce^{5x}+D.$$

We therefore see that the solution space of the equation $y''-5y'=0$ is 2-dimensional with basis $\{1, e^{5x}\}$.

27.   If we take the positive sign in Eq. (20) of the text, then we have $v^2 = y^2+a^2$ where $v(x) = y'(x)$. Then

$$\left(\frac{dy}{dx}\right)^2 = y^2+a^2, \text{ so } \frac{dx}{dy} = \frac{1}{\sqrt{y^2+a^2}}$$

(taking the positive square root as in the text). Then

$$x = \int \frac{dy}{\sqrt{y^2 + a^2}} = \int \frac{a\,du}{\sqrt{a^2u^2 + a^2}} \qquad (y = a\,u)$$

$$= \int \frac{du}{\sqrt{u^2 + 1}} = \sinh^{-1} u + b = \sinh^{-1} \frac{y}{a} + b.$$

It follows that

$$y(x) = a\sinh(x - b) = a\left(\sinh x \cosh b - \cosh x \sinh b\right)$$
$$= A\cosh x + B\sinh x.$$

29. **(a)** The verification in a component-wise manner that V is a vector space is the same as the verification that $\mathbf{R}^n$ is a vector space, except with vectors having infinitely many components rather than finitely many components. It boils down to the fact that a linear combination of infinite sequences of real numbers is itself such a sequence,

$$a \cdot \{x_n\}_1^\infty + b \cdot \{y_n\}_1^\infty = \{ax_n + by_n\}_1^\infty.$$

(b) If $\mathbf{e}_n = \{0, \cdots, 0, 1, 0, 0, \cdots\}$ is the indicated infinite sequence with 1 in the nth position, then the fact that

$$c_1\mathbf{e}_1 + c_2\mathbf{e}_2 + \cdots + c_k\mathbf{e}_k = \{c_1, c_2, \cdots, c_k, 0, 0, \cdots\}$$

evidently implies that any finite set $\mathbf{e}_1, \mathbf{e}_2, \cdots, \mathbf{e}_k$ of these vectors is linearly independent. Thus V contains "arbitrarily large" sets of linearly independent vectors, and therefore is infinite-dimensional.

31. **(a)** If $z_1 = a_1 + ib_1$ and $z_2 = a_2 + ib_2$, then direct computation shows that

$$T(c_1z_1 + c_2z_2) = c_1T(z_1) + c_2T(z_2) = \begin{bmatrix} c_1a_1 + c_2a_2 & -c_1b_1 - b_2a_2 \\ c_1b_1 + b_2a_2 & c_1a_1 + c_2a_2 \end{bmatrix}.$$

(b) If $z_1 = a_1 + ib_1$ and $z_2 = a_2 + ib_2$, then $z_1z_2 = (a_1a_2 - b_1b_2) + i(a_1b_2 + a_2b_1)$ and direct computation shows that

$$T(z_1z_2) = T(z_1)T(z_2) = \begin{bmatrix} a_1a_2 - b_1b_2 & -a_1b_2 - a_2b_1 \\ a_1b_2 + a_2b_1 & a_1a_2 - b_1b_2 \end{bmatrix}.$$

(b)   If $z = a + ib$ then

$$\frac{1}{z} = \frac{1}{a+bi} \cdot \frac{a-bi}{a-bi} = \frac{a-bi}{a^2+b^2}.$$

Therefore

$$T(z^{-1}) = \frac{1}{a^2+b^2}\begin{bmatrix} a & b \\ -b & a \end{bmatrix} = \begin{bmatrix} a & -b \\ b & a \end{bmatrix}^{-1} = T(z)^{-1}.$$

CHAPTER 5

LINEAR EQUATIONS OF HIGHER ORDER

SECTION 5.1

INTRODUCTION: SECOND-ORDER LINEAR EQUATIONS

In this section the central ideas of the theory of linear differential equations are introduced and illustrated concretely in the context of **second-order** equations. These key concepts include superposition of solutions (Theorem 1), existence and uniqueness of solutions (Theorem 2), linear independence, the Wronskian (Theorem 3), and general solutions (Theorem 4). This discussion of second-order equations serves as preparation for the treatment of nth order linear equations in Section 5.2. Although the concepts in this section may seem somewhat abstract to students, the problems set is quite tangible and largely computational.

In each of Problems 1–16 the verification that y_1 and y_2 satisfy the given differential equation is a routine matter. As in Example 2, we then impose the given initial conditions on the general solution $y = c_1 y_1 + c_2 y_2$. This yields two linear equations that determine the values of the constants c_1 and c_2.

1.   Imposition of the initial conditions $y(0) = 0$, $y'(0) = 5$ on the general solution

   $y(x) = c_1 e^x + c_2 e^{-x}$ yields the two equations $c_1 + c_2 = 0$, $c_1 - c_2 = 0$ with solution $c_1 = \dfrac{5}{2}$,

   $c_2 = -\dfrac{5}{2}$. Hence the desired particular solution is $y(x) = \dfrac{5}{2}\left(e^x - e^{-x}\right)$.

3.   Imposition of the initial conditions $y(0) = 3$, $y'(0) = 8$ on the general solution

   $y(x) = c_1 \cos 2x + c_2 \sin 2x$ yields the two equations $c_1 = 3$, $2c_2 = 8$ with solution $c_1 = 3$,

   $c_2 = 4$. Hence the desired particular solution is $y(x) = 3\cos 2x + 4\sin 2x$.

5.   Imposition of the initial conditions $y(0) = 1$, $y'(0) = 0$ on the general solution

   $y(x) = c_1 e^x + c_2 e^{2x}$ yields the two equations $c_1 + c_2 = 1$, $c_1 + 2c_2 = 0$ with solution

   $c_1 = 2$, $c_2 = -1$. Hence the desired particular solution is $y(x) = 2e^x - e^{2x}$.

7.   Imposition of the initial conditions $y(0) = -2$, $y'(0) = 8$ on the general solution
$y(x) = c_1 + c_2 e^{-x}$ yields the two equations $c_1 + c_2 = -2$, $-c_2 = 8$ with solution $c_1 = 6$,
$c_2 = -8$. Hence the desired particular solution is $y(x) = 6 - 8e^{-x}$.

9.   Imposition of the initial conditions $y(0) = 2$, $y'(0) = -1$ on the general solution
$y(x) = c_1 e^{-x} + c_2 x e^{-x}$ yields the two equations $c_1 = 2$, $-c_1 + c_2 = -1$ with solution
$c_1 = 2$ $c_2 = 1$. Hence the desired particular solution is $y(x) = 2e^{-x} + xe^{-x}$.

11.   Imposition of the initial conditions $y(0) = 0$, $y'(0) = 5$ on the general solution
$y(x) = c_1 e^x \cos x + c_2 e^x \sin x$ yields the two equations $c_1 = 0$, $c_1 + c_2 = 5$ with solution
$c_1 = 0$, $c_2 = 5$. Hence the desired particular solution is $y(x) = 5e^x \sin x$.

13.   Imposition of the initial conditions $y(1) = 3$, $y'(1) = 1$ on the general solution
$y(x) = c_1 x + c_2 x^2$ yields the two equations $c_1 + c_2 = 3$, $c_1 + 2c_2 = 1$ with solution $c_1 = 5$,
$c_2 = -2$. Hence the desired particular solution is $y(x) = 5x - 2x^2$.

15.   Imposition of the initial conditions $y(1) = 7$, $y'(1) = 2$ on the general solution
$y(x) = c_1 x + c_2 x \ln x$ yields the two equations $c_1 = 7$, $c_1 + c_2 = 2$ with solution $c_1 = 7$,
$c_2 = -5$. Hence the desired particular solution is $y(x) = 7x - 5x \ln x$.

17.   If $y = \dfrac{c}{x}$, then $y' + y^2 = -\dfrac{c}{x^2} + \dfrac{c^2}{x^2} = \dfrac{c(c-1)}{x^2} \neq 0$ unless either $c = 0$ or $c = 1$.

19.   If $y = 1 + \sqrt{x}$, then $yy'' + (y')^2 = \left(1 + \sqrt{x}\right)\left(-\dfrac{x^{-3/2}}{4}\right) + \left(\dfrac{x^{-1/2}}{2}\right)^2 = -\dfrac{x^{-3/2}}{4} \neq 0$.

21.   Linearly independent, because $x^3 = +x^2 |x|$ if $x > 0$, whereas $x^3 = -x^2 |x|$ if $x < 0$.

23.   Linearly independent, because $f(x) = +g(x)$ if $x > 0$, whereas $f(x) = -g(x)$ if $x < 0$.

25. $f(x) = e^x \sin x$ and $g(x) = e^x \cos x$ are linearly independent, because $f(x) = kg(x)$
would imply that $\sin x = k \cos x$, whereas $\sin x$ and $\cos x$ are linearly independent, as
noted in Example 3.

27. The operator notation used elsewhere in this chapter is convenient here. Let $L[y]$ denote $y'' + py' + qy$. Then $L[y_c] = 0$ and $L[y_p] = f$, so $L[y_c + y_p] = 0 + f = f$.

29. There is no contradiction, because if the given differential equation is divided by x^2 to get the form in Equation (8) in the text, then the resulting functions $p(x) = -\dfrac{4}{x}$ and $q(x) = \dfrac{6}{x^2}$ are not continuous at $x = 0$.

31. $W(y_1, y_2) = -2x$ vanishes at $x = 0$, whereas if y_1 and y_2 were (linearly independent) solutions of an equation $y'' + py' + qy = 0$ with p and q both continuous on an open interval I containing $x = 0$, then Theorem 3 would imply that $W \neq 0$ on I.

(c) Because the exponential factor is never zero.

In Problems 33–42 we give the characteristic equation, its roots, and the corresponding general solution.

33. $r^2 - 3r + 2 = 0$; $r = 1, 2$; $y(x) = c_1 e^x + c_2 e^{2x}$

35. $r^2 + 5r = 0$; $r = 0, -5$; $y(x) = c_1 + c_2 e^{-5x}$

37. $2r^2 - r - 2 = 0$; $r = 1, -\dfrac{1}{2}$; $y(x) = c_1 e^{-x/2} + c_2 e^x$

39. $4r^2 + 4r + 1 = 0$; $r = -\dfrac{1}{2}$ (repeated); $y(x) = (c_1 + c_2 x) e^{-x/2}$

41. $6r^2 - 7r - 20 = 0$; $r = -\dfrac{4}{3}, \dfrac{5}{2}$; $y(x) = c_1 e^{-4x/3} + c_2 e^{5x/2}$

In Problems 43–48 we first write and simplify the equation with the indicated characteristic roots, and then write the corresponding differential equation.

43. $(r - 0)(r + 10) = r^2 + 10r = 0$; $y'' + 10y' = 0$

45. $(r + 10)(r + 10) = r^2 + 20r + 100 = 0$; $y'' + 20y' + 100y = 0$

47. $(r - 0)(r - 0) = r^2 = 0$; $y'' = 0$

49.   The solution curve with $y(0)=1$, $y'(0)=6$ is $y(x)=8e^{-x}-7e^{-2x}$. We find that

$y'(x)=0$ when $x=\ln\dfrac{7}{4}$, so that $e^{-x}=\dfrac{4}{7}$ and $e^{-2x}=\dfrac{16}{49}$. It follows that $y\left(\ln\dfrac{7}{4}\right)=\dfrac{16}{7}$,

so the high point on the curve is $\left(\ln\dfrac{7}{4},\dfrac{16}{7}\right)\approx(0.56,2.29)$, which looks consistent with Fig. 5.1.6.

51.   **(a)** The substitution $v=\ln x$ gives

$$y'=\frac{dy}{dx}=\frac{dy}{dv}\frac{dv}{dx}=\frac{1}{x}\frac{dy}{dv}.$$

Then another differentiation using the chain and product rules gives

$$
\begin{aligned}
y''&=\frac{d^2y}{dx^2}\\
&=\frac{d}{dx}\left(\frac{dy}{dx}\right)\\
&=\frac{d}{dx}\left(\frac{1}{x}\cdot\frac{dy}{dv}\right)\\
&=-\frac{1}{x^2}\cdot\frac{dy}{dv}+\frac{1}{x}\cdot\frac{d}{dx}\left(\frac{dy}{dv}\right)\\
&=-\frac{1}{x^2}\cdot\frac{dy}{dv}+\frac{1}{x}\cdot\frac{d}{dv}\left(\frac{dy}{dv}\right)\cdot\frac{dv}{dx}\\
&=-\frac{1}{x^2}\cdot\frac{dy}{dv}+\frac{1}{x^2}\cdot\frac{d^2y}{dv^2}.
\end{aligned}
$$

Substitution of these expressions for y' and y'' into Eq. (21) in the text then yields immediately the desired Eq. (23):

$$a\frac{d^2y}{dv^2}+(b-a)\frac{dy}{dv}+cy=0.$$

(b) If the roots r_1 and r_2 of the characteristic equation of Eq. (23) are real and distinct, then a general solution of the original Euler equation is

$$y(x)=c_1e^{r_1v}+c_2e^{r_2v}=c_1\left(e^v\right)^{r_1}+c_2\left(e^v\right)^{r_2}=c_1x^{r_1}+c_2x^{r_2}.$$

53.   The substitution $v=\ln x$ yields the converted equation $\dfrac{dy^2}{dv^2}+\dfrac{dy}{dv}-12y=0$, whose char-

acteristic equation $r^2+r-12=0$ has roots $r_1=-4$ and $r_2=3$. Because $e^v=x$, the cor-

responding general solution is $y=c_1e^{-4v}+c_2e^{3v}=c_1x^{-4}+c_2x^3$.

55.   The substitution $v = \ln x$ yields the converted equation $\dfrac{dy^2}{dv^2} = 0$, whose characteristic equation $r^2 = 0$ has repeated roots $r_1, r_2 = 0$. Because $v = \ln x$, the corresponding general solution is $y = c_1 + c_2 v = c_1 + c_2 \ln x$.

SECTION 5.2

GENERAL SOLUTIONS OF LINEAR EQUATIONS

Students should check each of Theorems 1 through 4 in this section to see that, in the case $n = 2$, it reduces to the corresponding theorem in Section 5.1. Similarly, the computational problems for this section largely parallel those for the previous section. By the end of Section 5.2 students should understand that, although we do not prove the existence-uniqueness theorem now, it provides the basis for everything we do with linear differential equations.

The linear combinations listed in Problems 1–6 were discovered "by inspection"—that is, by trial and error.

1. $\dfrac{5}{2} \cdot 2x + \left(-\dfrac{8}{3} \right) \cdot 3x^2 + (-1)\left(5x - 8x^2 \right) = 0$ for all x.

3. $1 \cdot 0 + 0 \cdot \sin x + 0 \cdot e^x = 0$ for all x.

5. $1 \cdot 17 + (-34) \cdot \cos^2 x + 17 \cdot \cos 2x = 0$ for all x, because $2\cos^2 x = 1 + \cos 2x$.

7. $W = \begin{vmatrix} 1 & x & x^2 \\ 0 & 1 & 2x \\ 0 & 0 & 2 \end{vmatrix} = 2$ is nonzero everywhere.

9. $W = e^x \left(\cos^2 x + \sin^2 x \right) = e^x$ is never zero.

11. $W = x^3 e^{2x}$ is nonzero if $x \neq 0$.

In each of Problems 13-20 we first form the general solution
$y(x) = c_1 y_1(x) + c_2 y_2(x) + c_3 y_3(x)$, then calculate $y'(x)$ and $y''(x)$, and finally impose the
given initial conditions to determine the values of the coefficients c_1, c_2, c_3.

13. Imposition of the initial conditions $y(0) = 1$, $y'(0) = 2$, $y''(0) = 0$ on the general solution $y(x) = c_1 e^x + c_2 e^{-x} + c_3 e^{-2x}$ yields the three equations

$$c_1 + c_2 + c_3 = 1, \quad c_1 - c_2 - 2c_3 = 2, \quad c_1 + c_2 + 4c_3 = 0,$$

with solution $c_1 = \dfrac{4}{3}$, $c_2 = 0$, $c_3 = -\dfrac{1}{3}$. Hence the desired particular solution is given by

$$y(x) = \frac{1}{3}\left(4e^x - e^{-2x}\right).$$

15. Imposition of the initial conditions $y(0) = 2$, $y'(0) = 0$, $y''(0) = 0$ on the general solution $y(x) = c_1 e^x + c_2 x e^x + c_3 x^2 e^{3x}$ yields the three equations

$$c_1 = 2, \quad c_1 + c_2 = 0, \quad c_1 + 2c_2 + 2c_3 = 0,$$

with solution $c_1 = 2$, $c_2 = -2$, $c_3 = 1$. Hence the desired particular solution is given by
$$y(x) = \left(2 - 2x + x^2\right)e^x.$$

17. Imposition of the initial conditions $y(0) = 3$, $y'(0) = 1$, $y''(0) = 2$ on the general solution $y(x) = c_1 + c_2 \cos 3x + c_3 \sin 3x$ yields the three equations

$$c_1 + c_3 = 3, \quad 3c_3 = -1, \quad -9c_2 = 2$$

with solution $c_1 = \dfrac{29}{9}$, $c_2 = -\dfrac{2}{9}$, $c_3 = -\dfrac{1}{3}$. Hence the desired particular solution is given

by $y(x) = \dfrac{29}{9} - \dfrac{2}{9}\cos 3x - \dfrac{1}{3}\sin 3x$.

19. Imposition of the initial conditions $y(1) = 6$, $y'(1) = 14$, $y''(1) = 22$ on the general solution $y(x) = c_1 x + c_2 x^2 + c_3 x^3$ yields the three equations

$$c_1 + c_2 + c_3 = 6, \quad c_1 + 2c_2 + 3c_3 = 14, \quad 2c_2 + 6c_3 = 22,$$

with solution $c_1 = 1$, $c_2 = 2$, $c_3 = 3$. Hence the desired particular solution is given by
$$y(x) = x + 2x^2 + 3x^3.$$

In each of Problems 21-24 we first form the general solution

$$y(x) = y_c(x) + y_p(x) = c_1 y_1(x) + c_2 y_2(x) + y_p(x),$$

then calculate $y'(x)$, and finally impose the given initial conditions to determine the values of the coefficients c_1 and c_2.

21. Imposition of the initial conditions $y(0) = 2$, $y'(0) = -2$ on the general solution $y(x) = c_1 \cos x + c_2 \sin x + 3x$ yields the two equations $c_1 = 2$, $c_2 + 3 = -2$ with solution $c_1 = 2$, $c_2 = -5$. Hence the desired particular solution is given by $y(x) = 2\cos x - 5\sin x + 3x$.

23. Imposition of the initial conditions $y(0) = 3$, $y'(0) = 11$ on the general solution $y(x) = c_1 e^{-x} + c_2 e^{3x} - 2$ yields the two equations $c_1 + c_2 - 2 = 3$, $-c_1 + 3c_2 = 11$ with solution $c_1 = 1$, $c_2 = 4$. Hence the desired particular solution is given by $y(x) = e^{-x} + 4e^{3x} - 2$.

25. $Ly = L[y_1 + y_2] = Ly_1 + Ly_2 = f + g$

27. The equations

$$c_1 + c_2 x + c_3 x^2 = 10, \quad c_2 + 2c_3 x = 0, \quad 2c_3 = 0$$

(the latter two obtained by successive differentiation of the first one) evidently imply that $c_1 = c_2 = c_3 = 0$.

29. If $c_0 e^{rx} + c_1 x e^{rx} + \cdots + c_n x^n e^{rx} = 0$, then division by e^{rx} yields $c_0 + c_1 x + \cdots + c_n x^n = 0$, so the result of Problem 28 applies.

31. (a) Substitution of $x = a$ in the differential equation gives $y''(a) = -py'(a) - q(a)$.

(b) If $y(0) = 1$ and $y'(0) = 0$, then the equation $y'' - 2y' - 5y = 0$ implies that $y''(0) = 2y'(0) + 5y(0) = 5$.

33. This follows from the fact that

$$\begin{vmatrix} 1 & 1 & 1 \\ a & b & c \\ a^2 & b^2 & c^2 \end{vmatrix} = (b-a)(c-b)(c-a)$$

when a, b, and c are distinct, which can be verified by expanding both sides of the equation.

37.   When we substitute $y = vx^3$ in the given differential equation and simplify, we get the

separable equation $xv'' + v' = 0$, which we write as $\dfrac{v''}{v'} = -\dfrac{1}{x}$. Integrating gives

$\ln v' = -\ln x + \ln A$, and then solving for v' leads to $v' = \dfrac{A}{x}$, or finally $v(x) = A\ln x + B$.

With $A = 1$ and $B = 0$ we get $v(x) = \ln x$, and thus $y_2(x) = x^3 \ln x$.

39.   When we substitute $y = ve^{x/2}$ in the given differential equation and simplify, we eventually get the simple equation $v'' = 0$, with general solution $v(x) = Ax + B$. With $A = 1$ and $B = 0$ we get $v(x) = x$, and hence $y_2(x) = xe^{x/2}$.

41.   When we substitute $y = ve^x$ in the given differential equation and simplify, we get the

separable equation $(1 + x)v'' + xv' = 0$, which we write as $\dfrac{v''}{v'} = -\dfrac{x}{1+x} = -1 + \dfrac{1}{1+x}$. Inte-

grating gives $\ln v' = -x + \ln(1+x) + \ln A$, and then solving for v' leads to

$v' = A(1+x)e^{-x}$, or finally $v(x) = A\int(1+x)e^{-x}\,dx = -A(2+x)e^{-x} + B$. With $A = -1$

and $B = 0$ we get $v(x) = (2+x)e^{-x}$, and hence $y_2(x) = 2 + x$.

43.   When we substitute $y = vx$ in the given differential equation and simplify, we get the

separable equation $x(x^2 - 1)v'' = (2 - 4x^2)v'$, which we write using the method of partial

fractions as

$$\frac{v''}{v'} = \frac{2 - 4x^2}{x(x^2 - 1)} = -\frac{2}{x} - \frac{1}{1+x} + \frac{1}{1-x}.$$

Integrating gives

$$\ln v' = -2\ln x - \ln(1+x) - \ln(1-x) + \ln A,$$

and then solving for v' leads to

$$v' = \frac{A}{x^2(1-x^2)} = A\left[\frac{1}{x^2} + \frac{1}{2(1+x)} + \frac{1}{2(1-x)}\right],$$

or finally

$$v(x) = A\left[-\frac{1}{x} + \frac{1}{2}\ln(1+x) - \frac{1}{2}\ln(1-x)\right] + B.$$

With $A = -1$ and $B = 0$ we get $v(x) = \dfrac{1}{x} - \dfrac{1}{2}\ln(1+x) + \dfrac{1}{2}\ln(1-x)$, and hence

$$y_2(x) = 1 - \frac{x}{2}\ln\frac{1+x}{1-x}.$$

SECTION 5.3

HOMOGENEOUS EQUATIONS WITH CONSTANT COEFFICIENTS

This is a purely computational section devoted to the single most widely applicable type of higher order differential equations—linear ones with constant coefficients. In Problems 1–20, we first write the characteristic equation and list its roots, then give the corresponding general solution of the given differential equation. Explanatory comments are included only when the solution of the characteristic equation is not routine.

1. $r^2 - 4 = (r-2)(r+2) = 0$; $r = -2, 2$; $y(x) = c_1 e^{2x} + c_2 e^{-2x}$

3. $r^2 + 3r - 10 = (r+5)(r-2) = 0$; $r = -5, 2$; $y(x) = c_1 e^{2x} + c_2 e^{-5x}$

5. $r^2 + 6r + 9 = (r+3)^2 = 0$; $r = -3$ (repeated); $y(x) = c_1 e^{-3x} + c_2 x e^{-3x}$

7. $4r^2 - 12r + 9 = (2r-3)^2 = 0$; $r = -\dfrac{3}{2}$ (repeated); $y(x) = c_1 e^{3x/2} + c_2 x e^{3x/2}$

9. $r^2 + 8r + 25 = 0$; $r = \dfrac{-8 \pm \sqrt{-36}}{2} = -4 \pm 3i$; $y(x) = e^{-4x}\left(c_1 \cos 3x + c_2 \sin 3x\right)$

11. $r^4 - 8r^3 + 16r^2 = r^2(r-4)^2 = 0$; $r = 0, 0, 4, 4$; $y(x) = c_1 + c_2 x + c_3 e^{4x} + c_4 x e^{4x}$

13. $9r^3 + 12r^2 + 4r = r(3r+2)^2 = 0$; $r = 0, -\dfrac{2}{3}, -\dfrac{2}{3}$; $y(x) = c_1 + c_2 e^{-2x/3} + c_3 x e^{-2x/3}$

15. $4r^4 - 8r^2 + 16 = (r^2-4)^2 = (r-2)^2(r+2)^2 = 0$; $r = 2, 2, -2, -2$;

$$y(x) = c_1 e^{2x} + c_2 x e^{2x} + c_3 e^{-2x} + c_4 x e^{-2x}$$

17. $6r^4 + 11r^2 + 4 = (2r^2+1)(3r^2+4) = 0$; $r = \pm\dfrac{i}{\sqrt{2}}, \pm\dfrac{2i}{\sqrt{3}}$;

$$y(x) = c_1 \cos\frac{x}{\sqrt{2}} + c_2 \sin\frac{x}{\sqrt{2}} + c_3 \cos\frac{2x}{\sqrt{3}} + c_4 \sin\frac{2x}{\sqrt{3}}$$

19. Factoring by grouping gives $r^3 + r^2 - r - 1 = r(r^2 - 1) + (r^2 - 1) = (r - 1)(r + 1)^2 = 0$; $r = 1, -1, -1$; $y(x) = c_1 e^x + c_2 e^{-x} + c_3 x e^{-x}$.

21. Imposition of the initial conditions $y(0) = 7$, $y'(0) = 11$ on the general solution $y(x) = c_1 e^x + c_2 e^{3x}$ yields the two equations $c_1 + c_2 = 7$, $c_1 + 3c_2 = 11$ with solution $c_1 = 5$, $c_2 = 2$. Hence the desired particular solution is $y(x) = 5e^x + 2e^{3x}$.

23. Imposition of the initial conditions $y(0) = 3$, $y'(0) = 1$ on the general solution $y(x) = e^{3x}(c_1 \cos 4x + c_2 \sin 4x)$ yields the two equations $c_1 = 3$, $3c_1 + 4c_2 = 1$ with solution $c_1 = 3$, $c_2 = -2$. Hence the desired particular solution is
$$y(x) = e^{3x}(3\cos 4x - 2\sin 4x).$$

25. Imposition of the initial conditions $y(0) = -1$, $y'(0) = 0$, $y''(0) = 1$ on the general solution $y(x) = c_1 + c_2 x + c_3 e^{-2x/3}$ yields the three equations

$$c_1 + c_3 = -1, \quad c_2 - \frac{2c_3}{3} = 0, \quad \frac{4c_3}{9} = 1,$$

with solution $c_1 = -\frac{13}{4}$, $c_2 = \frac{3}{2}$, $c_3 = \frac{9}{4}$. Hence the desired particular solution is

$$y(x) = -\frac{13}{4} + \frac{3}{2}x + \frac{9}{4}e^{-2x/3}.$$

27. First we spot the root $r = 1$. Then long division of the polynomial $r^3 + 3r^2 - 4$ by $r - 1$ yields the quadratic factor $r^2 + 4r + 4 = (r + 2)^2$, with roots $r = -2, -2$. Hence the general solution is $y(x) = c_1 e^x + c_2 e^{-2x} + c_3 x e^{-2x}$.

29. First we spot the root $r = -3$. Then long division of the polynomial $r^3 + 27$ by $r + 3$ yields the quadratic factor $r^2 - 3r + 9$, with roots $r = \frac{3}{2} \pm i\frac{3\sqrt{3}}{2}$. Hence the general solution is $y(x) = c_1 e^{-3x} + e^{3x/2}\left(c_2 \cos\frac{3\sqrt{3}}{2}x + c_3 \sin\frac{3\sqrt{3}}{2}x\right)$.

31. The characteristic equation $r^3 + 3r^2 + 4r - 8 = 0$ has the evident root $r = 1$, and long division then yields the quadratic factor $r^2 + 4r + 8 = (r + 2)^2 + 4$, corresponding to the com-

plex conjugate roots $-2 \pm 2i$. Hence the general solution is
$$y(x) = c_1 e^x + e^{-2x}(c_2 \cos 2x + c_3 \sin 2x).$$

33. Knowing that $y = e^{3x}$ is one solution, we divide the characteristic polynomial $r^3 + 3r^2 - 54$ by $r - 3$ and get the quadratic factor $r^2 + 6r + 18 = (r + 3)^2 + 9$. Hence the general solution is $y(x) = c_1 e^{3x} + e^{-3x}(c_2 \cos 3x + c_3 \sin 3x)$.

35. The fact that $y = \cos 2x$ is one solution tells us that $r^2 + 4$ is a factor of the characteristic polynomial $6r^4 + 5r^3 + 25r^2 + 20r + 4$. Then long division yields the quadratic factor $6r^2 + 5r + 1 = (3r + 1)(2r + 1)$, with roots $r = -\dfrac{1}{2}, -\dfrac{1}{3}$. Hence the general solution is
$$y(x) = c_1 e^{-x/2} + c_2 e^{-x/3} + c_3 \cos 2x + c_4 \sin 2x.$$

37. The characteristic equation is $r^4 - r^3 = r^3(r - 1) = 0$, so the general solution is $y(x) = A + Bx + Cx^2 + De^x$. Imposition of the given initial conditions yields the equations
$$A + D = 18, \quad B + D = 12, \quad 2C + D = 13, \quad D = 7$$
with solution $A = 11$, $B = 5$, $C = 3$, $D = 7$. Hence the desired particular solution is $y(x) = 11 + 5x + 3x^2 + 7e^x$.

39. The characteristic polynomial is $(r - 2)^3 = r^3 - 6r^2 + 12r - 8$, so the differential equation is $y''' - 6y'' + 12y' - 8y = 0$.

41. The characteristic polynomial is $(r^2 + 4)(r^2 - 4) = r^4 - 16$, so the differential equation is $y^{(4)} - 16y = 0$.

45. The characteristic polynomial is the quadratic polynomial of Problem 44(b). Hence the general solution is
$$y(x) = c_1 e^{-ix} + c_2 e^{3ix} = c_1(\cos x - i \sin x) + c_2(\cos 3x + i \sin 3x).$$

47. The characteristic roots are $r = \pm\sqrt{-2 + 2i\sqrt{3}} = \pm(1 + i\sqrt{3})$, so the general solution is
$$y(x) = c_1 e^{(1 + i\sqrt{3})x} + c_2 e^{-(1 + i\sqrt{3})x} = c_1 e^x(\cos \sqrt{3}x + i \sin \sqrt{3}x) + c_2 e^{-x}(\cos \sqrt{3}x - i \sin \sqrt{3}x).$$

49. We adopt the same strategy as was used in Problem 48. The general solution is $y(x) = Ae^{2x} + Be^{-x} + C\cos x + D\sin x$. Imposition of the given initial conditions yields the equations

$$
\begin{aligned}
A + B + C &= 0 \\
2A - B + D &= 0 \\
4A + B - C &= 0 \\
8A - B - D &= 30
\end{aligned}
$$

that we solve for $A = 2$, $B = -5$, $C = 3$, and $D = -9$. Thus

$$y(x) = 2e^{2x} - 5e^{-x} + 3\cos x - 9\sin x.$$

51. In the solution of Problem 51 in Section 5.1 we showed that the substitution $v = \ln x$ gives $y' = \dfrac{dy}{dx} = \dfrac{1}{x}\dfrac{dy}{dv}$ and $y'' = \dfrac{d^2 y}{dx^2} = -\dfrac{1}{x^2}\cdot\dfrac{dy}{dv} + \dfrac{1}{x^2}\cdot\dfrac{d^2 y}{dv^2}$. A further differentiation using the chain rule gives

$$y''' = \frac{d^3 y}{dx^3} = \frac{2}{x^3}\cdot\frac{dy}{dv} - \frac{3}{x^3}\cdot\frac{d^2 y}{dv^2} + \frac{1}{x^3}\cdot\frac{d^3 y}{dv^3}.$$

Substitution of these expressions for y', y'', and y''' into the third-order Euler equation $ax^3 y''' + bx^2 y'' + cxy' + dy = 0$, together with collection of coefficients, yields the desired constant-coefficient equation

$$a\frac{d^3 y}{dv^3} + (b - 3a)\frac{d^2 y}{dv^2} + (c - b + 2a)\frac{dy}{dv} + d\cdot y = 0.$$

In Problems 52 through 58 we list first the transformed constant-coefficient equation, then its characteristic equation and roots, and finally the corresponding general solution with $v = \ln x$ and $e^v = x$.

53. $\dfrac{d^2 y}{dv^2} + 6\dfrac{dy}{dv} + 25y = 0$; $r^2 + 6r + 25 = 0$; $r = -3 \pm 4i$;

$$y(x) = e^{-3v}\left(c_1 \cos 4v + c_2 \sin 4v\right) = x^{-3}\left[c_1 \cos(4\ln x) + c_2 \sin(4\ln x)\right]$$

55. $\dfrac{d^3 y}{dv^3} - 4\dfrac{d^2 y}{dv^2} + 4\dfrac{dy}{dv} = 0$; $r^3 - 4r^2 + 4r = 0$; $r = 0, 2, 2$;

$$y(x) = c_1 + c_2 e^{2v} + c_3 v e^{2v} = c_1 + x^2\left(c_2 + c_3 \ln x\right)$$

57. $\dfrac{d^3 y}{dv^3} - 5\dfrac{d^2 y}{dv^2} + 5\dfrac{dy}{dv} = 0$; $r^3 - 4r^2 + 4r = 0$; $r = 0, 3 \pm \sqrt{3}$;

$$y(x) = c_1 + c_2 e^{(3-\sqrt{3})v} + c_3 v e^{(3+\sqrt{3})v} = c_1 + x^3\left(c_2 x^{-\sqrt{3}} + c_3 x^{+\sqrt{3}}\right)$$

SECTION 5.4

MECHANICAL VIBRATIONS

In this section we discuss four types of free motion of a mass on a spring—undamped, under-damped, critically damped, and overdamped. However, the undamped and underdamped cas-es—in which actual oscillations occur—are emphasized because they are both the most interest-ing and the most important cases for applications.

1. Frequency: $\omega_0 = \sqrt{\dfrac{k}{m}} = \sqrt{\dfrac{16}{4}} = 2\,\text{rad/sec} = \dfrac{1}{\pi}\,\text{Hz}$; period: $P = \dfrac{2\pi}{\omega_0} = \dfrac{2\pi}{2} = \pi\,\text{sec}$

3. The spring constant is $k = \dfrac{15\,N}{0.20\,\text{m}} = 75\,\text{N/m}$. The solution of $3x'' + 75x = 0$ with

$x(0) = 0$ and $x'(0) = -10$ is $x(t) = -2\sin 5t$. Thus the amplitude is 2 m, the frequency

is $\omega_0 = \sqrt{\dfrac{k}{m}} = \sqrt{\dfrac{75}{3}} = 5\,\text{rad/sec} = \dfrac{2.5}{\pi}\,\text{Hz}$, and the period is $\dfrac{2\pi}{5}\,\text{sec}$.

5. The gravitational acceleration at distance R from the center of the earth is $g = \dfrac{GM}{R^2}$. Ac-

cording to Equation (6) in the text, the (circular) frequency ω of a (linearized) pendulum

is given by $\omega^2 = \dfrac{g}{L} = \dfrac{GM}{R^2 L}$, so its period is $p = \dfrac{2\pi}{\omega} = 2\pi R\sqrt{\dfrac{L}{GM}}$.

7. The period equation $p = 3960\sqrt{100.10} = (3960 + x)\sqrt{100}$ yields

$x \approx 1.9795\,\text{mi} \approx 10.450\,\text{ft}$ for the altitude of the mountain.

9. Designating $x(t)$ as in the suggestion, we see that the mass is subject to a restorative

force $F_S = -kx$ together with the force of gravity $W = mg$. We also assume that the

mass is subject to a damping force $F_R = -cx'$. Applying Newton's law then gives

$mx'' = -kx + mg - cx'$, or $mx'' + cx' + kx = mg$. Finally, substituting $y = x - s_0$, so that

$x = y + s_0$ and thus $x' = y'$ and $x'' = y''$, yields $my'' + cy' + k(y + s_0) = mg$, or

$my'' + cy' + ky = mg - ks_0$, which is Equation (5) with $F(t)$ assuming the constant value

$mg - ks_0$.

11. The differential equation from Problem 10 must be modified to reflect the fact that the

weight density of water is $62.4\,\text{lb/ft}^3$ (as opposed to $1\,\text{g/cm}^3$ in the cgs system). Thus

the weight of water displaced by the buoy is given by $62.4\pi r^2 \cdot x$. Moreover, the mass and weight of the buoy are given to be 3.125 slugs and 100 lb, respectively. Applying $ma = F$ then gives $3.125x'' = 100 - 62.4\pi r^2 \cdot x$, or $x'' + \dfrac{62.4\pi}{3.125}r^2 \cdot x = 32$. The frequency

of the oscillations of the buoy is therefore $\dfrac{\omega_0}{2\pi}$, where $\omega_0 = \sqrt{\dfrac{62.4\pi}{3.125}} \cdot r$. Since the fre-

quency of the buoy's motion is observed to be $\dfrac{4\,\text{cycles}}{10\,\text{sec}} = 0.4\,\text{cycles/sec}$, we can equate

the two to conclude that $\dfrac{1}{2\pi}\sqrt{\dfrac{62.4\pi}{3.125}} \cdot r = 0.4$, which gives

$r = 0.8\sqrt{\dfrac{3.125\pi}{62.4}} \approx 0.3173\,\text{ft} \approx 3.8\,\text{in}$.

13. **(a)** The characteristic equation $10r^2 + 9r + 2 = (5r + 2)(2r + 1) = 0$ has roots $r = -\dfrac{2}{5}, -\dfrac{1}{2}$.

When we impose the initial conditions $x(0) = 0$, $x'(0) = 5$ on the general solution

$x(t) = c_1 e^{-2t/5} + c_2 e^{-t/2}$ we get the particular solution $x(t) = 50\left(e^{-2t/5} - e^{-t/2}\right)$.

(b) The derivative $x'(t) = 25e^{-t/2} - 20e^{-2t/5} = 5e^{-2t/5}\left(5e^{-t/10} - 4\right) = 0$ when

$t = 10\ln\dfrac{5}{4} \approx 2.23144$. Hence the mass's farthest distance to the right is given by

$x\left(10\ln\dfrac{5}{4}\right) = \dfrac{512}{125} = 4.096$.

In Problems 15-21 the graph of the damped motion $x(t)$, that is, with the dashpot attached, is shown as a solid line; the graph of the corresponding undamped motion $u(t)$ is dashed.

15. **With damping:** The characteristic equation $\dfrac{1}{2}r^2 + 3r + 4 = 0$ has roots $r = -2, -4$. When

we impose the initial conditions $x(0) = 2$, $x'(0) = 0$ on the general solution

$x(t) = c_1 e^{-2t} + c_2 e^{-4t}$ we get the particular solution $x(t) = 4e^{-2t} - 2e^{-4t}$ that describes overdamped motion.

Without damping: The characteristic equation $\dfrac{1}{2}r^2 + 4 = 0$ has roots $r = \pm 2i\sqrt{2}$. When

we impose the initial conditions $x(0) = 2$, $x'(0) = 0$ on the general solution

$u(t) = A\cos\left(2\sqrt{2}t\right) + B\sin\left(2\sqrt{2}t\right)$ we get the particular solution $u(t) = 2\cos\left(2\sqrt{2}t\right)$.

Problem 15

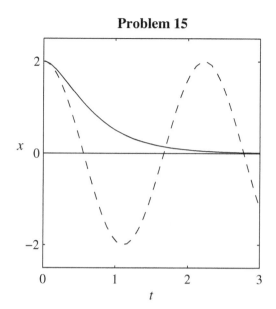

Problem 17

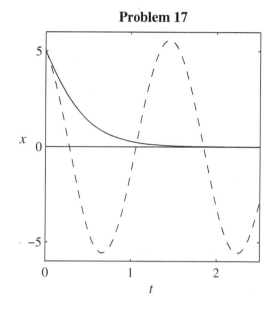

17. **With damping:** The characteristic equation $r^2 + 8r + 16 = 0$ has roots $r = -4, -4$. When we impose the initial conditions $x(0) = 5$, $x'(0) = -10$ on the general solution $x(t) = (c_1 + c_2 t)e^{-4t}$ we get the particular solution $x(t) = 5e^{-4t}(2t + 1)$ that describes critically damped motion.

Without damping: The characteristic equation $r^2 + 16 = 0$ has roots $r = \pm 4i$. When we impose the initial conditions $x(0) = 5$, $x'(0) = -10$ on the general solution $u(t) = A\cos 4t + B\sin 4t$ we get the particular solution

$$u(t) = 5\cos 4t + \frac{5}{2}\sin 4t \approx \frac{5}{2}\sqrt{5}\cos(4t - 5.8195).$$

19. The characteristic equation $4r^2 + 20r + 169 = 0$ has roots $r = -\frac{5}{2} \pm 6i$. When we impose the initial conditions $x(0) = 4$, $x'(0) = 16$ on the general solution $x(t) = e^{-5t/2}(A\cos 6t + B\sin 6t)$ we get the particular solution

$$x(t) = e^{-5t/2}\left(4\cos 6t + \frac{13}{3}\sin 6t\right) \approx \frac{1}{3}\sqrt{313}\, e^{-5t/2}\cos(6t - 0.8254)$$

that describes underdamped motion.

Without damping: The characteristic equation $4r^2 + 169 = 0$ has roots $r = \pm\frac{13}{2}i$. When we impose the initial conditions $x(0) = 4$, $x'(0) = 16$ on the general solution $u(t) = A\cos\left(\frac{13}{2}t\right) + B\sin\left(\frac{13}{2}t\right)$ we get the particular solution

$$u(t) = 4\cos\left(\frac{13t}{2}\right) + \frac{32}{13}\sin\left(\frac{13t}{2}\right) \approx \frac{4}{13}\sqrt{233}\cos\left(\frac{13}{2}t - 0.5517\right).$$

Problem 19

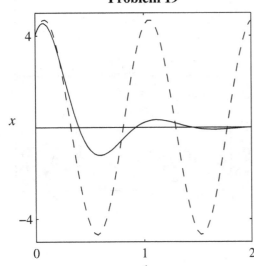

Problem 21

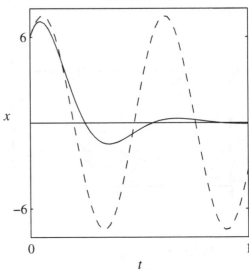

21. **With damping:** The characteristic equation $r^2 + 10r + 125 = 0$ has roots $r = -5 \pm 10i$. When we impose the initial conditions $x(0) = 6$, $x'(0) = 50$ on the general solution $x(t) = e^{-5t}(A\cos 10t + B\sin 10t)$ we get the particular solution

$$x(t) = e^{-5t}(6\cos 10t + 8\sin 10t) \approx 10e^{-5t}\cos(10t - 0.9273)$$

that describes underdamped motion.

Without damping: The characteristic equation $r^2 + 125 = 0$ has roots $r = \pm 5\sqrt{5}i$. When we impose the initial conditions $x(0) = 6$, $x'(0) = 50$ on the general solution $u(t) = A\cos\left(5\sqrt{5}t\right) + B\sin\left(5\sqrt{5}t\right)$ we get the particular solution

$$u(t) = 6\cos\left(5\sqrt{5}t\right) + 2\sqrt{5}\sin\left(5\sqrt{5}t\right) \approx 2\sqrt{14}\cos\left(5\sqrt{5}t - 0.6405\right).$$

23. **(a)** With $m = 100\,\text{slug}$ we get $\omega = \sqrt{\dfrac{k}{100}}$. But we are given that

$$\omega = (80\,\text{cycles/min}) \cdot 2\pi \cdot (1\,\text{min}/60\,\text{sec}) = \frac{8\pi}{3},$$

and equating the two values yields $k \approx 7018\,\text{lb/ft}$.

(b) With $\omega_1 = 2\pi \cdot \dfrac{78}{60}\,\text{cycles/sec}$, Equation (21) in the text yields $c \approx 372.31\,\text{lb}/(\text{ft/sec})$.

Hence $p = \dfrac{c}{2m} \approx 1.8615$. Finally $e^{-pt} = 0.01$ gives $t \approx 2.47\,\text{sec}$.

31. The binomial series

$$(1+x)^\alpha = 1 + \alpha x + \frac{\alpha(\alpha-1)}{2!}x^2 + \frac{\alpha(\alpha-1)(\alpha-2)}{3!}x^3 + \cdots$$

converges if $|x| < 1$. (See, for instance, Section 10.8 of Edwards and Penney, *Calculus: Early Transcendentals*, 7th edition, Pearson, 2008.) With $\alpha = \frac{1}{2}$ and $x = -\frac{c^2}{4mk}$ in Eq. (22) of Section 5.4 in the differential equations text, the binomial series gives

$$\omega_1 = \sqrt{\omega_0^2 - p^2} = \sqrt{\frac{k}{m} - \frac{c^2}{4m^2}} = \sqrt{\frac{k}{m}}\sqrt{1 - \frac{c^2}{4mk}} = \sqrt{\frac{k}{m}}\left(1 - \frac{c^2}{8mk} - \frac{c^4}{128m^2k^2} - \cdots\right) \approx \omega_0\left(1 - \frac{c^2}{8mk}\right).$$

33. If $x_1 = x(t_1)$ and $x_2 = x(t_2)$ are two successive local maxima, then $\omega_1 t_2 = \omega_1 t_1 + 2\pi$, and so $x_1 = Ce^{-pt_1}\cos(\omega_1 t_1 - \alpha)$ and $x_2 = Ce^{-pt_2}\cos(\omega_1 t_2 - \alpha) = Ce^{-pt_2}\cos(\omega_1 t_1 - \alpha)$. Hence

$\frac{x_1}{x_2} = e^{-p(t_1-t_2)}$ and therefore $\ln\left(\frac{x_1}{x_2}\right) = -p(t_1 - t_2) = \frac{2\pi p}{\omega_1}$.

35. The characteristic equation $r^2 + 2r + 1 = 0$ has roots $r = -1, -1$. When we impose the initial conditions $x(0) = 0$, $x'(1) = 0$ on the general solution $x(t) = (c_1 + c_2 t)e^{-t}$ we get the particular solution $x_1(t) = te^{-t}$.

37. The characteristic equation $r^2 + 2r + (1 + 10^{-2n}) = 0$ has roots $r = -1 \pm 10^{-n}i$. When we impose the initial conditions $x(0) = 0$, $x'(1) = 0$ on the general solution

$$x(t) = e^{-t}\left[A\cos(10^{-n}t) + B\sin(10^{-n}t)\right]$$

we get the equations $c_1 = 0$, $-c_1 + 10^{-n}c_2 = 1$ with solution $c_1 = 0$, $c_2 = 10^n$. This gives the particular solution $x_3(t) = 10^n e^{-t}\sin(10^{-n}t)$.

SECTION 5.5

NONHOMOGENEOUS EQUATIONS AND UNDETERMINED COEFFICIENTS

The method of undetermined coefficients is based on "educated guessing". If we can guess correctly the **form** of a particular solution of a nonhomogeneous linear equation with constant coefficients, then we can determine the particular solution explicitly by substitution in the given differential equation. It is pointed out at the end of Section 5.5 that this simple approach is not always successful—in which case the method of variation of parameters is available if a complementary function is known. However, undetermined coefficients *does* turn out to work well with a surprisingly large number of the nonhomogeneous linear differential equations that arise in elementary scientific applications.

In each of Problems 1-20 we give first the form of the trial solution y_{trial}, then the equations in the coefficients we get when we substitute y_{trial} into the differential equation and collect like terms, and finally the resulting particular solution y_p.

1. $y_{trial} = Ae^{3x}$; $25A = 1$; $y_p = \dfrac{1}{25}e^{3x}$.

3. $y_{trial} = A\cos 3x + B\sin 3x$; $-15A - 3B = 0$, $3A - 15B = 2$; $y_p = \dfrac{1}{39}\cos 3x - \dfrac{5}{39}\sin 3x$.

5.   First we substitute $\dfrac{1 - \cos 2x}{2}$ for $\sin^2 x$ on the right-hand side of the differential equation, leading to $y_{trial} = A + B\cos 2x + C\sin 2x$, and then

$$A = \frac{1}{2}, \quad -3B + 2C = -\frac{1}{2}, \quad -2B - 3C = 0;$$

$$y_p = \frac{1}{2} + \frac{3}{26}\cos 2x - \frac{1}{13}\sin 2x.$$

7.   First we substitute $\dfrac{e^x - e^{-x}}{2}$ for $\sinh x$ on the right-hand side of the differential equatio,

leading to $y_{trial} = Ae^x + Be^{-x}$; $-3A = \dfrac{1}{2}$, $-3B = -\dfrac{1}{2}$; $y_p = \dfrac{1}{6}e^{-x} - \dfrac{1}{6}e^x = -\dfrac{1}{3}\sinh x$. (Note that according to Rule 1 in the text, we could also have started with $y_{trial} = A\cosh x + B\sinh x$.)

9.   First we note that e^x is part of the complementary function $y_c = c_1 e^x + c_2 e^{-3x}$. Then $y_{trial} = A + x(B + Cx)e^x$, and then

$$-3A = 1 \quad 4B + 2C = 0 \quad 8C = 1;$$

$$y_p = -\frac{1}{3} + x\left(-\frac{1}{16} + \frac{1}{8}x\right)e^x.$$

11. First we note the duplication with the complementary function
$y_c = c_1 + c_2 \cos 2x + c_3 \sin 2x$. Then $y_{\text{trial}} = x(A + Bx)$; $4A = -1$, $8B = 3$;

$$y_p = x\left(-\frac{1}{4} + \frac{3}{8}x\right) = \frac{1}{8}(3x^2 - 2x).$$

13. $y_{\text{trial}} = e^x(A \cos x + B \sin x)$; $7A + 4B = 0$, $-4A + 7B = 1$; $y_p = \frac{1}{65}e^x(7 \sin x - 4 \cos x)$.

15. This is something of a trick problem. We cannot solve the characteristic equation $r^5 + 5r^4 - 1 = 0$ to find the complementary function, but we can see that the complementary function contains no constant term (why?). Hence we can take $y_{\text{trial}} = A$, leading immediately to the particular solution $y_p = -17$.

17. First we note the duplication with the complementary function $y_c = c_1 \cos x + c_2 \sin x$.
Then $y_{\text{trial}} = x\left[(A + Bx)\cos x + (C + Dx)\sin x\right]$;

$$2B + 2C = 0 \quad 4D = 1 \quad -2A + 2D = 1 \quad -4B = 0;$$

$$y_p = x\left(-\frac{1}{4}\cos x + \frac{1}{4}x \sin x\right) = \frac{1}{4}(x^2 \sin x - x \cos x).$$

19. First we note the duplication with the part $c_1 + c_2 x$ of the complementary function (which corresponds to the factor r^2 of the characteristic polynomial). Then
$y_{\text{trial}} = x^2(A + Bx + Cx^2)$;

$$4A + 12B = -1, \quad 12B + 48C = 0, \quad 24C = 3;$$

$$y_p = x^2\left(\frac{5}{4} - \frac{1}{2}x + \frac{1}{8}x^2\right) = \frac{1}{8}(10x^2 - 4x^3 + x^4).$$

In Problems 21-30 we list first the complementary function y_c, then the initially proposed trial function y_i, and finally the actual trial function y_p, in which duplication with the complementary function has been eliminated.

21. $y_c = e^x(c_1 \cos x + c_2 \sin x)$; $y_i = e^x(A \cos x + B \sin x)$; $y_p = x \cdot e^x(A \cos x + B \sin x)$

23. $y_c = c_1 \cos x + c_2 \sin x$; $y_i = (A + Bx)\cos 2x + (C + Dx)\sin 2x$;
$y_p = x \cdot \left[(A + Bx)\cos 2x + (C + Dx)\sin 2x\right]$

25. $y_c = c_1 e^{-x} + c_2 e^{-2x}$; $y_i = (A + Bx)e^{-x} + (C + Dx)e^{-2x}$;

$y_p = x \cdot (A + Bx)e^{-x} + x \cdot (C + Dx)e^{-2x}$

27. $y_c = (c_1 \cos x + c_2 \sin x) + (c_3 \cos 2x + c_4 \sin 2x)$;

$y_i = (A\cos x + B\sin x) + (C\cos 2x + D\sin 2x)$;

$y_p = x \cdot \left[(A\cos x + B\sin x) + (C\cos 2x + D\sin 2x) \right]$

29. $y_c = (c_1 + c_2 x + c_3 x^2)e^x + c_4 e^{2x} + c_5 e^{-2x}$; $y_i = (A + Bx)e^x + Ce^{2x} + De^{-2x}$;

$y_p = x^3 \cdot (A + Bx)e^x + x \cdot (Ce^{2x}) + x \cdot (De^{-2x})$

In Problems 31-40 we list first the complementary function y_c, the trial solution y_{tr} for the method of undetermined coefficients, and the corresponding general solution $y_g = y_c + y_p$, where y_p results from determining the coefficients in y_{tr} so as to satisfy the given nonhomogeneous differential equation. Then we list the linear equations obtained by imposing the given initial conditions, and finally the resulting particular solution $y(x)$.

31. $y_c = c_1 \cos 2x + c_2 \sin 2x$; $y_{tr} = A + Bx$; $y_g = c_1 \cos 2x + c_2 \sin 2x + \dfrac{x}{2}$; $c_1 = 1$, $2c_2 + \dfrac{1}{2} = 2$;

$y(x) = \cos 2x + (3/4)\sin 2x + x/2$

33. $y_c = c_1 \cos 3x + c_2 \sin 3x$; $y_{tr} = A\cos 2x + B\sin 2x$; $y_g = c_1 \cos 3x + c_2 \sin 3x + \dfrac{1}{5}\sin 2x$;

$c_1 = 1$, $3c_2 + \dfrac{2}{5} = 0$, $y(x) = \cos 3x - \dfrac{2}{15}\sin 3x + \dfrac{1}{5}\sin 2x$

35. $y_c = e^x (c_1 \cos x + c_2 \sin x)$; $y_{tr} = A + Bx$; $y_g = e^x (c_1 \cos x + c_2 \sin x) + 1 + \dfrac{x}{2}$; $c_1 + 1 = 3$,

$c_1 + c_2 + \dfrac{1}{2} = 0$; $y(x) = e^x \left(2\cos x - \dfrac{5}{2}\sin x \right) + 1 + \dfrac{x}{2}$

37. $y_c = c_1 + c_2 e^x + c_3 x e^x$; $y_{tr} = x \cdot (A) + x^2 \cdot (B + Cx)e^x$

$y_g = c_1 + c_2 e^x + c_3 x e^x + x - \dfrac{1}{2}x^2 e^x + \dfrac{1}{6}x^3 e^x$

$c_1 + c_2 = 0$, $c_2 + c_3 + 1 = 0$, $c_2 + 2c_3 - 1 = 1$

$y(x) = 4 + x + e^x \left(-4 + 3x - \dfrac{1}{2}x^2 + \dfrac{1}{6}x^3 \right)$

39. $y_c = c_1 + c_2 x + c_3 e^{-x}$; $y_{tr} = x^2 \cdot (A + Bx) + x \cdot (Ce^{-x})$

$$y_g = c_1 + c_2 x + c_3 e^{-x} - \frac{x^2}{2} + \frac{x^3}{6} + xe^{-x}$$

$$c_1 + c_3 = 1, \quad c_2 - c_3 + 1 = 0, \quad c_3 - 3 = 1$$

$$y(x) = \frac{1}{6}\left(-18 + 18x - 3x^2 + x^3\right) + (4 + x)e^{-x}$$

41. The trial solution $y_{tr} = A + Bx + Cx^2 + Dx^3 + Ex^4 + Fx^5$ leads to the equations

$$
\begin{array}{rcrcrcrcrcrcl}
2A & - & B & - & 2C & - & 6D & + & 24E & & & = & 0 \\
 & & -2B & - & 2C & - & 6D & - & 24E & + & 120F & = & 0 \\
 & & & & -2C & - & 3D & - & 12E & - & 60F & = & 0 \\
 & & & & & & -2D & - & 4E & - & 20F & = & 0 \\
 & & & & & & & & -2E & - & 5F & = & 0 \\
 & & & & & & & & & & -2F & = & 8
\end{array}
$$

that are readily solved by back-substitution. The resulting particular solution is

$$y(x) = -255 - 450x + 30x^2 + 20x^3 + 10x^4 - 4x^5.$$

43. **(a)** Applying Euler's formula gives

$$\cos 3x + i\sin 3x = (\cos x + i\sin x)^3 = \cos^3 x + 3i\cos^2 x \sin x - 3\cos x \sin^2 x - i\sin^3 x.$$

When we equate real parts we get the equation

$$\cos^3 x - 3(\cos x)(1 - \cos^2 x) = 4\cos^3 x - 3\cos x$$

and readily solve for $\cos^3 x = \frac{3}{4}\cos x + \frac{1}{4}\cos 3x$. The formula for $\sin^3 x$ is derived similarly by equating imaginary parts in the first equation above.

(b) Upon substituting the trial solution $y_p = A\cos x + B\sin x + C\cos 3x + D\sin 3x$ in the differential equation $y'' + 4y = \frac{3}{4}\cos x + \frac{1}{4}\cos 3x$, we find that

$$A = \frac{1}{4}, \quad B = 0, \quad C = -\frac{1}{20}, \quad D = 0.$$

The resulting general solution is

$$y(x) = c_1 \cos 2x + c_2 \sin 2x + \frac{1}{4}\cos x - \frac{1}{20}\cos 3x.$$

45. We substitute

$$\sin^4 x = \frac{1}{4}(1 - \cos 2x)^2 = \frac{1}{4}\left(1 - 2\cos 2x + \cos^2 2x\right) = \frac{1}{8}\left(3 - 4\cos 2x + \cos 4x\right)$$

on the right-hand side of the differential equation, and then substitute the trial solution

$$y_p = A\cos 2x + B\sin 2x + C\cos 4x + D\sin 4x + E.$$

We find that

$$A = -\frac{1}{10}, \quad B = 0, \quad C = -\frac{1}{56}, \quad D = 0, \quad E = \frac{1}{24}.$$

The resulting general solution is

$$y = c_1 \cos 3x + c_2 \sin 3x + \frac{1}{24} - \frac{1}{10}\cos 2x - \frac{1}{56}\cos 4x.$$

In Problems 47–49 we list the independent solutions y_1 and y_2 of the associated homogeneous equation, their Wronskian $W = W(y_1, y_2)$, the coefficient functions

$$u_1(x) = -\int \frac{y_2(x)f(x)}{W(x)}dx \quad \text{and} \quad u_2(x) = \int \frac{y_1(x)f(x)}{W(x)}dx$$

in the particular solution $y_p = u_1 y_1 + u_2 y_2$ of Eq. (32) in the text, and finally y_p itself.

47. $y_1 = e^{-2x}$, $y_2 = e^{-x}$, $W = e^{-3x}$, $u_1 = -\frac{4}{3}e^{3x}$, $u_2 = 2e^{2x}$, $y_p = \frac{2}{3}e^{x}$

49. $y_1 = e^{2x}$, $y_2 = xe^{2x}$, $W = e^{4x}$, $u_1 = -x^2$, $u_2 = 2x$, $y_p = x^2 e^{2x}$.

51. $y_1 = \cos 2x$, $y_2 = \sin 2x$, $W = 2$. Liberal use of trigonometric sum and product identities yields

$$u_1 = \frac{1}{20}(\cos 5x - 5\cos x) \quad \text{and} \quad u_2 = \frac{1}{20}(\sin 5x + 5\sin x).$$

Thus

$$y_p = \frac{1}{20}(\cos 5x - 5\cos x)\cos 2x + \frac{1}{20}(\sin 5x + 5\sin x)\sin 2x$$

$$= \frac{1}{20}\left[(\cos 5x \cos 2x + \sin 5x \sin 2x) - 5(\cos x \cos 2x - \sin x \sin 2x)\right]$$

$$= \frac{1}{20}(\cos 3x - 5\cos 3x)$$

$$= -\frac{1}{5}\cos 3x \; (!)$$

53. $y_1 = \cos 3x$, $y_2 = \sin 3x$, $W = 3$; $u_1' = -\dfrac{2}{3}\tan 3x$, so $u_1 = \dfrac{2}{9}\ln|\cos 3x|$; $u_2' = \dfrac{2}{3}$, so

$u_2 = \dfrac{2}{3}x$. Thus

$$y_p = \left(\cos 3x\right)\cdot\frac{2}{9}\ln|\cos 3x| + \left(\sin 3x\right)\cdot\frac{2}{3}x$$

$$= \frac{2}{9}\left[3x\sin 3x + \left(\cos 3x\right)\ln|\cos 3x|\right].$$

55. $y_1 = \cos 2x$, $y_2 = \sin 2x$, $W = 2$;

$$u_1' = -\frac{1}{2}\sin^2 x \sin 2x = -\frac{1}{2}\cdot\frac{1-\cos 2x}{2}\cdot\sin 2x = -\frac{1}{4}\left(\sin 2x - \cos 2x \sin 2x\right),$$

so $u_1 = \dfrac{1}{16}\left(2\cos 2x - \cos^2 2x\right)$;

$$u_2' = \frac{1}{2}\sin^2 x \cos 2x = \frac{1}{2}\cdot\frac{1-\cos 2x}{2}\cdot\cos 2x = \frac{1}{4}\left(\cos 2x - \cos^2 2x\right) = \frac{1}{8}\left[2\cos 2x - \left(1+\cos 4x\right)\right],$$

so $u_2 = \dfrac{1}{8}\left(\sin 2x - x - \dfrac{\sin 4x}{4}\right)$. Thus

$$y_p = \frac{1}{16}\left(2\cos 2x - \cos^2 2x\right)\cos 2x + \frac{1}{8}\left(\sin 2x - x - \frac{\sin 4x}{4}\right)\sin 2x$$

$$= \frac{1}{16}\left(\underline{2\cos^2 2x} - \cos^3 2x + \underline{2\sin^2 2x} - 2x\sin 2x - \frac{1}{2}\sin 4x \sin 2x\right)$$

$$= \frac{1}{16}\left(2 - 2x\sin 2x \underline{\underline{-\cos^3 2x - \frac{1}{2}\sin 4x \sin 2x}}\right),$$

because the single-underlined terms sum to 2. The double-underlined terms reduce to

$$-\cos^3 2x - \frac{1}{2}\sin 4x \sin 2x = -\cos^3 2x - \frac{1}{2}\left(2\sin 2x \cos 2x\right)\sin 2x$$

$$= -\cos^3 2x - \sin^2 2x \cos 2x$$

$$= -\cos^3 2x - \left(1 - \cos^2 2x\right)\cos 2x$$

$$= -\cos 2x.$$

Therefore we can take $y_p = \dfrac{1}{16}\left(2 - 2x\sin 2x - \cos 2x\right)$. However, because $\cos 2x$ is a solution of the associated homogeneous equation $y'' + y = 0$—that is, because $\cos 2x$ is part of the complimentary function y_c—we can in fact omit the $\cos 2x$ term from y_p, leading to the simpler version $y_p = \dfrac{1}{8}\left(1 - x\sin 2x\right)$.

57.   With $y_1 = x$, $y_2 = x^{-1}$, and $f(x) = 72x^3$, Equations (31) in the text take the form

$$xu_1' + x^{-1}u_2' = 0,$$
$$u_1' - x^{-2}u_2' = 72x^3.$$

Upon multiplying the second equation by x and then adding, we readily solve first for $u_1' = 36x^3$, so $u_1 = 9x^4$; then $u_2' = -x^2 u_1'$, so $u_2 - -6x^6$. It follows that

$$y_p = y_1 u_1 + y_2 u_2 = x(9x^4) + (x^{-1})(-6x^6) = 3x^5.$$

59. $y_1 = x^2$, $y_2 = x^2 \ln x$, $W = x^3$, $f(x) = x^2$; $u_1' = -x \ln x$, $u_2' = x$; $y_p = \dfrac{1}{4}x^4$.

61. $y_1 = \cos(\ln x)$, $y_2 = \sin(\ln x)$, $W = \dfrac{1}{x}$, $f(x) = \dfrac{\ln x}{x^2}$; from $u_1' = -\dfrac{(\ln x)\sin(\ln x)}{x}$ and $u_2' = \dfrac{(\ln x)\cos(\ln x)}{x}$ integration by parts yields

$$u_1 = -\int \frac{(\ln x)\sin(\ln x)}{x}dx = -\int \frac{\sin(\ln x)}{x} \cdot \ln x\, dx$$
$$= \cos(\ln x) \cdot \ln x - \int \frac{\cos(\ln x)}{x}dx = \cos(\ln x) \cdot \ln x - \sin(\ln x)$$

and

$$u_2 = \int \frac{(\ln x)\cos(\ln x)}{x}dx = \int \frac{\cos(\ln x)}{x} \cdot \ln x\, dx$$
$$= \sin(\ln x) \cdot \ln x - \int \frac{\sin(\ln x)}{x}dx = \sin(\ln x) \cdot \ln x + \cos(\ln x).$$

Thus

$$y_p = u_1 y_1 + u_2 y_2$$
$$= \left[\cos(\ln x) \cdot \ln x - \sin(\ln x)\right]\cos(\ln x) + \left[\sin(\ln x) \cdot \ln x + \cos(\ln x)\right]\sin(\ln x)$$
$$= \left[\cos^2(\ln x) + \sin^2(\ln x)\right]\ln x - \cancel{\sin(\ln x)\cos(\ln x)} + \cancel{\cos(\ln x)\sin(\ln x)}$$
$$= \ln x \ (!)$$

63.   This is simply a matter of solving the equations in (31) for the derivatives

$$u_1' = -\frac{y_2(x)f(x)}{W(x)} \quad \text{and} \quad u_2' = \frac{y_1(x)f(x)}{W(x)},$$

integrating each, and then substituting the results in (32).

SECTION 5.6

FORCED OSCILLATIONS AND RESONANCE

1. Trial of $x = A\cos 2t$ yields the particular solution $x_p = 2\cos 2t$. (Can you see that because the differential equation contains no first-derivative term, there is no need to include a $\sin 2t$ term in the trial solution?) Hence the general solution is

$$x(t) = c_1 \cos 3t + c_2 \sin 3t + 2\cos 2t.$$

The initial conditions imply that $c_1 = -2$ and $c_2 = 0$, so $x(t) = 2\cos 2t - 2\cos 3t$. The figure shows the graph of $x(t)$.

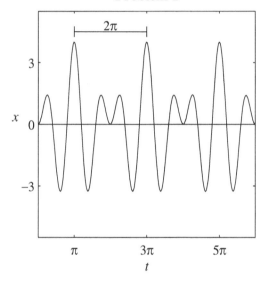

Problem 1

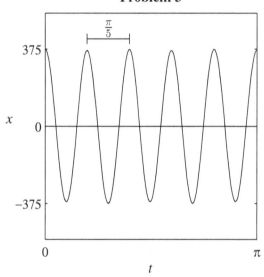

Problem 3

3. First we apply the method of undetermined coefficients with trial solution $x = A\cos 5t + B\sin 5t$ to find the particular solution

$$x_p = 3\cos 5t + 4\sin 5t = 5\left(\frac{3}{5}\cos 5t + \frac{4}{5}\sin 5t\right) = 5\cos(5t - \beta),$$

where $\beta = \tan^{-1}\dfrac{4}{3} \approx 0.9273$. Hence the general solution is

$$x(t) = c_1 \cos 10t + c_2 \sin 10t + 3\cos 5t + 4\sin 5t.$$

The initial conditions $x(0) = 375$, $x'(0) = 0$ now yield $c_1 = 372$ and $c_2 = -2$, so the part of the solution with frequency $\omega = 10$ is

$$x_c = 372\cos 10t - 2\sin 10t$$

$$= \sqrt{138388}\left(\frac{372}{\sqrt{138388}}\cos 10t - \frac{2}{\sqrt{138388}}\sin 10t\right)$$

$$= \sqrt{138388}\cos(10t - \alpha),$$

where $\alpha = 2\pi - \tan^{-1}\dfrac{1}{186} \approx 6.2778$ is a fourth-quadrant angle. The figure shows the graph of $x(t)$.

5. Substitution of the trial solution $x = C\cos\omega t$ gives $C = \dfrac{F_0}{k - m\omega^2}$. Then imposition of the initial conditions $x(0) = x_0$, $x'(0) = 0$ on the general solution

$$x(t) = c_1\cos\omega_0 t + c_2\sin\omega_0 t + C\cos\omega t$$

(where $\omega_0 = \sqrt{k/m}$) gives the particular solution $x(t) = (x_0 - C)\cos\omega_0 t + C\cos\omega t$.

In Problems 7–10 we give first the trial solution x_p involving undetermined coefficients A and B, then the equations that determine these coefficients, and finally the resulting steady periodic solution x_{sp}. In each case the figure shows the graphs of $x_{sp}(t)$ and the adjusted forcing function $F_1(t) = F(t)/m\omega$.

7. $x_p = A\cos 3t + B\sin 3t$; $-5A + 12B = 10$, $12A + 5B = 0$.

$$x_{sp}(t) = -\frac{50}{169}\cos 3t + \frac{120}{169}\sin 3t = \frac{10}{13}\left(-\frac{5}{13}\cos 3t + \frac{12}{13}\sin 3t\right) = \frac{10}{13}\cos(3t - \alpha),$$

where $\alpha = \pi - \tan^{-1}\dfrac{12}{5} \approx 1.9656$, a 2nd-quadrant angle.

Problem 7

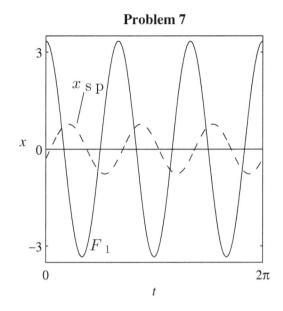

Problem 9

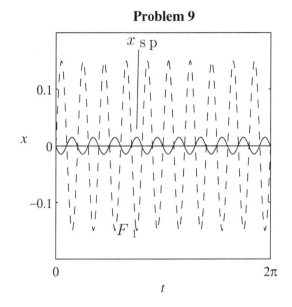

9. $x_p = A\cos 10t + 10\sin 5t$; $-199A + 20B = 0$, $20A + 199B = -3$.

$$x_{sp}(t) = -\frac{60}{40001}\cos 10t - \frac{597}{40001}\sin 10t$$

$$= \frac{3}{\sqrt{40001}}\left(-\frac{20}{\sqrt{40001}}\cos 10t - \frac{199}{\sqrt{40001}}\sin 10t\right)$$

$$= \frac{3}{\sqrt{40001}}\cos(10t - \alpha),$$

where $\alpha = \pi + \tan^{-1}\dfrac{199}{20} \approx 4.6122$, a 3rd-quadrant angle.

Each solution in Problems 11–14 has two parts. For the first part, we give first the trial solution x_p involving undetermined coefficients A and B, then the equations that determine these coefficients, and finally the resulting steady periodic solution x_{sp}. For the second part, we give first the general solution $x(t)$ involving the coefficients c_1 and c_2 in the transient solution, then the equations that determine these coefficients, and finally the resulting transient solution x_{tr}, so that $x(t) = x_{tr}(t) + x_{sp}(t)$ satisfies the given initial conditions. For each problem, the graph shows the graphs of both $x(t)$ and $x_{sp}(t)$.

11. $x_p = A\cos 3t + B\sin 3t$; $-4A + 12B = 10$, $12A + 4B = 0$.

$$x_{sp}(t) = -\frac{1}{4}\cos 3t + \frac{3}{4}\sin 3t = \frac{\sqrt{10}}{4}\left(-\frac{1}{\sqrt{10}}\cos 3t + \frac{3}{\sqrt{10}}\sin 3t\right) = \frac{\sqrt{10}}{4}\cos(3t - \alpha),$$

where $\alpha = \pi - \tan^{-1} 3 \approx 1.8925$, a 2nd-quadrant angle.

$$x(t) = e^{-2t}\left(c_1 \cos t + c_2 \sin t\right) + x_{sp}(t); \ c_1 - \frac{1}{4} = 0, \ -2c_1 + c_2 + \frac{9}{4} = 0.$$

$$x_{tr}(t) = e^{-2t}\left(\frac{1}{4}\cos t - \frac{7}{4}\sin t\right) = \frac{\sqrt{50}}{4} e^{-2t}\left(\frac{1}{\sqrt{50}}\cos t - \frac{7}{\sqrt{50}}\sin t\right) = \frac{5}{4}\sqrt{2}e^{-2t}\cos(t-\beta),$$

where $\beta = 2\pi - \tan^{-1} 7 \approx 4.8543$, a 4th-quadrant angle.

Problem 11	**Problem 13**
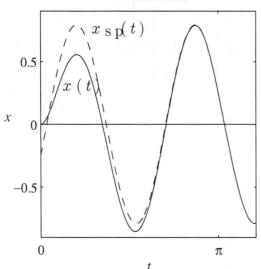	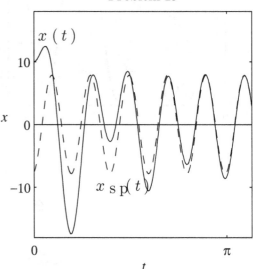

13. $x_p = A\cos 10t + B\sin 10t$; $74A + 20B = 600$, $20A + 74B = 0$.

$$x_{sp}(t) = -\frac{11100}{1469}\cos 10t + \frac{3000}{1469}\sin 10t$$

$$= \frac{300}{\sqrt{1469}}\left(-\frac{37}{\sqrt{1469}}\cos 10t + \frac{10}{\sqrt{1469}}\sin 10t\right)$$

$$= \frac{300}{\sqrt{1469}}\cos(10t - \alpha),$$

where $\alpha = \pi - \tan^{-1}\frac{10}{37} \approx 2.9320$, a 2nd-quadrant angle.

$$x(t) = e^{-t}\left(c_1 \cos 5t + c_2 \sin 5t\right) + x_{sp}(t); \ c_1 - \frac{11100}{1469} = 10, \ -c_1 + 5c_2 = -\frac{30000}{1469}.$$

$$x_{tr}(t) = \frac{e^{-t}}{1469}\left(25790\cos 5t - 842\sin 5t\right)$$

$$= \frac{2\sqrt{166458266}}{1469}e^{-t}\left(\frac{12895}{\sqrt{166458266}}\cos 5t - \frac{421}{\sqrt{166458266}}\sin 5t\right)$$

$$= 2\sqrt{\frac{113314}{1469}}e^{-t}\cos(5t - \beta),$$

where $\beta = 2\pi - \tan^{-1}\dfrac{421}{12895} \approx 6.2505$, a 4th-quadrant angle.

In Problems 15-18 we substitute $x(t) = A(\omega)\cos\omega t + B(\omega)\sin\omega t$ into the differential equation $mx'' + cx' + kx = F_0\cos\omega t$ with the given numerical values of m, c, k, and F_0. We give first the equations in A and B that result upon collection of coefficients of $\cos\omega t$ and $\sin\omega t$, followed by the values of $A(\omega)$ and $B(\omega)$ that we get by solving these equations. Finally, $C = \sqrt{A^2 + B^2}$ gives the amplitude of the resulting forced oscillations as a function of the forcing frequency ω, and we show the graph of the function $C(\omega)$.

15. $(2 - \omega^2)A + 2\omega B = 2$, $-2\omega A + (2 - \omega^2)B = 0$; $A = \dfrac{2(2 - \omega^2)}{4 + \omega^4}$, $B = \dfrac{4\omega}{4 + \omega^4}$;

$C(\omega) = \dfrac{2}{\sqrt{4 + \omega^4}}$ begins with $C(0) = 1$ and steadily decreases as ω increases. Hence there is no practical resonance frequency.

Problem 15

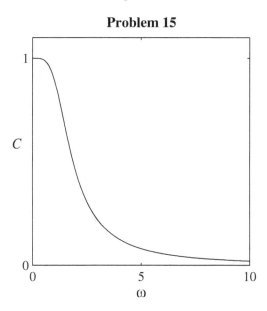

Problem 17

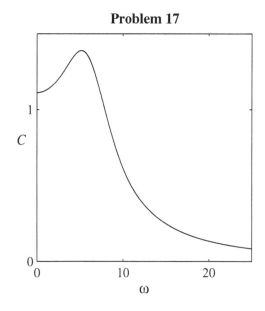

17. $(45 - \omega^2)A + 6\omega B = 50$, $-6\omega A + (45 - \omega^2)B = 0$; $A = \dfrac{50(45 - \omega^2)}{2025 - 54\omega^2 + \omega^4}$,

$B = \dfrac{300\omega}{2025 - 54\omega^2 + \omega^4}$; $C(\omega) = \dfrac{50}{\sqrt{2025 - 54\omega^2 + \omega^4}}$. So, to find the maximum value of $C(\omega)$, we calculate its derivative

$$C'(\omega) = \dfrac{-100\omega(-27 + \omega^2)}{(2025 - 54\omega^2 + \omega^4)^{3/2}}.$$

Hence the practical resonance frequency (where the derivative vanishes) is $\omega = \sqrt{27} = 3\sqrt{3}$.

19. $m = 100/32$ slugs and $k = 1200$ lb/ft, so the critical frequency is

$$\omega_0 = \sqrt{\frac{k}{m}} = \sqrt{384} \text{ rad/sec} = \frac{\sqrt{384}}{2\pi} \text{ Hz} \approx 3.12 \text{ Hz}.$$

21.   If θ is the angular displacement from the vertical, then the (essentially horizontal) displacement of the mass is $x = L\theta$, so twice its total energy (KE + PE) is

$$m(x')^2 + kx^2 + 2mgh = mL^2(\theta')^2 + kL^2\theta^2 + 2mgL(1 - \cos\theta) = C.$$

Differentiation, substitution of $\theta \approx \sin\theta$, and simplification yield $\theta'' + \left(\dfrac{k}{m} + \dfrac{g}{L}\right)\theta = 0$, so

$$\omega_0 = \sqrt{\frac{k}{m} + \frac{g}{L}}.$$

23.   **(a)** In units of ft-lb-sec we have $m = 1000$ and $k = 10000$, so
$\omega_0 = \sqrt{10}$ rad/sec ≈ 0.50 Hz.

(b) We are given that $\omega = 2\pi/2.25 \approx 2.79$ rad/sec, and the equation $mx'' + kx = F(t)$

simplifies to $x'' + 10x = \dfrac{1}{4}\omega^2 \sin\omega t$. When we substitute $x(t) = A\sin\omega t$ we find that the

amplitude is $A = \dfrac{\omega^2}{4(10 - \omega^2)} \approx 0.8854$ ft ≈ 10.63 in.

25.   Substitution of the trial solution $x = A\cos\omega t + B\sin\omega t$ in the differential equation followed by collection of coefficients as usual yields the equations

$$(k - m\omega^2)A + (c\omega)B = 0, \quad -(c\omega)A + (k - m\omega^2)B = F_0$$

with coefficient determinant $\Delta = (k - m\omega^2)^2 + (c\omega)^2$ and solution

$$A = \frac{1}{\Delta}(-c\omega)F_0, \quad B = \frac{1}{\Delta}(k - m\omega^2)F_0.$$

Hence

$$x(t) = \frac{F_0}{\sqrt{\Delta}}\left(\frac{k - m\omega^2}{\sqrt{\Delta}}\sin\omega t - \frac{c\omega}{\sqrt{\Delta}}\cos\omega t\right) = C\sin(\omega t - \alpha),$$

where $C = \dfrac{1}{\sqrt{\Delta}}F_0$, $\sin\alpha = \dfrac{c\omega}{\sqrt{\Delta}}$, and $\cos\alpha = \dfrac{1}{\sqrt{\Delta}}(k - m\omega^2)$.

27. The derivative of $C(\omega) = \dfrac{F_0}{\sqrt{\left(k - m\omega^2\right)^2 + \left(c\omega\right)^2}}$ is given by

$$C'(\omega) = -\frac{\omega F_0}{2} \cdot \frac{\left(c^2 - 2km\right) + 2\left(m\omega\right)^2}{\left[\left(k - m\omega^2\right)^2 + \left(c\omega\right)^2\right]^{3/2}}.$$

(a) Therefore, if $c^2 \geq \left(c_{\text{cr}}/\sqrt{2}\right)^2 = 2km$, it is clear from the numerator that $C'(\omega) < 0$ for all ω, so that $C(\omega)$ steadily decreases as ω increases.

(b) If instead $c^2 < 2km$, however, then the numerator (and hence $C'(\omega)$) vanishes when

$$\omega = \omega_m = \sqrt{\frac{k}{m} - \frac{c^2}{2m^2}} < \sqrt{\frac{k}{m}} = \omega_0.$$

Calculation then shows that

$$C''(\omega_m) = \frac{16 F_0 m^3 \left(c^2 - 2km\right)}{c^3 \left(4km - c^2\right)^{3/2}} < 0,$$

so it follows from the second-derivative test that $C(\omega_m)$ is a local maximum value.

29. We need only substitute $E_0 = ac\omega$ and $F_0 = ak$ in the result of Problem 26

CHAPTER 6

EIGENVALUES AND EIGENVECTORS

SECTION 6.1

INTRODUCTION TO EIGENVALUES

In each of Problems 1–32 we first list the characteristic polynomial $p(\lambda) = \left| \mathbf{A} - \lambda \mathbf{I} \right|$ of the given matrix $\mathbf{A}$, and then the roots of $p(\lambda)$ — which are the eigenvalues of $\mathbf{A}$. All of the eigenvalues that appear in Problems 1–26 are integers, so each characteristic polynomial factors readily. For each eigenvalue λ_j of the matrix $\mathbf{A}$, we determine the associate eigenvector(s) by finding a basis for the solution space of the linear system $\left(\mathbf{A} - \lambda_j \mathbf{I} \right) \mathbf{v} = \mathbf{0}$. We write this linear system in scalar form in terms of the components of $\mathbf{v} = \begin{bmatrix} a & b & \cdots \end{bmatrix}^T$. In most cases an associated eigenvector is then apparent. If $\mathbf{A}$ is a 2×2 matrix, for instance, then our two scalar equations will be multiples one of the other, so we can substitute a convenient numerical value for the first component a of $\mathbf{v}$ and then solve either equation for the second component b (or vice versa).

1. Characteristic polynomial: $p(\lambda) = \lambda^2 - 5\lambda + 6 = (\lambda - 2)(\lambda - 3)$

Eigenvalues: $\lambda_1 = 2, \quad \lambda_2 = 3$

With $\lambda_1 = 2$:
$\left. \begin{array}{r} 2a - 2b = 0 \\ a - b = 0 \end{array} \right\}$ $\qquad \mathbf{v}_1 = \begin{bmatrix} 1 \\ 1 \end{bmatrix}$

With $\lambda_2 = 3$:
$\left. \begin{array}{r} a - 2b = 0 \\ a - 2b = 0 \end{array} \right\}$ $\qquad \mathbf{v}_2 = \begin{bmatrix} 2 \\ 1 \end{bmatrix}$

3. Characteristic polynomial: $p(\lambda) = \lambda^2 - 7\lambda + 10 = (\lambda - 2)(\lambda - 5)$

Eigenvalues: $\lambda_1 = 2, \quad \lambda_2 = 5$

With $\lambda_1 = 2$:
$\left. \begin{array}{r} 6a - 6b = 0 \\ 3a - 3b = 0 \end{array} \right\}$ $\qquad \mathbf{v}_1 = \begin{bmatrix} 1 \\ 1 \end{bmatrix}$

With $\lambda_2 = 5$:
$\left. \begin{array}{r} 3a - 6b = 0 \\ 3a - 6b = 0 \end{array} \right\}$ $\qquad \mathbf{v}_2 = \begin{bmatrix} 2 \\ 1 \end{bmatrix}$

5.   Characteristic polynomial: $p(\lambda) = \lambda^2 - 5\lambda + 4 = (\lambda - 1)(\lambda - 4)$

        Eigenvalues: $\lambda_1 = 1, \quad \lambda_2 = 4$

         With $\lambda_1 = 1$: $\left.\begin{array}{r} 9a - 9b = 0 \\ 6a - 6b = 0 \end{array}\right\}$ $\mathbf{v}_1 = \begin{bmatrix} 1 \\ 1 \end{bmatrix}$

         With $\lambda_2 = 4$: $\left.\begin{array}{r} 6a - 9b = 0 \\ 6a - 9b = 0 \end{array}\right\}$ $\mathbf{v}_2 = \begin{bmatrix} 3 \\ 2 \end{bmatrix}$

7.   Characteristic polynomial: $p(\lambda) = \lambda^2 - 6\lambda + 8 = (\lambda - 2)(\lambda - 4)$

        Eigenvalues: $\lambda_1 = 2, \quad \lambda_2 = 4$

         With $\lambda_1 = 2$: $\left.\begin{array}{r} 8a - 8b = 0 \\ 6a - 6b = 0 \end{array}\right\}$ $\mathbf{v}_1 = \begin{bmatrix} 1 \\ 1 \end{bmatrix}$

         With $\lambda_2 = 4$: $\left.\begin{array}{r} 6a - 8b = 0 \\ 6a - 8b = 0 \end{array}\right\}$ $\mathbf{v}_2 = \begin{bmatrix} 4 \\ 3 \end{bmatrix}$

9.   Characteristic polynomial: $p(\lambda) = \lambda^2 - 7\lambda + 12 = (\lambda - 3)(\lambda - 4)$

        Eigenvalues: $\lambda_1 = 3, \quad \lambda_2 = 4$

         With $\lambda_1 = 3$: $\left.\begin{array}{r} 5a - 10b = 0 \\ 2a - 4b = 0 \end{array}\right\}$ $\mathbf{v}_1 = \begin{bmatrix} 2 \\ 1 \end{bmatrix}$

         With $\lambda_2 = 4$: $\left.\begin{array}{r} 4a - 10b = 0 \\ 2a - 5b = 0 \end{array}\right\}$ $\mathbf{v}_2 = \begin{bmatrix} 5 \\ 2 \end{bmatrix}$

11.   Characteristic polynomial: $p(\lambda) = \lambda^2 - 9\lambda + 20 = (\lambda - 4)(\lambda - 5)$

        Eigenvalues: $\lambda_1 = 4, \quad \lambda_2 = 5$

         With $\lambda_1 = 4$: $\left.\begin{array}{r} 15a - 10b = 0 \\ 21a - 14b = 0 \end{array}\right\}$ $\mathbf{v}_1 = \begin{bmatrix} 2 \\ 3 \end{bmatrix}$

         With $\lambda_2 = 5$: $\left.\begin{array}{r} 14a - 10b = 0 \\ 21a - 15b = 0 \end{array}\right\}$ $\mathbf{v}_2 = \begin{bmatrix} 5 \\ 7 \end{bmatrix}$

13.   Characteristic polynomial: $p(\lambda) = -\lambda^3 + 3\lambda^2 - 2\lambda = -\lambda(\lambda - 1)(\lambda - 2)$

        Eigenvalues: $\lambda_1 = 0, \quad \lambda_2 = 1, \quad \lambda_3 = 2$

With $\lambda_1 = 0$:
$$\left.\begin{array}{r} 2a = 0 \\ 2a - 2b - c = 0 \\ -2a + 6b + 3c = 0 \end{array}\right\}$$
$\mathbf{v}_1 = \begin{bmatrix} 0 \\ -1 \\ 2 \end{bmatrix}$

With $\lambda_2 = 1$:
$$\left.\begin{array}{r} a = 0 \\ 2a - 3b - c = 0 \\ -2a + 6b + 2c = 0 \end{array}\right\}$$
$\mathbf{v}_2 = \begin{bmatrix} 0 \\ -1 \\ 3 \end{bmatrix}$

With $\lambda_3 = 2$:
$$\left.\begin{array}{r} 0 = 0 \\ 2a - 4b - c = 0 \\ -2a + 6b + c = 0 \end{array}\right\}$$
$\mathbf{v}_3 = \begin{bmatrix} 1 \\ 0 \\ 2 \end{bmatrix}$

15. Characteristic polynomial: $p(\lambda) = -\lambda^3 + 3\lambda^2 - 2\lambda = -\lambda(\lambda - 1)(\lambda - 2)$

Eigenvalues: $\lambda_1 = 0, \quad \lambda_2 = 1, \quad \lambda_3 = 2$

With $\lambda_1 = 0$:
$$\left.\begin{array}{r} 2a - 2b = 0 \\ 2a - 2b - c = 0 \\ -2a + 2b + 3c = 0 \end{array}\right\}$$
$\mathbf{v}_1 = \begin{bmatrix} 1 \\ 1 \\ 0 \end{bmatrix}$

With $\lambda_2 = 1$:
$$\left.\begin{array}{r} a - 2b = 0 \\ 2a - 3b - c = 0 \\ -2a + 2b + 2c = 0 \end{array}\right\}$$
$\mathbf{v}_2 = \begin{bmatrix} 2 \\ 1 \\ 1 \end{bmatrix}$

With $\lambda_3 = 2$:
$$\left.\begin{array}{r} -2b = 0 \\ 2a - 4b - c = 0 \\ -2a + 2b + c = 0 \end{array}\right\}$$
$\mathbf{v}_3 = \begin{bmatrix} 1 \\ 0 \\ 2 \end{bmatrix}$

17. Characteristic polynomial: $p(\lambda) = -\lambda^3 + 6\lambda^2 - 11\lambda + 6 = -(\lambda - 1)(\lambda - 2)(\lambda - 3)$

Eigenvalues: $\lambda_1 = 1, \quad \lambda_2 = 2, \quad \lambda_3 = 3$

With $\lambda_1 = 1$:
$$\left.\begin{array}{r} 2a + 5b - 2c = 0 \\ b = 0 \\ 2b = 0 \end{array}\right\}$$
$\mathbf{v}_1 = \begin{bmatrix} 1 \\ 0 \\ 1 \end{bmatrix}$

With $\lambda_2 = 2$:
$$\left.\begin{array}{r} a + 5b - 2c = 0 \\ 0 = 0 \\ 2b - c = 0 \end{array}\right\}$$
$\mathbf{v}_2 = \begin{bmatrix} -1 \\ 1 \\ 2 \end{bmatrix}$

$$\text{With } \lambda_3 = 3: \quad \left. \begin{array}{r} 5b - 2c = 0 \\ -b = 0 \\ 2b - 2c = 0 \end{array} \right\} \qquad \mathbf{v}_3 = \begin{bmatrix} 1 \\ 0 \\ 0 \end{bmatrix}$$

19. Characteristic polynomial: $p(\lambda) = -\lambda^3 + 5\lambda^2 - 7\lambda + 3 = -(\lambda - 1)^2(\lambda - 3)$

Eigenvalues: $\lambda_1 = \lambda_2 = 1, \quad \lambda_3 = 3$

$$\text{With } \lambda_1 = 1: \quad \left. \begin{array}{r} 2a + 6b - 2c = 0 \\ 0 = 0 \\ 0 = 0 \end{array} \right\} \qquad \mathbf{v}_1 = \begin{bmatrix} 1 \\ 0 \\ 1 \end{bmatrix}, \quad \mathbf{v}_2 = \begin{bmatrix} -3 \\ 1 \\ 0 \end{bmatrix}$$

The eigenspace of $\lambda_1 = 1$ is 2-dimensional. We get the eigenvector $\mathbf{v}_1$ with $b = 0$, $c = 1$, and the eigenvector $\mathbf{v}_2$ with $b = 1$, $c = 0$.

$$\text{With } \lambda_3 = 3: \quad \left. \begin{array}{r} 6b - 2c = 0 \\ -2b = 0 \\ -2c = 0 \end{array} \right\} \qquad \mathbf{v}_3 = \begin{bmatrix} 1 \\ 0 \\ 0 \end{bmatrix}$$

21. Characteristic polynomial: $p(\lambda) = -\lambda^3 + 5\lambda^2 - 8\lambda + 4 = -(\lambda - 1)(\lambda - 2)^2$

Eigenvalues: $\lambda_1 = 1, \quad \lambda_2 = \lambda_3 = 2$

$$\text{With } \lambda_1 = 1: \quad \left. \begin{array}{r} 3a - 3b + c = 0 \\ 2a - 2b + c = 0 \\ c = 0 \end{array} \right\} \qquad \mathbf{v}_1 = \begin{bmatrix} 1 \\ 1 \\ 0 \end{bmatrix}$$

$$\text{With } \lambda_2 = 2: \quad \left. \begin{array}{r} 2a - 3b + c = 0 \\ 2a - 3b + c = 0 \\ 0 = 0 \end{array} \right\} \qquad \mathbf{v}_2 = \begin{bmatrix} 3 \\ 2 \\ 0 \end{bmatrix} \quad \mathbf{v}_3 = \begin{bmatrix} -1 \\ 0 \\ 2 \end{bmatrix}$$

The eigenspace of $\lambda_2 = 2$ is 2-dimensional. We get the eigenvector $\mathbf{v}_2$ with $b = 2$, $c = 0$, and the eigenvector $\mathbf{v}_3$ with $b = 0$, $c = 2$.

23. Characteristic polynomial: $p(\lambda) = (\lambda - 1)(\lambda - 2)(\lambda - 3)(\lambda - 4)$

Eigenvalues: $\lambda_1 = 1, \quad \lambda_2 = 2, \quad \lambda_3 = 3, \quad \lambda_4 = 4$

$$\text{With } \lambda_1 = 1: \quad \left. \begin{array}{r} 2b + 2c + 2d = 0 \\ b + 2c + 2d = 0 \\ 2c + 2d = 0 \\ 3d = 0 \end{array} \right\} \qquad \mathbf{v}_1 = \begin{bmatrix} 1 \\ 0 \\ 0 \\ 0 \end{bmatrix}$$

With $\lambda_2 = 2$:
$$\left.\begin{array}{r} -a+2b+2c+2d = 0 \\ 2c+2d = 0 \\ c+2d = 0 \\ 2d = 0 \end{array}\right\} \qquad \mathbf{v}_2 = \begin{bmatrix} 2 \\ 1 \\ 0 \\ 0 \end{bmatrix}$$

With $\lambda_3 = 3$:
$$\left.\begin{array}{r} -2a+2b+2c+2d = 0 \\ b+2c+2d = 0 \\ 2d = 0 \\ d = 0 \end{array}\right\} \qquad \mathbf{v}_3 = \begin{bmatrix} 3 \\ 2 \\ 1 \\ 0 \end{bmatrix}$$

With $\lambda_4 = 4$:
$$\left.\begin{array}{r} -3a+2b+2c+2d = 0 \\ -2b+2c+2d = 0 \\ -c+2d = 0 \\ 0 = 0 \end{array}\right\} \qquad \mathbf{v}_4 = \begin{bmatrix} 4 \\ 3 \\ 2 \\ 1 \end{bmatrix}$$

25.   Characteristic polynomial: $p(\lambda) = (\lambda-1)^2(\lambda-2)^2$

Eigenvalues: $\lambda_1 = \lambda_2 = 1, \quad \lambda_3 = \lambda_4 = 2$

With $\lambda_1 = 1$:
$$\left.\begin{array}{r} c = 0 \\ c = 0 \\ c = 0 \\ d = 0 \end{array}\right\} \qquad \mathbf{v}_1 = \begin{bmatrix} 1 \\ 0 \\ 0 \\ 0 \end{bmatrix}, \qquad \mathbf{v}_2 = \begin{bmatrix} 0 \\ 1 \\ 0 \\ 0 \end{bmatrix}$$

The eigenspace of $\lambda_1 = 1$ is 2-dimensional. We note that $c = d = 0$, but a and b are arbitrary.

With $\lambda_3 = 2$:
$$\left.\begin{array}{r} -a+c = 0 \\ -b+c = 0 \\ 0 = 0 \\ 0 = 0 \end{array}\right\} \qquad \mathbf{v}_3 = \begin{bmatrix} 0 \\ 0 \\ 0 \\ 1 \end{bmatrix}, \qquad \mathbf{v}_4 = \begin{bmatrix} 1 \\ 1 \\ 1 \\ 0 \end{bmatrix}$$

The eigenspace of $\lambda_3 = 2$ is 2-dimensional. We get the eigenvector $\mathbf{v}_3$ with $b = 0$, $c = 1$, and the eigenvector $\mathbf{v}_4$ with $b = 1$, $c = 0$.

27.   Characteristic polynomial: $p(\lambda) = \lambda^2 + 1$

Eigenvalues: $\lambda_1 = -i, \quad \lambda_2 = +i$

With $\lambda_1 = -i$:
$$\left.\begin{array}{r} ia+b = 0 \\ -a+ib = 0 \end{array}\right\} \qquad \mathbf{v}_1 = \begin{bmatrix} i \\ 1 \end{bmatrix}$$

With $\lambda_2 = +i$: $\left.\begin{array}{r} -ia+b = 0 \\ -a-ib = 0 \end{array}\right\}$ $\mathbf{v}_2 = \begin{bmatrix} -i \\ 1 \end{bmatrix}$

29. Characteristic polynomial: $p(\lambda) = \lambda^2 + 36$

Eigenvalues: $\lambda_1 = -6i, \quad \lambda_2 = +6i$

With $\lambda_1 = -6i$: $\left.\begin{array}{r} 6ia-3b = 0 \\ 12a+6ib = 0 \end{array}\right\}$ $\mathbf{v}_1 = \begin{bmatrix} -i \\ 2 \end{bmatrix}$

With $\lambda_2 = +6i$: $\left.\begin{array}{r} -6ia-3b = 0 \\ 12a-6ib = 0 \end{array}\right\}$ $\mathbf{v}_2 = \begin{bmatrix} i \\ 2 \end{bmatrix}$

31. Characteristic polynomial: $p(\lambda) = \lambda^2 + 144$

Eigenvalues: $\lambda_1 = -12i, \quad \lambda_2 = +12i$

With $\lambda_1 = -12i$: $\left.\begin{array}{r} 12ia+24b = 0 \\ -6a+12ib = 0 \end{array}\right\}$ $\mathbf{v}_1 = \begin{bmatrix} 2i \\ 1 \end{bmatrix}$

With $\lambda_2 = +12i$: $\left.\begin{array}{r} -12ia+24b = 0 \\ -6a-12ib = 0 \end{array}\right\}$ $\mathbf{v}_2 = \begin{bmatrix} -2i \\ 1 \end{bmatrix}$

33. If $\mathbf{Av} = \lambda\mathbf{v}$ and we assume that $\mathbf{A}^{n-1}\mathbf{v} = \lambda^{n-1}\mathbf{v}$ — meaning that λ^{n-1} is an eigenvalue of $\mathbf{A}^{n-1}$ with associated eigenvector $\mathbf{v}$, then multiplication by $\mathbf{A}$ yields

$$\mathbf{A}^n\mathbf{v} = \mathbf{A}\cdot\mathbf{A}^{n-1}\mathbf{v} = \mathbf{A}\cdot\lambda^{n-1}\mathbf{v} = \lambda^{n-1}\cdot\mathbf{Av} = \lambda^{n-1}\cdot\lambda\mathbf{v} = \lambda^n\mathbf{v}.$$

Thus λ^n is an eigenvalue of the matrix $\mathbf{A}^n$ with associated eigenvector $\mathbf{v}$.

35. **(a)** Note first that $(\mathbf{A}-\lambda\mathbf{I})^T = (\mathbf{A}^T-\lambda\mathbf{I})$ because $\mathbf{I}^T = \mathbf{I}$. Since the determinant of a square matrix equals the determinant of its transpose, it follows that

$$|\mathbf{A}-\lambda\mathbf{I}| = |\mathbf{A}^T-\lambda\mathbf{I}|.$$

Thus the matrices $\mathbf{A}$ and $\mathbf{A}^T$ have the same characteristic polynomial, and therefore have the same eigenvalues.

(b) Consider the matrix $\mathbf{A} = \begin{bmatrix} 1 & 0 \\ 1 & 1 \end{bmatrix}$ with characteristic equation $(\lambda-1)^2 = 0$ and the single eigenvalue $\lambda = 1$. Then $\mathbf{A}-\mathbf{I} = \begin{bmatrix} 0 & 0 \\ 1 & 0 \end{bmatrix}$ and it follows that the only associated eigenvector is a multiple of $\begin{bmatrix} 0 & 1 \end{bmatrix}^T$. The transpose $\mathbf{A}^T = \begin{bmatrix} 1 & 1 \\ 0 & 1 \end{bmatrix}$ has the same characteristic equation and eigenvalue, but we see similarly that its only eigenvector is a multiple of $\begin{bmatrix} 1 & 0 \end{bmatrix}^T$. Thus $\mathbf{A}$ and $\mathbf{A}^T$ have the same eigenvalue but different eigenvectors.

37. If $|\mathbf{A} - \lambda\mathbf{I}| = (-1)^n \lambda^n + c_{n-1}\lambda^{n-1} + \cdots + c_1\lambda + c_0$, then substitution of $\lambda = 0$ yields $c_0 = |\mathbf{A} - 0\mathbf{I}| = |\mathbf{A}|$ for the constant term in the characteristic polynomial.

39. If the characteristic equation of the $n \times n$ matrix $\mathbf{A}$ with eigenvalues $\lambda_1, \lambda_2, \cdots, \lambda_n$ (not necessarily distinct) is written in the factored form

$$(\lambda - \lambda_1)(\lambda - \lambda_2) \cdots (\lambda - \lambda_n) = 0,$$

then it should be clear that upon multiplying out the factors the coefficient of λ^{n-1} will be $-(\lambda_1 + \lambda_2 + \cdots + \lambda_n)$. But according to Problem 38, this coefficient also equals $-(\text{trace}\,\mathbf{A})$. Therefore

$$\lambda_1 + \lambda_2 + \cdots + \lambda_n = \text{trace}\,\mathbf{A} = a_{11} + a_{22} + \cdots a_{nn}.$$

41. We find that $\text{trace}\,\mathbf{A} = 8$ and $\det\mathbf{A} = -60$, so the characteristic polynomial of the given matrix $\mathbf{A}$ is

$$p(\lambda) = \lambda^4 - 8\lambda^3 + c_2\lambda^2 + c_1\lambda - 60.$$

Substitution of

$$\lambda = 1, \ p(1) = \det(\mathbf{A} - \mathbf{I}) = -24 \text{ and } \lambda = -1, \ p(-1) = \det(\mathbf{A} + \mathbf{I}) = -72$$

yields the equations

$$c_2 + c_1 = 43, \quad c_2 - c_1 = -21$$

that we solve readily for $c_1 = 32$, $c_2 = 11$. Hence the characteristic equation of $\mathbf{A}$ is

$$\lambda^4 - 8\lambda^3 + 11\lambda^2 + 32\lambda - 60 = 0.$$

Trying $\lambda = \pm 1, \pm 2$ in turn, we discover the eigenvalues $\lambda_1 = -2$ and $\lambda_2 = 2$. Then division of the quartic by $(\lambda^2 - 4)$ yields

$$\lambda^2 - 8\lambda + 15 = (\lambda - 3)(\lambda - 5),$$

so the other two eigenvalues are $\lambda_3 = 3$ and $\lambda_4 = 5$. We proceed to find the eigenvectors associated with these four eigenvalues.

With $\lambda_1 = -2$:

$$\left.\begin{array}{r}24a-9b-8c-8d = 0\\10a-5b-14c+2d = 0\\10a+10c-10d = 0\\29a-9b-3c-13d = 0\end{array}\right\} \rightarrow \left.\begin{array}{r}a-(1/2)d = 0\\b = 0\\c-(1/2)d = 0\\0 = 0\end{array}\right\} \quad \mathbf{v}_1 = \begin{bmatrix}1\\0\\1\\2\end{bmatrix}$$

With $\lambda_2 = 2$:

$$\left.\begin{array}{r}20a-9b-8c-8d = 0\\10a-9b-14c+2d = 0\\10a+6c-10d = 0\\29a-9b-3c-17d = 0\end{array}\right\} \rightarrow \left.\begin{array}{r}a-d = 0\\b-(4/3)d = 0\\c = 0\\0 = 0\end{array}\right\} \quad \mathbf{v}_2 = \begin{bmatrix}3\\4\\0\\3\end{bmatrix}$$

With $\lambda_3 = 3$:

$$\left.\begin{array}{r}19a-9b-8c-8d = 0\\10a-10b-14c+2d = 0\\10a+5c-10d = 0\\29a-9b-3c-18d = 0\end{array}\right\} \rightarrow \left.\begin{array}{r}a-(3/4)d = 0\\b-(1/4)d = 0\\c-(1/2)d = 0\\0 = 0\end{array}\right\} \quad \mathbf{v}_3 = \begin{bmatrix}3\\1\\2\\4\end{bmatrix}$$

With $\lambda_4 = 5$:

$$\left.\begin{array}{r}17a-9b-8c-8d = 0\\10a-12b-14c+2d = 0\\10a+3c-10d = 0\\29a-9b-3c-20d = 0\end{array}\right\} \rightarrow \left.\begin{array}{r}a-d = 0\\b-d = 0\\c = 0\\0 = 0\end{array}\right\} \quad \mathbf{v}_4 = \begin{bmatrix}1\\1\\0\\1\end{bmatrix}$$

SECTION 6.2

DIAGONALIZATION OF MATRICES

In Problems 1–28 we first find the eigenvalues and associated eigenvectors of the given $n \times n$ matrix $\mathbf{A}$. If $\mathbf{A}$ has n linearly independent eigenvectors, then we can proceed to set up the desired diagonalizing matrix $\mathbf{P} = \begin{bmatrix}\mathbf{v}_1 & \mathbf{v}_2 & \cdots & \mathbf{v}_n\end{bmatrix}$ and diagonal matrix $\mathbf{D}$ such that $\mathbf{P}^{-1}\mathbf{AP} = \mathbf{D}$. If you write the eigenvalues in a different order on the diagonal of $\mathbf{D}$, then naturally the eigenvector columns of $\mathbf{P}$ must be rearranged in the same order.

1.   Characteristic polynomial: $p(\lambda) = \lambda^2 - 4\lambda + 3 = (\lambda - 1)(\lambda - 3)$

Eigenvalues: $\lambda_1 = 1, \quad \lambda_2 = 3$

With $\lambda_1 = 1$: $\qquad\left.\begin{array}{l} 4a - 4b = 0 \\ 2a - 2b = 0 \end{array}\right\} \qquad \mathbf{v}_1 = \begin{bmatrix} 1 \\ 1 \end{bmatrix}$

With $\lambda_2 = 3$: $\qquad\left.\begin{array}{l} 2a - 4b = 0 \\ 2a - 4b = 0 \end{array}\right\} \qquad \mathbf{v}_2 = \begin{bmatrix} 2 \\ 1 \end{bmatrix}$

$\mathbf{P} = \begin{bmatrix} 1 & 2 \\ 1 & 1 \end{bmatrix}, \qquad \mathbf{D} = \begin{bmatrix} 1 & 0 \\ 0 & 3 \end{bmatrix}$

3.   Characteristic polynomial: $\qquad p(\lambda) = \lambda^2 - 5\lambda + 6 = (\lambda - 2)(\lambda - 3)$

Eigenvalues: $\quad \lambda_1 = 2, \quad \lambda_2 = 3$

With $\lambda_1 = 2$: $\qquad\left.\begin{array}{l} 3a - 3b = 0 \\ 2a - 2b = 0 \end{array}\right\} \qquad \mathbf{v}_1 = \begin{bmatrix} 1 \\ 1 \end{bmatrix}$

With $\lambda_2 = 3$: $\qquad\left.\begin{array}{l} 2a - 3b = 0 \\ 2a - 3b = 0 \end{array}\right\} \qquad \mathbf{v}_2 = \begin{bmatrix} 3 \\ 2 \end{bmatrix}$

$\mathbf{P} = \begin{bmatrix} 1 & 3 \\ 1 & 2 \end{bmatrix}, \qquad \mathbf{D} = \begin{bmatrix} 2 & 0 \\ 0 & 3 \end{bmatrix}$

5.   Characteristic polynomial: $\qquad p(\lambda) = \lambda^2 - 4\lambda + 3 = (\lambda - 1)(\lambda - 3)$

Eigenvalues: $\quad \lambda_1 = 1, \quad \lambda_2 = 3$

With $\lambda_1 = 1$: $\qquad\left.\begin{array}{l} 8a - 8b = 0 \\ 6a - 6b = 0 \end{array}\right\} \qquad \mathbf{v}_1 = \begin{bmatrix} 1 \\ 1 \end{bmatrix}$

With $\lambda_2 = 3$: $\qquad\left.\begin{array}{l} 6a - 8b = 0 \\ 6a - 8b = 0 \end{array}\right\} \qquad \mathbf{v}_2 = \begin{bmatrix} 4 \\ 3 \end{bmatrix}$

$\mathbf{P} = \begin{bmatrix} 1 & 4 \\ 1 & 3 \end{bmatrix}, \qquad \mathbf{D} = \begin{bmatrix} 1 & 0 \\ 0 & 3 \end{bmatrix}$

7.   Characteristic polynomial: $\qquad p(\lambda) = \lambda^2 - 3\lambda + 2 = (\lambda - 1)(\lambda - 2)$

Eigenvalues: $\quad \lambda_1 = 1, \quad \lambda_2 = 2$

With $\lambda_1 = 1$: $\qquad\left.\begin{array}{l} 5a - 10b = 0 \\ 2a - 4b = 0 \end{array}\right\} \qquad \mathbf{v}_1 = \begin{bmatrix} 2 \\ 1 \end{bmatrix}$

With $\lambda_2 = 2$:
$$\left.\begin{array}{r} 4a - 10b = 0 \\ 2a - 5b = 0 \end{array}\right\}$$
$$\mathbf{v}_2 = \begin{bmatrix} 5 \\ 2 \end{bmatrix}$$

$$\mathbf{P} = \begin{bmatrix} 2 & 5 \\ 1 & 2 \end{bmatrix}, \qquad \mathbf{D} = \begin{bmatrix} 1 & 0 \\ 0 & 2 \end{bmatrix}$$

9. Characteristic polynomial: $p(\lambda) = \lambda^2 - 2\lambda + 1 = (\lambda - 1)^2$

Eigenvalues: $\lambda_1 = 1, \quad \lambda_2 = 1$

With $\lambda_1 = 1$:
$$\left.\begin{array}{r} 4a - 2b = 0 \\ 2a - b = 0 \end{array}\right\}$$
$$\mathbf{v}_1 = \begin{bmatrix} 2 \\ 1 \end{bmatrix}$$

Because the given matrix $\mathbf{A}$ has only the single eigenvector $\mathbf{v}_1$, it is not diagonalizable.

11. Characteristic polynomial: $p(\lambda) = \lambda^2 - 4\lambda + 4 = (\lambda - 2)^2$

Eigenvalues: $\lambda_1 = 2, \quad \lambda_2 = 2$

With $\lambda_1 = 2$:
$$\left.\begin{array}{r} 3a + b = 0 \\ -9a - 3b = 0 \end{array}\right\}$$
$$\mathbf{v}_1 = \begin{bmatrix} -1 \\ 3 \end{bmatrix}$$

Because the given matrix $\mathbf{A}$ has only the single eigenvector $\mathbf{v}_1$, it is not diagonalizable.

13. Characteristic polynomial: $p(\lambda) = -\lambda^3 + 5\lambda^2 - 8\lambda + 4 = -(\lambda - 1)(\lambda - 2)^2$

Eigenvalues: $\lambda_1 = 1, \quad \lambda_2 = 2, \quad \lambda_3 = 2$

With $\lambda_1 = 1$:
$$\left.\begin{array}{r} 3b = 0 \\ b = 0 \\ c = 0 \end{array}\right\}$$
$$\mathbf{v}_1 = \begin{bmatrix} 1 \\ 0 \\ 0 \end{bmatrix}$$

With $\lambda_2 = 2$:
$$\left.\begin{array}{r} 3b - a = 0 \\ 0 = 0 \\ 0 = 0 \end{array}\right\}$$
$$\mathbf{v}_2 = \begin{bmatrix} 0 \\ 0 \\ 1 \end{bmatrix}, \quad \mathbf{v}_3 = \begin{bmatrix} 3 \\ 1 \\ 0 \end{bmatrix}$$

The eigenspace of $\lambda_2 = 2$ is 2-dimensional. We get the eigenvector $\mathbf{v}_2$ with $b = 0$, $c = 1$, and the eigenvector $\mathbf{v}_3$ with $b = 1$, $c = 0$.

$$\mathbf{P} = \begin{bmatrix} 1 & 0 & 3 \\ 0 & 0 & 1 \\ 0 & 1 & 0 \end{bmatrix}, \qquad \mathbf{D} = \begin{bmatrix} 1 & 0 & 0 \\ 0 & 2 & 0 \\ 0 & 0 & 2 \end{bmatrix}$$

15.   Characteristic polynomial: $p(\lambda) = -\lambda^3 + 2\lambda^2 - \lambda = -\lambda(\lambda - 1)^2$

Eigenvalues: $\lambda_1 = 0$, $\lambda_2 = 1$, $\lambda_3 = 1$

With $\lambda_1 = 0$: $\left.\begin{array}{r} 3a - 3b + c = 0 \\ 2a - 2b + c = 0 \\ c = 0 \end{array}\right\}$ $\mathbf{v}_1 = \begin{bmatrix} 1 \\ 1 \\ 0 \end{bmatrix}$

With $\lambda_2 = 1$: $\left.\begin{array}{r} 2a - 3b + c = 0 \\ 2a - 3b + c = 0 \\ 0 = 0 \end{array}\right\}$ $\mathbf{v}_2 = \begin{bmatrix} -1 \\ 0 \\ 2 \end{bmatrix}$, $\mathbf{v}_3 = \begin{bmatrix} 3 \\ 2 \\ 0 \end{bmatrix}$

The eigenspace of $\lambda_2 = 2$ is 2-dimensional. We get the eigenvector $\mathbf{v}_2$ with $b = 0$, $c = 2$, and the eigenvector $\mathbf{v}_3$ with $b = 2$, $c = 0$.

$$\mathbf{P} = \begin{bmatrix} 1 & -1 & 3 \\ 1 & 0 & 2 \\ 0 & 2 & 0 \end{bmatrix}, \qquad \mathbf{D} = \begin{bmatrix} 0 & 0 & 0 \\ 0 & 1 & 0 \\ 0 & 0 & 1 \end{bmatrix}$$

17.   Characteristic polynomial: $p(\lambda) = -\lambda^3 + 2\lambda^2 + \lambda - 2 = -(\lambda + 1)(\lambda - 1)(\lambda - 2)$

Eigenvalues: $\lambda_1 = -1$, $\lambda_2 = 1$, $\lambda_3 = 2$

With $\lambda_1 = -1$: $\left.\begin{array}{r} 8a - 8b + 3c = 0 \\ 6a - 6b + 3c = 0 \\ 2a - 2b + 3c = 0 \end{array}\right\}$ $\mathbf{v}_1 = \begin{bmatrix} 1 \\ 1 \\ 0 \end{bmatrix}$

With $\lambda_2 = 1$: $\left.\begin{array}{r} 6a - 8b + 3c = 0 \\ 6a - 8b + 3c = 0 \\ 2a - 2b + c = 0 \end{array}\right\}$ $\mathbf{v}_2 = \begin{bmatrix} -1 \\ 0 \\ 2 \end{bmatrix}$

With $\lambda_3 = 2$: $\left.\begin{array}{r} 5a - 8b + 3c = 0 \\ 6a - 9b + 3c = 0 \\ 2a - 2b = 0 \end{array}\right\}$ $\mathbf{v}_3 = \begin{bmatrix} 1 \\ 1 \\ 1 \end{bmatrix}$

$$\mathbf{P} = \begin{bmatrix} 1 & -1 & 1 \\ 1 & 0 & 1 \\ 0 & 2 & 1 \end{bmatrix}, \qquad \mathbf{D} = \begin{bmatrix} -1 & 0 & 0 \\ 0 & 1 & 0 \\ 0 & 0 & 2 \end{bmatrix}$$

19.   Characteristic polynomial: $p(\lambda) = -\lambda^3 + 6\lambda^2 - 11\lambda + 6 = -(\lambda - 1)(\lambda - 2)(\lambda - 3)$

Eigenvalues: $\lambda_1 = 1$, $\lambda_2 = 2$, $\lambda_3 = 3$

With $\lambda_1 = 1$:
$$\left.\begin{array}{r} b - c = 0 \\ -2a + 3b - c = 0 \\ -4a + 4b = 0 \end{array}\right\}$$
$$\mathbf{v}_1 = \begin{bmatrix} 1 \\ 1 \\ 1 \end{bmatrix}$$

With $\lambda_2 = 2$:
$$\left.\begin{array}{r} -a + b - c = 0 \\ -2a + 2b - c = 0 \\ -4a + 4b - c = 0 \end{array}\right\}$$
$$\mathbf{v}_2 = \begin{bmatrix} 1 \\ 1 \\ 0 \end{bmatrix}$$

With $\lambda_3 = 3$:
$$\left.\begin{array}{r} -2a + b - c = 0 \\ -2a + b - c = 0 \\ -4a + 4b - 2c = 0 \end{array}\right\}$$
$$\mathbf{v}_3 = \begin{bmatrix} -1 \\ 0 \\ 2 \end{bmatrix}$$

$$\mathbf{P} = \begin{bmatrix} 1 & 1 & -1 \\ 1 & 1 & 0 \\ 1 & 0 & 2 \end{bmatrix}, \qquad \mathbf{D} = \begin{bmatrix} 1 & 0 & 0 \\ 0 & 2 & 0 \\ 0 & 0 & 3 \end{bmatrix}$$

21. Characteristic polynomial: $\quad p(\lambda) = -\lambda^3 + 3\lambda^2 - 3\lambda + 1 = -(\lambda - 1)^3$

Eigenvalues: $\quad \lambda_1 = 1, \quad \lambda_2 = 1, \quad \lambda_3 = 1$

With $\lambda_1 = 1$:
$$\left.\begin{array}{r} b - a = 0 \\ b - a = 0 \\ b - a = 0 \end{array}\right\}$$
$$\mathbf{v}_1 = \begin{bmatrix} 0 \\ 0 \\ 1 \end{bmatrix}, \quad \mathbf{v}_2 = \begin{bmatrix} 1 \\ 1 \\ 0 \end{bmatrix}$$

The eigenspace of $\lambda_1 = 1$ is 2-dimensional. We get the eigenvector $\mathbf{v}_1$ with $b = 0$, $c = 1$, and the eigenvector $\mathbf{v}_2$ with $b = 1$, $c = 0$. Because the given matrix $\mathbf{A}$ has only two linearly independent eigenvectors, it is not diagonalizable.

23. Characteristic polynomial: $\quad p(\lambda) = -\lambda^3 + 4\lambda^2 - 5\lambda + 2 = -(\lambda - 1)^2(\lambda - 2)$

Eigenvalues: $\quad \lambda_1 = 1, \quad \lambda_2 = 1, \quad \lambda_3 = 2$

With $\lambda_1 = 1$:
$$\left.\begin{array}{r} -3a + 4b - c = 0 \\ -3a + 4b - c = 0 \\ -a + b = 0 \end{array}\right\}$$
$$\mathbf{v}_1 = \begin{bmatrix} 1 \\ 1 \\ 1 \end{bmatrix}$$

With $\lambda_3 = 2$:
$$\left.\begin{array}{r} -4a + 4b - c = 0 \\ -3a + 3b - c = 0 \\ -a + b - c = 0 \end{array}\right\}$$
$$\mathbf{v}_2 = \begin{bmatrix} 1 \\ 1 \\ 0 \end{bmatrix}$$

The given matrix $\mathbf{A}$ has only the two linearly independent eigenvectors $\mathbf{v}_1$ and $\mathbf{v}_2$, and therefore is not diagonalizable.

25. Characteristic polynomial: $p(\lambda) = (\lambda+1)^2(\lambda-1)^2$

Eigenvalues: $\lambda_1 = -1,\quad \lambda_2 = -1,\quad \lambda_3 = 1,\quad \lambda_4 = 1$

With $\lambda_1 = -1$:
$$\left.\begin{matrix} 2a - 2c = 0 \\ 2b - 2c = 0 \\ 0 = 0 \\ 0 = 0 \end{matrix}\right\}$$
$$\mathbf{v}_1 = \begin{bmatrix} 0 \\ 0 \\ 0 \\ 1 \end{bmatrix}, \quad \mathbf{v}_2 = \begin{bmatrix} 1 \\ 1 \\ 1 \\ 0 \end{bmatrix}$$

The eigenspace of $\lambda_1 = -1$ is 2-dimensional. We get the eigenvector $\mathbf{v}_1$ with $c = 0,\ d = 1,$ and the eigenvector $\mathbf{v}_2$ with $c = 1,\ d = 0.$

With $\lambda_3 = 1$:
$$\left.\begin{matrix} -2c = 0 \\ -2c = 0 \\ -2c = 0 \\ -2d = 0 \end{matrix}\right\}$$
$$\mathbf{v}_3 = \begin{bmatrix} 0 \\ 1 \\ 0 \\ 0 \end{bmatrix}, \quad \mathbf{v}_4 = \begin{bmatrix} 1 \\ 0 \\ 0 \\ 0 \end{bmatrix}$$

The eigenspace of $\lambda_3 = 1$ is also 2-dimensional. We get the eigenvector $\mathbf{v}_3$ with $a = 0,\ b = 1,$ and the eigenvector $\mathbf{v}_4$ with $a = 1,\ b = 0.$

$$\mathbf{P} = \begin{bmatrix} 0 & 1 & 0 & 1 \\ 0 & 1 & 1 & 0 \\ 0 & 1 & 0 & 0 \\ 1 & 0 & 0 & 0 \end{bmatrix}, \qquad \mathbf{D} = \begin{bmatrix} -1 & 0 & 0 & 0 \\ 0 & -1 & 0 & 0 \\ 0 & 0 & 1 & 0 \\ 0 & 0 & 0 & 1 \end{bmatrix}$$

27. Characteristic polynomial: $p(\lambda) = (\lambda-1)^3(\lambda-2)$

Eigenvalues: $\lambda_1 = 1,\quad \lambda_2 = 1,\quad \lambda_3 = 1,\quad \lambda_4 = 2$

With $\lambda_1 = 1$:
$$\left.\begin{matrix} b = 0 \\ c = 0 \\ d = 0 \\ d = 0 \end{matrix}\right\}$$
$$\mathbf{v}_1 = \begin{bmatrix} 1 \\ 0 \\ 0 \\ 0 \end{bmatrix}$$

The eigenspace of $\lambda_1 = 1$ is 1-dimensional, with only a single associated eigenvector.

With $\lambda_4 = 2$:
$$\left.\begin{matrix} b - a = 0 \\ c - b = 0 \\ d - c = 0 \\ 0 = 0 \end{matrix}\right\}$$
$$\mathbf{v}_4 = \begin{bmatrix} 1 \\ 1 \\ 1 \\ 1 \end{bmatrix}$$

The given matrix $\mathbf{A}$ has only the two linearly independent eigenvectors $\mathbf{v}_1$ and $\mathbf{v}_4$, and therefore is not diagonalizable.

29.   If $\mathbf{A}$ is similar to $\mathbf{B}$ and $\mathbf{B}$ is similar to $\mathbf{C}$, so $\mathbf{A} = \mathbf{P}^{-1}\mathbf{BP}$ and $\mathbf{B} = \mathbf{Q}^{-1}\mathbf{CQ}$, then

$$\mathbf{A} = \mathbf{P}^{-1}(\mathbf{Q}^{-1}\mathbf{CQ})\mathbf{P} = (\mathbf{P}^{-1}\mathbf{Q}^{-1})\mathbf{C}(\mathbf{QP}) = (\mathbf{QP})^{-1}\mathbf{C}(\mathbf{QP}) = \mathbf{R}^{-1}\mathbf{CR}$$

with $\mathbf{R} = \mathbf{QP}$, so $\mathbf{A}$ is similar to $\mathbf{C}$.

31.   If $\mathbf{A}$ is similar to $\mathbf{B}$ so $\mathbf{A} = \mathbf{P}^{-1}\mathbf{BP}$ then $\mathbf{A}^{-1} = (\mathbf{P}^{-1}\mathbf{BP})^{-1} = \mathbf{P}^{-1}\mathbf{B}^{-1}\mathbf{P}$, so $\mathbf{A}^{-1}$ is similar to $\mathbf{B}^{-1}$.

33.   If $\mathbf{A}$ and $\mathbf{B}$ are similar with $\mathbf{A} = \mathbf{P}^{-1}\mathbf{BP}$, then

$$\left|\mathbf{A}\right| = \left|\mathbf{P}^{-1}\mathbf{BP}\right| = \left|\mathbf{P}^{-1}\right|\left|\mathbf{B}\right|\left|\mathbf{P}\right| = \left|\mathbf{P}\right|^{-1}\left|\mathbf{B}\right|\left|\mathbf{P}\right| = \left|\mathbf{B}\right|.$$

Moreover, by Problem 32 the two matrices have the same eigenvalues, and by Problem 39 in Section 6.1, the trace of a square matrix with real eigenvalues is equal to the sum of those eigenvalues. Therefore trace $\mathbf{A} = $ (eigenvalue sum) $= $ trace $\mathbf{B}$.

35.   Three eigenvectors associated with three distinct eigenvalues can be arranged in six different orders as the column vectors of the diagonalizing matrix $\mathbf{P} = \begin{bmatrix} \mathbf{v}_1 & \mathbf{v}_2 & \mathbf{v}_3 \end{bmatrix}^T$.

37.   If $\mathbf{A} = \mathbf{PDP}^{-1}$ with $\mathbf{P}$ the eigenvector matrix of $\mathbf{A}$ and $\mathbf{D}$ its diagonal matrix of eigenvalues, then $\mathbf{A}^2 = (\mathbf{PDP}^{-1})(\mathbf{PDP}^{-1}) = \mathbf{PD}(\mathbf{P}^{-1}\mathbf{P})\mathbf{DP}^{-1} = \mathbf{PD}^2\mathbf{P}^{-1}$. Thus the same (eigenvector) matrix $\mathbf{P}$ diagonalizes $\mathbf{A}^2$, but the resulting diagonal (eigenvalue) matrix $\mathbf{D}^2$ is the square of the one for $\mathbf{A}$. The diagonal elements of $\mathbf{D}^2$ are the eigenvalues of $\mathbf{A}^2$ and the diagonal elements of $\mathbf{D}$ are the eigenvalues of $\mathbf{A}$, so the former are the squares of the latter.

39.   Let the $n \times n$ matrix $\mathbf{A}$ have $k \leq n$ distinct eigenvalues $\lambda_1, \lambda_2, \cdots, \lambda_k$. Then the definition of algebraic multiplicity and the fact that all solutions of the nth degree polynomial equation $\left|\mathbf{A} - \lambda\mathbf{I}\right| = 0$ are real imply that the sum of the multiplicities of the eigenvalues equals n,

$$p_1 + p_2 + \cdots + p_k = n.$$

Now Theorem 4 in this section implies that $\mathbf{A}$ is diagonalizable if and only if

$$q_1 + q_2 + \cdots + q_k = n$$

where q_i denotes the geometric multiplicity of λ_i $(i = 1, 2, \cdots, k)$. But, because $p_i \geq q_i$ for each $i = 1, 2, \cdots, k$, the two equations displayed above can both be satisfied if and only if $p_i = q_i$ for each i.

SECTION 6.3

APPLICATIONS INVOLVING POWERS OF MATRICES

In Problems 1–10 we first find the eigensystem of the given matrix **A** so as to determine its eigenvector matrix **P** and its diagonal eigenvalue matrix **D**. Then we calculate the matrix power $\mathbf{A}^5 = \mathbf{P}\mathbf{D}^5\mathbf{P}^{-1}$.

1.   Characteristic polynomial: $p(\lambda) = \lambda^2 - 3\lambda + 2 = (\lambda - 1)(\lambda - 2)$

         Eigenvalues: $\lambda_1 = 1, \quad \lambda_2 = 2$

         With $\lambda_1 = 1$: $\left.\begin{array}{r} 2a - 2b = 0 \\ a - b = 0 \end{array}\right\}$ $\mathbf{v}_1 = \begin{bmatrix} 1 \\ 1 \end{bmatrix}$

         With $\lambda_2 = 2$: $\left.\begin{array}{r} a - 2b = 0 \\ a - 2b = 0 \end{array}\right\}$ $\mathbf{v}_2 = \begin{bmatrix} 2 \\ 1 \end{bmatrix}$

$$\mathbf{P} = \begin{bmatrix} 1 & 2 \\ 1 & 1 \end{bmatrix}, \quad \mathbf{D} = \begin{bmatrix} 1 & 0 \\ 0 & 2 \end{bmatrix}, \quad \mathbf{P}^{-1} = \begin{bmatrix} -1 & 2 \\ 1 & -1 \end{bmatrix}$$

$$\mathbf{A}^5 = \begin{bmatrix} 1 & 2 \\ 1 & 1 \end{bmatrix}\begin{bmatrix} 1 & 0 \\ 0 & 32 \end{bmatrix}\begin{bmatrix} -1 & 2 \\ 1 & -1 \end{bmatrix} = \begin{bmatrix} 63 & -62 \\ 31 & -30 \end{bmatrix}$$

3.   Characteristic polynomial: $p(\lambda) = \lambda^2 - 2\lambda = \lambda(\lambda - 2)$

         Eigenvalues: $\lambda_1 = 0, \quad \lambda_2 = 2$

         With $\lambda_1 = 0$: $\left.\begin{array}{r} 6a - 6b = 0 \\ 4a - 4b = 0 \end{array}\right\}$ $\mathbf{v}_1 = \begin{bmatrix} 1 \\ 1 \end{bmatrix}$

         With $\lambda_2 = 2$: $\left.\begin{array}{r} 4a - 6b = 0 \\ 4a - 6b = 0 \end{array}\right\}$ $\mathbf{v}_2 = \begin{bmatrix} 3 \\ 2 \end{bmatrix}$

$$\mathbf{P} = \begin{bmatrix} 1 & 3 \\ 1 & 2 \end{bmatrix}, \quad \mathbf{D} = \begin{bmatrix} 0 & 0 \\ 0 & 2 \end{bmatrix}, \quad \mathbf{P}^{-1} = \begin{bmatrix} -2 & 3 \\ 1 & -1 \end{bmatrix}$$

$$\mathbf{A}^5 = \begin{bmatrix} 1 & 3 \\ 1 & 2 \end{bmatrix}\begin{bmatrix} 0 & 0 \\ 0 & 32 \end{bmatrix}\begin{bmatrix} -2 & 3 \\ 1 & -1 \end{bmatrix} = \begin{bmatrix} 96 & -96 \\ 64 & -64 \end{bmatrix}$$

5.   Characteristic polynomial: $p(\lambda) = \lambda^2 - 3\lambda + 2 = (\lambda - 1)(\lambda - 2)$

         Eigenvalues: $\lambda_1 = 1, \quad \lambda_2 = 2$

With $\lambda_1 = 1$:
$$\left.\begin{array}{r} 4a - 4b = 0 \\ 3a - 3b = 0 \end{array}\right\}$$
$$\mathbf{v}_1 = \begin{bmatrix} 1 \\ 1 \end{bmatrix}$$

With $\lambda_2 = 2$:
$$\left.\begin{array}{r} 3a - 4b = 0 \\ 3a - 4b = 0 \end{array}\right\}$$
$$\mathbf{v}_2 = \begin{bmatrix} 4 \\ 3 \end{bmatrix}$$

$$\mathbf{P} = \begin{bmatrix} 1 & 4 \\ 1 & 3 \end{bmatrix}, \quad \mathbf{D} = \begin{bmatrix} 1 & 0 \\ 0 & 2 \end{bmatrix}, \quad \mathbf{P}^{-1} = \begin{bmatrix} -3 & 4 \\ 1 & -1 \end{bmatrix}$$

$$\mathbf{A}^5 = \begin{bmatrix} 1 & 4 \\ 1 & 3 \end{bmatrix}\begin{bmatrix} 1 & 0 \\ 0 & 32 \end{bmatrix}\begin{bmatrix} -3 & 4 \\ 1 & -1 \end{bmatrix} = \begin{bmatrix} 125 & -124 \\ 93 & -92 \end{bmatrix}$$

7.   Characteristic polynomial: $p(\lambda) = -(\lambda - 1)(\lambda - 2)^2$

Eigenvalues: $\lambda_1 = 1, \quad \lambda_2 = 2, \quad \lambda_3 = 2$

With $\lambda_1 = 1$:
$$\left.\begin{array}{r} 3b = 0 \\ b = 0 \\ c = 0 \end{array}\right\}$$
$$\mathbf{v}_1 = \begin{bmatrix} 1 \\ 0 \\ 0 \end{bmatrix}$$

With $\lambda_2 = 2$:
$$\left.\begin{array}{r} 3b - a = 0 \\ 0 = 0 \\ 0 = 0 \end{array}\right\}$$
$$\mathbf{v}_2 = \begin{bmatrix} 0 \\ 0 \\ 1 \end{bmatrix}, \quad \mathbf{v}_3 = \begin{bmatrix} 3 \\ 1 \\ 0 \end{bmatrix}$$

$$\mathbf{P} = \begin{bmatrix} 1 & 0 & 3 \\ 0 & 0 & 1 \\ 0 & 1 & 0 \end{bmatrix}, \quad \mathbf{D} = \begin{bmatrix} 1 & 0 & 0 \\ 0 & 2 & 0 \\ 0 & 0 & 2 \end{bmatrix}, \quad \mathbf{P}^{-1} = \begin{bmatrix} 1 & -3 & 0 \\ 0 & 0 & 1 \\ 0 & 1 & 0 \end{bmatrix}$$

$$\mathbf{A}^5 = \begin{bmatrix} 1 & 0 & 3 \\ 0 & 0 & 1 \\ 0 & 1 & 0 \end{bmatrix}\begin{bmatrix} 1 & 0 & 0 \\ 0 & 32 & 0 \\ 0 & 0 & 32 \end{bmatrix}\begin{bmatrix} 1 & -3 & 0 \\ 0 & 0 & 1 \\ 0 & 1 & 0 \end{bmatrix} = \begin{bmatrix} 1 & 93 & 0 \\ 0 & 32 & 0 \\ 0 & 0 & 32 \end{bmatrix}$$

9.   Characteristic polynomial: $p(\lambda) = -(\lambda - 1)(\lambda - 2)^2$

Eigenvalues: $\lambda_1 = 1, \quad \lambda_2 = 2, \quad \lambda_3 = 2$

With $\lambda_1 = 1$:
$$\left.\begin{array}{r} c - 3b = 0 \\ b = 0 \\ c = 0 \end{array}\right\}$$
$$\mathbf{v}_1 = \begin{bmatrix} 1 \\ 0 \\ 0 \end{bmatrix}$$

With $\lambda_2 = 2$:
$$\left.\begin{array}{r} -a-3b+c = 0 \\ 0 = 0 \\ 0 = 0 \end{array}\right\} \quad \mathbf{v}_2 = \begin{bmatrix} 1 \\ 0 \\ 1 \end{bmatrix}, \quad \mathbf{v}_3 = \begin{bmatrix} -3 \\ 1 \\ 0 \end{bmatrix}$$

$$\mathbf{P} = \begin{bmatrix} 1 & 1 & -3 \\ 0 & 0 & 1 \\ 0 & 1 & 0 \end{bmatrix}, \quad \mathbf{D} = \begin{bmatrix} 1 & 0 & 0 \\ 0 & 2 & 0 \\ 0 & 0 & 2 \end{bmatrix}, \quad \mathbf{P}^{-1} = \begin{bmatrix} 1 & 3 & -1 \\ 0 & 0 & 1 \\ 0 & 1 & 0 \end{bmatrix}$$

$$\mathbf{A}^5 = \begin{bmatrix} 1 & 1 & -3 \\ 0 & 0 & 1 \\ 0 & 1 & 0 \end{bmatrix} \begin{bmatrix} 1 & 0 & 0 \\ 0 & 32 & 0 \\ 0 & 0 & 32 \end{bmatrix} \begin{bmatrix} 1 & 3 & -1 \\ 0 & 0 & 1 \\ 0 & 1 & 0 \end{bmatrix} = \begin{bmatrix} 1 & -93 & 31 \\ 0 & 32 & 0 \\ 0 & 0 & 32 \end{bmatrix}$$

11. Characteristic polynomial: $\quad p(\lambda) = -\lambda^3 + \lambda = -\lambda(\lambda+1)(\lambda-1)$

Eigenvalues: $\quad \lambda_1 = -1, \quad \lambda_2 = 0, \quad \lambda_3 = 1$

With $\lambda_1 = -1$:
$$\left.\begin{array}{r} 2a = 0 \\ 6a+6b+2c = 0 \\ 21a-15b-5c = 0 \end{array}\right\} \quad \mathbf{v}_1 = \begin{bmatrix} 0 \\ -1 \\ 3 \end{bmatrix}$$

With $\lambda_2 = 0$:
$$\left.\begin{array}{r} a = 0 \\ 6a+5b+2c = 0 \\ 21a-15b-6c = 0 \end{array}\right\} \quad \mathbf{v}_2 = \begin{bmatrix} 0 \\ -2 \\ 5 \end{bmatrix}$$

With $\lambda_3 = 1$:
$$\left.\begin{array}{r} 0 = 0 \\ 6a+4b+2c = 0 \\ 21a-15b-7c = 0 \end{array}\right\} \quad \mathbf{v}_3 = \begin{bmatrix} -1 \\ -42 \\ 87 \end{bmatrix}$$

$$\mathbf{P} = \begin{bmatrix} 0 & 0 & -1 \\ -1 & -2 & -42 \\ 3 & 5 & 87 \end{bmatrix}, \quad \mathbf{D} = \begin{bmatrix} -1 & 0 & 0 \\ 0 & 0 & 0 \\ 0 & 0 & 1 \end{bmatrix}, \quad \mathbf{P}^{-1} = \begin{bmatrix} -36 & 5 & 2 \\ 39 & -3 & -1 \\ -1 & 0 & 0 \end{bmatrix}$$

$$\mathbf{A}^{10} = \begin{bmatrix} 0 & 0 & -1 \\ -1 & -2 & -42 \\ 3 & 5 & 87 \end{bmatrix} \begin{bmatrix} 1 & 0 & 0 \\ 0 & 0 & 0 \\ 0 & 0 & 1 \end{bmatrix} \begin{bmatrix} -36 & 5 & 2 \\ 39 & -3 & -1 \\ -1 & 0 & 0 \end{bmatrix} = \begin{bmatrix} 1 & 0 & 0 \\ 78 & -5 & -2 \\ -195 & 15 & 6 \end{bmatrix}$$

13. Characteristic polynomial: $\quad p(\lambda) = -\lambda^3 + \lambda = -\lambda(\lambda+1)(\lambda-1)$

Eigenvalues: $\quad \lambda_1 = -1, \quad \lambda_2 = 0, \quad \lambda_3 = 1$

With $\lambda_1 = -1$: $\left.\begin{array}{r} 2a-b+c = 0 \\ 2a-b+c = 0 \\ 4a-4b+2c = 0 \end{array}\right\}$ $\mathbf{v}_1 = \begin{bmatrix} -1 \\ 0 \\ 2 \end{bmatrix}$

With $\lambda_2 = 0$: $\left.\begin{array}{r} a-b+c = 0 \\ 2a-2b+c = 0 \\ 4a-4b+c = 0 \end{array}\right\}$ $\mathbf{v}_2 = \begin{bmatrix} 1 \\ 1 \\ 0 \end{bmatrix}$

With $\lambda_3 = 1$: $\left.\begin{array}{r} -b+c = 0 \\ 2a-3b+c = 0 \\ 4a-4b = 0 \end{array}\right\}$ $\mathbf{v}_3 = \begin{bmatrix} 1 \\ 1 \\ 1 \end{bmatrix}$

$$\mathbf{P} = \begin{bmatrix} -1 & 1 & 1 \\ 0 & 1 & 1 \\ 2 & 0 & 1 \end{bmatrix}, \quad \mathbf{D} = \begin{bmatrix} -1 & 0 & 0 \\ 0 & 0 & 0 \\ 0 & 0 & 1 \end{bmatrix}, \quad \mathbf{P}^{-1} = \begin{bmatrix} -1 & 1 & 0 \\ -2 & 3 & -1 \\ 2 & -2 & 1 \end{bmatrix}$$

$$\mathbf{A}^{10} = \begin{bmatrix} -1 & 1 & 1 \\ 0 & 1 & 1 \\ 2 & 0 & 1 \end{bmatrix}\begin{bmatrix} 1 & 0 & 0 \\ 0 & 0 & 0 \\ 0 & 0 & 1 \end{bmatrix}\begin{bmatrix} -1 & 1 & 0 \\ -2 & 3 & -1 \\ 2 & -2 & 1 \end{bmatrix} = \begin{bmatrix} 3 & -3 & 1 \\ 2 & -2 & 1 \\ 0 & 0 & 1 \end{bmatrix}$$

15. $p(\lambda) = \lambda^2 - 3\lambda + 2$ so $\mathbf{A}^2 - 3\mathbf{A} + 2\mathbf{I} = \mathbf{0}$

$$\mathbf{A}^2 = 3\mathbf{A} - 2\mathbf{I} = 3\begin{bmatrix} 5 & -4 \\ 3 & -2 \end{bmatrix} - 2\begin{bmatrix} 1 & 0 \\ 0 & 1 \end{bmatrix} = \begin{bmatrix} 13 & -12 \\ 9 & -8 \end{bmatrix}$$

$$\mathbf{A}^3 = 3\mathbf{A}^2 - 2\mathbf{A} = 3\begin{bmatrix} 13 & -12 \\ 9 & -8 \end{bmatrix} - 2\begin{bmatrix} 5 & -4 \\ 3 & -2 \end{bmatrix} = \begin{bmatrix} 29 & -28 \\ 21 & -20 \end{bmatrix}$$

$$\mathbf{A}^4 = 3\mathbf{A}^3 - 2\mathbf{A}^2 = 3\begin{bmatrix} 29 & -28 \\ 21 & -20 \end{bmatrix} - 2\begin{bmatrix} 13 & -12 \\ 9 & -8 \end{bmatrix} = \begin{bmatrix} 61 & -60 \\ 45 & -44 \end{bmatrix}$$

$$\mathbf{A}^{-1} = \frac{1}{2}(-\mathbf{A} + 3\mathbf{I}) = \frac{1}{2}\left(-\begin{bmatrix} 5 & -4 \\ 3 & -2 \end{bmatrix} + 3\begin{bmatrix} 1 & 0 \\ 0 & 1 \end{bmatrix}\right) = \frac{1}{2}\begin{bmatrix} -2 & 4 \\ -3 & 5 \end{bmatrix}$$

17. $p(\lambda) = -\lambda^3 + 5\lambda^2 - 8\lambda + 4$ so $-\mathbf{A}^3 + 5\mathbf{A}^2 - 8\mathbf{A} + 4\mathbf{I} = \mathbf{0}$

$$\mathbf{A}^2 = \begin{bmatrix} 1 & 9 & 0 \\ 0 & 4 & 0 \\ 0 & 0 & 4 \end{bmatrix}, \qquad \mathbf{A}^3 = 5\mathbf{A}^2 - 8\mathbf{A} + 4\mathbf{I} = \begin{bmatrix} 1 & 21 & 0 \\ 0 & 8 & 0 \\ 0 & 0 & 8 \end{bmatrix}$$

$$\mathbf{A}^4 = 5\mathbf{A}^3 - 8\mathbf{A}^2 + 4\mathbf{A} = \begin{bmatrix} 1 & 45 & 0 \\ 0 & 16 & 0 \\ 0 & 0 & 16 \end{bmatrix}$$

$$\mathbf{A}^{-1} = \frac{1}{4}\left(\mathbf{A}^2 - 5\mathbf{A} + 8\mathbf{I}\right) = \frac{1}{2}\begin{bmatrix} 2 & -3 & 0 \\ 0 & 1 & 0 \\ 0 & 0 & 1 \end{bmatrix}$$

19. $p(\lambda) = -\lambda^3 + 5\lambda^2 - 8\lambda + 4$ so $-\mathbf{A}^3 + 5\mathbf{A}^2 - 8\mathbf{A} + 4\mathbf{I} = \mathbf{0}$

$$\mathbf{A}^2 = \begin{bmatrix} 1 & -9 & 3 \\ 0 & 4 & 0 \\ 0 & 0 & 4 \end{bmatrix}, \qquad \mathbf{A}^3 = 5\mathbf{A}^2 - 8\mathbf{A} + 4\mathbf{I} = \begin{bmatrix} 1 & -21 & 7 \\ 0 & 8 & 0 \\ 0 & 0 & 8 \end{bmatrix}$$

$$\mathbf{A}^4 = 5\mathbf{A}^3 - 8\mathbf{A}^2 + 4\mathbf{A} = \begin{bmatrix} 1 & -45 & 15 \\ 0 & 16 & 0 \\ 0 & 0 & 16 \end{bmatrix}$$

$$\mathbf{A}^{-1} = \frac{1}{4}\left(\mathbf{A}^2 - 5\mathbf{A} + 8\mathbf{I}\right) = \frac{1}{2}\begin{bmatrix} 2 & 3 & -1 \\ 0 & 1 & 0 \\ 0 & 0 & 1 \end{bmatrix}$$

21. $p(\lambda) = -\lambda^3 + \lambda$ so $-\mathbf{A}^3 + \mathbf{A} = \mathbf{0}$

$$\mathbf{A}^2 = \begin{bmatrix} 1 & 0 & 0 \\ 78 & -5 & -2 \\ -195 & 15 & 6 \end{bmatrix}, \qquad \mathbf{A}^3 = \mathbf{A} = \begin{bmatrix} 1 & 0 & 0 \\ 6 & 5 & 2 \\ 21 & -15 & -6 \end{bmatrix}$$

$$\mathbf{A}^4 = \mathbf{A}^2 = \begin{bmatrix} 1 & 0 & 0 \\ 78 & -5 & -2 \\ -195 & 15 & 6 \end{bmatrix}$$

Because $\lambda = 0$ is an eigenvalue, $\mathbf{A}$ is singular and $\mathbf{A}^{-1}$ does not exist.

23. $p(\lambda) = -\lambda^3 + \lambda$ so $-\mathbf{A}^3 + \mathbf{A} = \mathbf{0}$

$$\mathbf{A}^2 = \begin{bmatrix} 3 & -3 & 1 \\ 2 & -2 & 1 \\ 0 & 0 & 1 \end{bmatrix}, \qquad \mathbf{A}^3 = \mathbf{A} = \begin{bmatrix} 1 & -1 & 1 \\ 2 & -2 & 1 \\ 4 & -4 & 1 \end{bmatrix}$$

$$A^4 = A^2 = \begin{bmatrix} 3 & -3 & 1 \\ 2 & -2 & 1 \\ 0 & 0 & 1 \end{bmatrix}$$

Because $\lambda = 0$ is an eigenvalue, A is singular and A^{-1} does not exist.

In Problems 25–30 we first find the eigensystem of the given transition matrix A so as to determine its eigenvector matrix P and its diagonal eigenvalue matrix D. Then we determine how the matrix power $A^k = PD^kP^{-1}$ behaves as $k \to \infty$. For simpler calculations of eigenvalues and eigenvectors, we write the entries of A in fractional rather than decimal form.

25. Characteristic polynomial: $p(\lambda) = \lambda^2 - \dfrac{9}{5}\lambda + \dfrac{4}{5} = \dfrac{1}{5}(\lambda - 1)(5\lambda - 4)$

Eigenvalues: $\lambda_1 = 1, \quad \lambda_2 = \dfrac{4}{5}$

With $\lambda_1 = 1$:
$\left. \begin{array}{l} -\dfrac{1}{10}a + \dfrac{1}{10}b = 0 \\[2mm] \dfrac{1}{10}a - \dfrac{1}{10}b = 0 \end{array} \right\}$
$\qquad \mathbf{v}_1 = \begin{bmatrix} 1 \\ 1 \end{bmatrix}$

With $\lambda_2 = \dfrac{4}{5}$:
$\left. \begin{array}{l} \dfrac{1}{10}a + \dfrac{1}{10}b = 0 \\[2mm] \dfrac{1}{10}a + \dfrac{1}{10}b = 0 \end{array} \right\}$
$\qquad \mathbf{v}_2 = \begin{bmatrix} -1 \\ 1 \end{bmatrix}$

$$P = \begin{bmatrix} 1 & -1 \\ 1 & 1 \end{bmatrix}, \quad D = \begin{bmatrix} 1 & 0 \\ 0 & 4/5 \end{bmatrix}, \quad P^{-1} = \dfrac{1}{2}\begin{bmatrix} 1 & 1 \\ -1 & 1 \end{bmatrix}$$

$$\mathbf{x}_k = A^k\mathbf{x}_0 = \begin{bmatrix} 1 & -1 \\ 1 & 1 \end{bmatrix}\begin{bmatrix} 1 & 0 \\ 0 & 4/5 \end{bmatrix}^k \dfrac{1}{2}\begin{bmatrix} 1 & 1 \\ -1 & 1 \end{bmatrix}\mathbf{x}_0$$

$$\to \begin{bmatrix} 1 & -1 \\ 1 & 1 \end{bmatrix}\begin{bmatrix} 1 & 0 \\ 0 & 0 \end{bmatrix}\dfrac{1}{2}\begin{bmatrix} 1 & 1 \\ -1 & 1 \end{bmatrix}\mathbf{x}_0 = \dfrac{1}{2}\begin{bmatrix} 1 & 1 \\ 1 & 1 \end{bmatrix}\begin{bmatrix} C_0 \\ S_0 \end{bmatrix} = (C_0 + S_0)\begin{bmatrix} 1/2 \\ 1/2 \end{bmatrix}$$

as $k \to \infty$. Thus the long-term distribution of population is 50% city, 50% suburban.

27. Characteristic polynomial: $p(\lambda) = \lambda^2 - \dfrac{8}{5}\lambda + \dfrac{3}{5} = \dfrac{1}{5}(\lambda - 1)(5\lambda - 3)$

Eigenvalues: $\lambda_1 = 1, \quad \lambda_2 = \dfrac{3}{5}$

With $\lambda_1 = 1$:
$$\left.\begin{array}{r} -\dfrac{1}{4}a + \dfrac{3}{20}b = 0 \\[2mm] \dfrac{1}{4}a - \dfrac{3}{20}b = 0 \end{array}\right\} \qquad \mathbf{v}_1 = \begin{bmatrix} 3 \\ 5 \end{bmatrix}$$

With $\lambda_2 = \dfrac{3}{5}$:
$$\left.\begin{array}{r} \dfrac{3}{20}a + \dfrac{3}{20}b = 0 \\[2mm] \dfrac{1}{4}a + \dfrac{1}{4}b = 0 \end{array}\right\} \qquad \mathbf{v}_2 = \begin{bmatrix} -1 \\ 1 \end{bmatrix}$$

$$\mathbf{P} = \begin{bmatrix} 3 & -1 \\ 5 & 1 \end{bmatrix}, \quad \mathbf{D} = \begin{bmatrix} 1 & 0 \\ 0 & 3/5 \end{bmatrix}, \quad \mathbf{P}^{-1} = \frac{1}{8}\begin{bmatrix} 1 & 1 \\ -5 & 3 \end{bmatrix}$$

$$\mathbf{x}_k = \mathbf{A}^k \mathbf{x}_0 = \begin{bmatrix} 3 & -1 \\ 5 & 1 \end{bmatrix}\begin{bmatrix} 1 & 0 \\ 0 & 3/5 \end{bmatrix}^k \frac{1}{8}\begin{bmatrix} 1 & 1 \\ -5 & 3 \end{bmatrix}\mathbf{x}_0$$

$$\rightarrow \begin{bmatrix} 3 & -1 \\ 5 & 1 \end{bmatrix}\begin{bmatrix} 1 & 0 \\ 0 & 0 \end{bmatrix}\frac{1}{8}\begin{bmatrix} 1 & 1 \\ -5 & 3 \end{bmatrix}\mathbf{x}_0 = \frac{1}{8}\begin{bmatrix} 3 & 3 \\ 5 & 5 \end{bmatrix}\begin{bmatrix} C_0 \\ S_0 \end{bmatrix} = (C_0 + S_0)\begin{bmatrix} 3/8 \\ 5/8 \end{bmatrix}$$

as $k \rightarrow \infty$. Thus the long-term distribution of population is 3/8 city, 5/8 suburban.

29. Characteristic polynomial: $\quad p(\lambda) = \lambda^2 - \dfrac{37}{20}\lambda + \dfrac{17}{20} = \dfrac{1}{20}(\lambda - 1)(20\lambda - 17)$

Eigenvalues: $\quad \lambda_1 = 1, \quad \lambda_2 = \dfrac{17}{20}$

With $\lambda_1 = 1$:
$$\left.\begin{array}{r} -\dfrac{1}{10}a + \dfrac{1}{20}b = 0 \\[2mm] \dfrac{1}{10}a - \dfrac{1}{20}b = 0 \end{array}\right\} \qquad \mathbf{v}_1 = \begin{bmatrix} 1 \\ 2 \end{bmatrix}$$

With $\lambda_2 = \dfrac{17}{20}$:
$$\left.\begin{array}{r} \dfrac{1}{20}a + \dfrac{1}{20}b = 0 \\[2mm] \dfrac{1}{10}a + \dfrac{1}{10}b = 0 \end{array}\right\} \qquad \mathbf{v}_2 = \begin{bmatrix} -1 \\ 1 \end{bmatrix}$$

$$\mathbf{P} = \begin{bmatrix} 1 & -1 \\ 2 & 1 \end{bmatrix}, \quad \mathbf{D} = \begin{bmatrix} 1 & 0 \\ 0 & 17/20 \end{bmatrix}, \quad \mathbf{P}^{-1} = \frac{1}{3}\begin{bmatrix} 1 & 1 \\ -2 & 1 \end{bmatrix}$$

$$\mathbf{x}_k = \mathbf{A}^k \mathbf{x}_0 = \begin{bmatrix} 1 & -1 \\ 2 & 1 \end{bmatrix}\begin{bmatrix} 1 & 0 \\ 0 & 17/20 \end{bmatrix}^k \frac{1}{3}\begin{bmatrix} 1 & 1 \\ -2 & 1 \end{bmatrix}\mathbf{x}_0$$

$$\rightarrow \begin{bmatrix} 1 & -1 \\ 2 & 1 \end{bmatrix}\begin{bmatrix} 1 & 0 \\ 0 & 0 \end{bmatrix}\frac{1}{3}\begin{bmatrix} 1 & 1 \\ -2 & 1 \end{bmatrix}\mathbf{x}_0 = \frac{1}{3}\begin{bmatrix} 1 & 1 \\ 2 & 2 \end{bmatrix}\begin{bmatrix} C_0 \\ S_0 \end{bmatrix} = (C_0 + S_0)\begin{bmatrix} 1/3 \\ 2/3 \end{bmatrix}$$

as $k \to \infty$. Thus the long-term distribution of population is 1/3 city, 2/3 suburban.

31.   Characteristic polynomial: $p(\lambda) = \lambda^2 - \dfrac{9}{5}\lambda + \dfrac{4}{5} = \dfrac{1}{5}(\lambda - 1)(5\lambda - 4)$

Eigenvalues: $\lambda_1 = 1,\quad \lambda_2 = \dfrac{4}{5}$

With $\lambda_1 = 1$:
$$\left.\begin{aligned} -\frac{2}{5}a + \frac{1}{2}b &= 0 \\ -\frac{4}{25}a + \frac{1}{5}b &= 0 \end{aligned}\right\} \qquad \mathbf{v}_1 = \begin{bmatrix} 5 \\ 4 \end{bmatrix}$$

With $\lambda_2 = \dfrac{4}{5}$:
$$\left.\begin{aligned} -\frac{1}{5}a + \frac{1}{2}b &= 0 \\ -\frac{4}{25}a + \frac{2}{5}b &= 0 \end{aligned}\right\} \qquad \mathbf{v}_2 = \begin{bmatrix} 5 \\ 2 \end{bmatrix}$$

$$\mathbf{P} = \begin{bmatrix} 5 & 5 \\ 4 & 2 \end{bmatrix}, \quad \mathbf{D} = \begin{bmatrix} 1 & 0 \\ 0 & 4/5 \end{bmatrix}, \quad \mathbf{P}^{-1} = \frac{1}{10}\begin{bmatrix} -2 & 5 \\ 4 & -5 \end{bmatrix}$$

$$\mathbf{x}_k = \mathbf{A}^k \mathbf{x}_0 = \begin{bmatrix} 5 & 5 \\ 4 & 2 \end{bmatrix}\begin{bmatrix} 1 & 0 \\ 0 & 4/5 \end{bmatrix}^k \frac{1}{10}\begin{bmatrix} -2 & 5 \\ 4 & -5 \end{bmatrix}\mathbf{x}_0$$

$$\rightarrow \begin{bmatrix} 5 & 5 \\ 4 & 2 \end{bmatrix}\begin{bmatrix} 1 & 0 \\ 0 & 0 \end{bmatrix}\frac{1}{10}\begin{bmatrix} -2 & 5 \\ 4 & -5 \end{bmatrix}\mathbf{x}_0 = \frac{1}{10}\begin{bmatrix} -10 & 25 \\ -8 & 20 \end{bmatrix}\begin{bmatrix} F_0 \\ R_0 \end{bmatrix} = \begin{bmatrix} 2.5R_0 - F_0 \\ 2R_0 - 0.8F_0 \end{bmatrix}$$

as $k \to \infty$. Thus the fox-rabbit population approaches a stable situation with $2.5R_0 - F_0$ foxes and $2R_0 - 0.8F_0$ rabbits.

33.   Characteristic polynomial: $p(\lambda) = \lambda^2 - \dfrac{9}{5}\lambda + \dfrac{63}{80} = \dfrac{1}{80}(20\lambda - 21)(4\lambda - 3)$

Eigenvalues: $\lambda_1 = \dfrac{19}{20},\quad \lambda_2 = \dfrac{17}{20}$

With $\lambda_1 = \dfrac{21}{20}$:
$$\left.\begin{aligned} -\frac{9}{20}a + \frac{1}{2}b &= 0 \\ -\frac{27}{200}a + \frac{3}{20}b &= 0 \end{aligned}\right\} \qquad \mathbf{v}_1 = \begin{bmatrix} 10 \\ 9 \end{bmatrix}$$

With $\lambda_2 = \dfrac{3}{4}$:
$$\left.\begin{array}{l} -\dfrac{3}{20}a + \dfrac{1}{2}b = 0 \\[2mm] -\dfrac{27}{200}a + \dfrac{9}{20}b = 0 \end{array}\right\} \qquad \mathbf{v}_2 = \begin{bmatrix} 10 \\ 3 \end{bmatrix}$$

$$\mathbf{P} = \begin{bmatrix} 10 & 10 \\ 9 & 3 \end{bmatrix}, \quad \mathbf{D} = \begin{bmatrix} 21/20 & 0 \\ 0 & 3/4 \end{bmatrix}, \quad \mathbf{P}^{-1} = \dfrac{1}{60}\begin{bmatrix} -3 & 10 \\ 9 & -10 \end{bmatrix}$$

$$\mathbf{x}_k = \mathbf{A}^k\mathbf{x}_0 = \begin{bmatrix} 10 & 10 \\ 9 & 3 \end{bmatrix}\begin{bmatrix} 21/20 & 0 \\ 0 & 3/4 \end{bmatrix}^k \dfrac{1}{60}\begin{bmatrix} -3 & 10 \\ 9 & -10 \end{bmatrix}\mathbf{x}_0$$

$$\approx \dfrac{1}{60}\left(\dfrac{21}{20}\right)^k\begin{bmatrix} 10 & 10 \\ 9 & 3 \end{bmatrix}\begin{bmatrix} 1 & 0 \\ 0 & 0 \end{bmatrix}\begin{bmatrix} -3 & 10 \\ 9 & -10 \end{bmatrix}\mathbf{x}_0 = \dfrac{1}{60}(1.05)^k\begin{bmatrix} -30 & 100 \\ -27 & 90 \end{bmatrix}\begin{bmatrix} F_0 \\ R_0 \end{bmatrix}$$

$$= \dfrac{1}{60}(1.05)^k\,(10R_0 - 3F_0)\begin{bmatrix} 10 \\ 9 \end{bmatrix}$$

when k is sufficiently large. Thus the fox and rabbit populations are both increasing at 5% per year, with 10 foxes for each 9 rabbits.

35. The fact that each $|\lambda| = 1$, so $\lambda = \pm 1$, implies that $\mathbf{D}^n = \mathbf{I}$ if n is even, in which case $\mathbf{A}^n = \mathbf{PD}^n\mathbf{P}^{-1} = \mathbf{PIP}^{-1} = \mathbf{I}$.

37. We find immediately that $\mathbf{A}^2 = -\mathbf{I}$, so $\mathbf{A}^3 = \mathbf{A}^2\mathbf{A} = -\mathbf{IA} = -\mathbf{A}$, $\mathbf{A}^4 = \mathbf{A}^3\mathbf{A} = -\mathbf{A}^2 = \mathbf{I}$, and so forth.

39. The characteristic equation of $\mathbf{A}$ is
$$\begin{aligned} (p - \lambda)(q - \lambda) &= (1 - p)(1 - q) = \lambda^2 - (p + q)\lambda + (p + q - 1) \\ &= (\lambda - 1)[\lambda - (p + q - 1)], \end{aligned}$$

so the eigenvalues of A are $\lambda_1 = 1$ and $\lambda_2 = p + q - 1$.

CHAPTER 7

LINEAR SYSTEMS OF DIFFERENTIAL EQUATIONS

SECTION 7.1

FIRST-ORDER SYSTEMS AND APPLICATIONS

1. Let $x_1 = x$ and $x_2 = x_1' = x'$, so $x_2' = x'' = -7x - 3x' + t^2$.

Equivalent system:

$$x_1' = x_2, \qquad x_2' = -7x_1 - 3x_2 + t^2$$

3. Let $x_1 = x$ and $x_2 = x_1' = x'$, so $x_2' = x'' = \left[\left(1 - t^2\right)x - tx'\right]/t^2$.

Equivalent system:

$$x_1' = x_2, \qquad t^2 x_2' = (1 - t^2)x_1 - tx_2$$

5. Let $x_1 = x$, $x_2 = x_1' = x'$, $x_3 = x_2' = x''$, so $x_3' = x''' = \left(x'\right)^2 + \cos x$.

Equivalent system:

$$x_1' = x_2, \qquad x_2' = x_3, \qquad x_3' = x_2^2 + \cos x_1$$

7. Let $x_1 = x$, $x_2 = x_1' = x'$, $y_1 = y$, $y_2 = y_1' = y'$ so $x_2' = x'' = -kx/(x^2 + y^2)^{3/2}$,
$y_2' = y'' = -ky/(x^2 + y^2)^{3/2}$.

Equivalent system:

$$x_1' = x_2, \qquad x_2' = -kx_1/\left(x_1^2 + y_1^2\right)^{3/2}$$
$$y_1' = y_2, \qquad y_2' = -ky_1/\left(x_1^2 + y_1^2\right)^{3/2}$$

9. Let $x_1 = x$, $x_2 = x_1' = x'$, $y_1 = y$, $y_2 = y_1' = y'$, $z_1 = z$, $z_2 = z_1' = z'$, so
$x_2' = x'' = 3x - y + 2z$, $y_2' = y'' = x + y - 4z$, $z_2' = z'' = 5x - y - z$.

Equivalent system:

$$x_1' = x_2, \qquad x_2' = 3x_1 - y_1 + 2z_1$$

$$y_1' = y_2, \qquad y_2' = x_1 + y_1 - 4z_1$$

$$z_1' = z_2, \qquad z_2' = 5x_1 - y_1 - z_1$$

11. The computation $x'' = y' = -x$ yields the single linear second-order equation $x'' + x = 0$ with characteristic equation $r^2 + 1 = 0$ and general solution

$$x(t) = A \cos t + B \sin t.$$

Then the original first equation $y = x'$ gives

$$y(t) = B \cos t - A \sin t.$$

The figure below shows a direction field and typical solution curves (obviously circles?) for the given system.

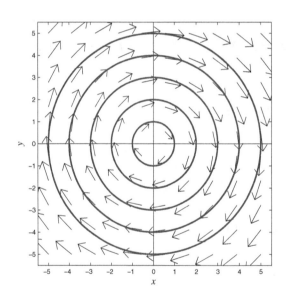

13. The computation $x'' = -2y' = -4x$ yields the single linear second-order equation $x'' + 4x = 0$ with characteristic equation $r^2 + 4 = 0$ and general solution

$$x(t) = A \cos 2t + B \sin 2t.$$

Then the original first equation $y = -x'/2$ gives

$$y(t) = -B \cos 2t + A \sin 2t.$$

Finally, the condition $x(0) = 1$ implies that $A = 1$, and then the condition $y(0) = 0$ gives $B = 0$. Hence the desired particular solution is given by

$$x(t) = \cos 2t, \qquad y(t) = \sin 2t.$$

The figure on the left below shows a direction field and some typical circular solution curves for the given system.

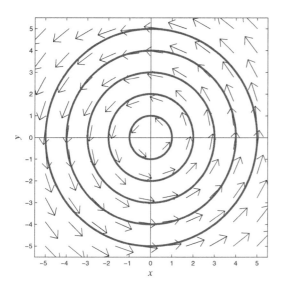   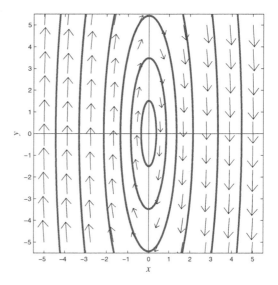

15.   The computation $x'' = y'/2 = -4x$ yields the single linear second-order equation $x'' + 4x = 0$ with characteristic equation $r^2 + 4 = 0$ and general solution

$$x(t) = A \cos 2t + B \sin 2t.$$

Then the original first equation $y = 2x'$ gives

$$y(t) = 4B \cos 2t - 4A \sin 2t.$$

The figure on the right above shows a direction field and some typical elliptical solution curves.

17.   The computation $x'' = y' = 6x - y = 6x - x'$ yields the single linear second-order equation $x'' + x' - 6x = 0$ with characteristic equation $r^2 + r - 6 = 0$ and characteristic roots $r = -3$ and 2, so the general solution

$$x(t) = A e^{-3t} + B e^{2t}.$$

Then the original first equation $y = x'$ gives

$$y(t) = -3A e^{-3t} + 2B e^{2t}.$$

Finally, the initial conditions

$$x(0) = A + B = 1, \quad y(0) = -3A + 2B = 2$$

imply that $A = 0$ and $B = 1$, so the desired particular solution is given by

$$x(t) = e^{-3t}, \qquad y(t) = 2 e^{2t}.$$

The figure below shows a direction field and some typical solution curves.

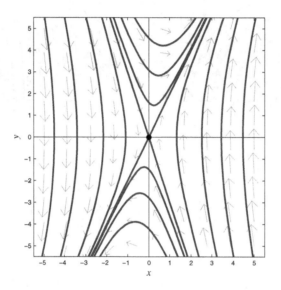

19. The computation $x'' = -y' = -13x - 4y = -13x + 4x'$ yields the single linear second-order equation $x'' - 4x' + 13x = 0$ with characteristic equation $r^2 - 4r + 13 = 0$ and characteristic roots $r = 2 \pm 3i$, hence the general solution is

$$x(t) = e^{2t}(A \cos 3t + B \sin 3t).$$

The initial condition $x(0) = 0$ then gives $A = 0$, so $x(t) = B e^{2t} \sin 3t$. Then the original first equation $y = -x'$ gives

$$y(t) = -e^{2t}(3B \cos 3t + 2B \sin 3t).$$

Finally, the initial condition $y(0) = 3$ gives $B = -1$, so the desired particular solution is given by

$$x(t) = -e^{2t} \sin 3t, \qquad y(t) = e^{2t}(3 \cos 3t + 2 \sin 3t).$$

The figure at the top of the next page shows a direction field and some typical solution curves.

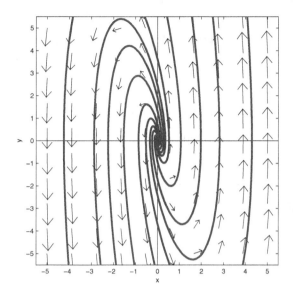

21.   **(a)**   Substituting the general solution found in Problem 11 we get

$$x^2 + y^2 = (A \cos t + B \sin t)^2 + (B \cos t - A \sin t)^2$$
$$= (A^2 + B^2)(\cos^2 t + \sin^2 t) = A^2 + B^2$$
$$x^2 + y^2 = C^2,$$

the equation of a circle of radius $C = (A^2 + B^2)^{1/2}$.

(b)   Substituting the general solution found in Problem 12 we get

$$x^2 - y^2 = (Ae^t + Be^{-t})^2 - (Ae^t - Be^{-t})^2 = 4AB,$$

the equation of a hyperbola.

23.   When we solve Equations (20) and (21) in the text for e^{-t} and e^{2t} we get

$$2x - y = 3Ae^{-t} \text{and} \qquad x + y = 3Be^{2t}.$$

Hence

$$(2x - y)^2(x + y) = (3Ae^{-t})^2(3Be^{2t}) = 27A^2B = C.$$

Clearly $y = 2x$ or $y = -x$ if $C = 0$, and expansion gives the equation $4x^3 - 3xy^2 + y^3 = C$.

25.   Looking at Fig. 7.1.12 in the text, we see that

$$my_1'' = -T \sin\theta_1 + T \sin\theta_2 \approx -T \tan\theta_1 + T \tan\theta_2 = -Ty_1 / L + T(y_2 - y_1)/L,$$

$$my_2'' = -T\sin\theta_2 - T\sin\theta_3 \approx -T\tan\theta_2 - T\tan\theta_3 = -T(y_2 - y_1)/L - Ty_2/L.$$

We get the desired equations when we multiply each of these equations by L/T and set $k = mL/T$.

27.   If θ is the polar angular coordinate of the point (x, y) and we write
$F = k/(x^2 + y^2) = k/r^2$, then Newton's second law gives

$$mx'' = -F\cos\theta = -(k/r^2)(x/r) = -kx/r^3,$$
$$my'' = -F\sin\theta = -(k/r^2)(y/r) = -ky/r^3,$$

29.   If $\mathbf{r} = (x, y, z)$ is the particle's position vector, then Newton's law $m\mathbf{r}'' = \mathbf{F}$ gives

$$m\mathbf{r}'' = q\mathbf{v}\times\mathbf{B} = q\begin{vmatrix} \mathbf{i} & \mathbf{j} & \mathbf{k} \\ x' & y' & z' \\ 0 & 0 & B \end{vmatrix} = +qBy'\mathbf{i} - qBx'\mathbf{j} = qB(-y', x', 0).$$

SECTION 7.2

MATRICES AND LINEAR SYSTEMS

1. $$(\mathbf{AB})' = \begin{bmatrix} t - 4t^2 + 6t^3 & t + t^2 - 4t^3 + 8t^4 \\ 3t + t^3 - t^4 & 4t^2 + t^3 + t^4 \end{bmatrix}' = \begin{bmatrix} 1 - 8t + 18t^2 & 1 + 2t - 12t^2 + 32t^3 \\ 3 + 3t^2 - 4t^3 & 8t + 3t^2 + 4t^3 \end{bmatrix}$$

$$\mathbf{A'B} + \mathbf{AB'} = \begin{bmatrix} 1 & 2 \\ 3t^2 & -\dfrac{1}{t^2} \end{bmatrix}\begin{bmatrix} 1-t & 1+t \\ 3t^2 & 4t^3 \end{bmatrix} + \begin{bmatrix} t & 2t-1 \\ t^3 & \dfrac{1}{t} \end{bmatrix}\begin{bmatrix} -1 & 1 \\ 6t & 12t^2 \end{bmatrix}$$

$$= \begin{bmatrix} 1-t+6t^2 & 1+t+8t^3 \\ -3+3t^2-3t^3 & -4t+3t^2+3t^3 \end{bmatrix} + \begin{bmatrix} -7t+12t^2 & t-12t^2+24t^3 \\ 6-t^3 & 12t+t^3 \end{bmatrix}$$

$$= \begin{bmatrix} 1-8t+18t^2 & 1+2t-12t^2+32t^3 \\ 3+3t^2-4t^3 & 8t+3t^2+4t^3 \end{bmatrix}$$

3. $$\mathbf{x} = \begin{bmatrix} x \\ y \end{bmatrix}, \qquad \mathbf{P}(t) = \begin{bmatrix} 0 & -3 \\ 3 & 0 \end{bmatrix}, \qquad \mathbf{f}(t) = \begin{bmatrix} 0 \\ 0 \end{bmatrix}$$

5. $\mathbf{x} = \begin{bmatrix} x \\ y \end{bmatrix}$, $\mathbf{P}(t) = \begin{bmatrix} 2 & 4 \\ 5 & -1 \end{bmatrix}$, $\mathbf{f}(t) = \begin{bmatrix} 3e^t \\ -t^2 \end{bmatrix}$

7. $\mathbf{x} = \begin{bmatrix} x \\ y \\ z \end{bmatrix}$, $\mathbf{P}(t) = \begin{bmatrix} 0 & 1 & 1 \\ 1 & 0 & 1 \\ 1 & 1 & 0 \end{bmatrix}$, $\mathbf{f}(t) = \begin{bmatrix} 0 \\ 0 \\ 0 \end{bmatrix}$

9. $\mathbf{x} = \begin{bmatrix} x \\ y \\ z \end{bmatrix}$, $\mathbf{P}(t) = \begin{bmatrix} 3 & -4 & 1 \\ 1 & 0 & -3 \\ 0 & 6 & -7 \end{bmatrix}$, $\mathbf{f}(t) = \begin{bmatrix} t \\ t^2 \\ t^3 \end{bmatrix}$

11. $\mathbf{x} = \begin{bmatrix} x_1 \\ x_2 \\ x_3 \\ x_4 \end{bmatrix}$, $\mathbf{P}(t) = \begin{bmatrix} 0 & 1 & 0 & 0 \\ 0 & 0 & 2 & 0 \\ 0 & 0 & 0 & 3 \\ 4 & 0 & 0 & 0 \end{bmatrix}$, $\mathbf{f}(t) = \begin{bmatrix} 0 \\ 0 \\ 0 \\ 0 \end{bmatrix}$

13. $W(t) = \begin{vmatrix} 2e^t & e^{2t} \\ -3e^t & -e^{2t} \end{vmatrix} = e^{3t} \neq 0$

$$\mathbf{x}_1' = \begin{bmatrix} 2e^t \\ -3e^t \end{bmatrix}' = \begin{bmatrix} 2e^t \\ -3e^t \end{bmatrix} = \begin{bmatrix} 4 & 2 \\ -3 & -1 \end{bmatrix} \begin{bmatrix} 2e^t \\ -3e^t \end{bmatrix} = \mathbf{A}\mathbf{x}_1$$

$$\mathbf{x}_2' = \begin{bmatrix} e^{2t} \\ -e^{2t} \end{bmatrix}' = \begin{bmatrix} 2e^{2t} \\ -2e^{2t} \end{bmatrix} = \begin{bmatrix} 4 & 2 \\ -3 & -1 \end{bmatrix} \begin{bmatrix} e^{2t} \\ -e^{2t} \end{bmatrix} = \mathbf{A}\mathbf{x}_2$$

$$\mathbf{x}(t) = c_1\mathbf{x}_1 + c_2\mathbf{x}_2 = c_1 \begin{bmatrix} 2e^t \\ -3e^t \end{bmatrix} + c_2 \begin{bmatrix} e^{2t} \\ -e^{2t} \end{bmatrix} = \begin{bmatrix} 2c_1e^t + c_2e^{2t} \\ -3c_1e^t - c_2e^{2t} \end{bmatrix}$$

In most of Problems 14–22, we omit the verifications of the given solutions. In each case, this is simply a matter of calculating both the derivative $\mathbf{x}_i'$ of the given solution vector and the product $\mathbf{A}\mathbf{x}_i$ (where $\mathbf{A}$ is the coefficient matrix in the given differential equation) to verify that $\mathbf{x}_i' = \mathbf{A}\mathbf{x}_i$ (just as in the verification of the solutions $\mathbf{x}_1$ and $\mathbf{x}_2$ in Problem 13 above).

15. $W(t) = \begin{vmatrix} e^{2t} & e^{-2t} \\ e^{2t} & 5e^{-2t} \end{vmatrix} = 4 \neq 0$

$$\mathbf{x}(t) = c_1\mathbf{x}_1 + c_2\mathbf{x}_2 = c_1 \begin{bmatrix} 1 \\ 1 \end{bmatrix} e^{2t} + c_2 \begin{bmatrix} 1 \\ 5 \end{bmatrix} e^{-2t} = \begin{bmatrix} c_1e^{2t} + c_2e^{-2t} \\ c_1e^{2t} + 5c_2e^{-2t} \end{bmatrix}$$

17. $W(t) = \begin{vmatrix} 3e^{2t} & e^{-5t} \\ 2e^{2t} & 3e^{-5t} \end{vmatrix} = 7e^{-3t} \neq 0$

$$\mathbf{x}(t) = c_1\mathbf{x}_1 + c_2\mathbf{x}_2 = c_1\begin{bmatrix} 3e^{2t} \\ 2e^{2t} \end{bmatrix} + c_2\begin{bmatrix} e^{-5t} \\ 3e^{-5t} \end{bmatrix} = \begin{bmatrix} 3c_1e^{2t} + c_2e^{-5t} \\ 2c_1e^{2t} + 3c_2e^{-5t} \end{bmatrix}$$

19. $W(t) = \begin{vmatrix} e^{2t} & e^{-t} & 0 \\ e^{2t} & 0 & e^{-t} \\ e^{2t} & -e^{-t} & -e^{-t} \end{vmatrix} = 3 \neq 0$

$$\mathbf{x}(t) = c_1\mathbf{x}_1 + c_2\mathbf{x}_2 + c_3\mathbf{x}_3 = c_1\begin{bmatrix} 1 \\ 1 \\ 1 \end{bmatrix}e^{2t} + c_2\begin{bmatrix} 1 \\ 0 \\ -1 \end{bmatrix}e^{-t} + c_3\begin{bmatrix} 0 \\ 1 \\ -1 \end{bmatrix}e^{-t} = \begin{bmatrix} c_1e^{2t} + c_2e^{-t} \\ c_1e^{2t} + c_3e^{-t} \\ c_1e^{2t} - c_2e^{-t} - c_3e^{-t} \end{bmatrix}$$

$$\mathbf{x}_1' = \begin{bmatrix} 2 \\ 2 \\ 2 \end{bmatrix}e^{t} = \begin{bmatrix} 0 & 1 & 1 \\ 1 & 0 & 1 \\ 1 & 1 & 0 \end{bmatrix}\begin{bmatrix} 1 \\ 1 \\ 1 \end{bmatrix}e^{t} = \mathbf{A}\mathbf{x}_1$$

$$\mathbf{x}_2' = \begin{bmatrix} -1 \\ 0 \\ 1 \end{bmatrix}e^{-t} = \begin{bmatrix} 0 & 1 & 1 \\ 1 & 0 & 1 \\ 1 & 1 & 0 \end{bmatrix}\begin{bmatrix} 1 \\ 0 \\ -1 \end{bmatrix}e^{-t} = \mathbf{A}\mathbf{x}_2$$

$$\mathbf{x}_3' = \begin{bmatrix} 0 \\ -1 \\ 1 \end{bmatrix}e^{t} = \begin{bmatrix} 0 & 1 & 1 \\ 1 & 0 & 1 \\ 1 & 1 & 0 \end{bmatrix}\begin{bmatrix} 0 \\ 1 \\ -1 \end{bmatrix}e^{-t} = \mathbf{A}\mathbf{x}_3$$

21. $W(t) = \begin{vmatrix} 3e^{-2t} & e^{t} & e^{3t} \\ -2e^{-2t} & -e^{t} & -e^{3t} \\ 2e^{-2t} & e^{t} & 0 \end{vmatrix} = e^{2t} \neq 0$

$$\mathbf{x}(t) = c_1\mathbf{x}_1 + c_2\mathbf{x}_2 + c_3\mathbf{x}_3 = c_1\begin{bmatrix} 3 \\ -2 \\ 2 \end{bmatrix}e^{-2t} + c_2\begin{bmatrix} 1 \\ -1 \\ 1 \end{bmatrix}e^{t} + c_3\begin{bmatrix} 1 \\ -1 \\ 0 \end{bmatrix}e^{3t} = \begin{bmatrix} 3c_1e^{-2t} + c_2e^{t} + c_3e^{3t} \\ -2c_1e^{-2t} - c_2e^{t} - c_3e^{3t} \\ 2c_1e^{-2t} + c_2e^{t} \end{bmatrix}$$

In Problems 23–26 (and similarly in Problems 27–32) we give first the scalar components $x_1(t)$ and $x_2(t)$ of a general solution, then the equations in the coefficients c_1 and c_2 that are obtained when the given initial conditions are imposed, and finally the resulting particular solution of the given system.

23. $\quad x_1(t) = c_1 e^{3t} + 2c_2 e^{-2t}, \quad x_2(t) = 3c_1 e^{3t} + c_2 e^{-2t}$

$\quad c_1 + 2c_2 = 0, \qquad 3c_1 + c_2 = 5$

$\quad x_1(t) = 2e^{3t} - 2e^{-2t}, \quad x_2(t) = 6e^{3t} - e^{-2t}$

25. $\quad x_1(t) = c_1 e^{3t} + c_2 e^{2t}, \quad x_2(t) = -c_1 e^{3t} - 2c_2 e^{2t}$

$\quad c_1 + c_2 = 11, \qquad -c_1 - 2c_2 = -7$

$\quad x_1(t) = 15e^{3t} - 4e^{2t}, \quad x_2(t) = -15e^{3t} + 8e^{2t}$

27. $\quad x_1(t) = 2c_1 e^{t} - 2c_2 e^{3t} + 2c_3 e^{5t}, \quad x_2(t) = 2c_1 e^{t} - 2c_3 e^{5t}, \quad x_3(t) = c_1 e^{t} + c_2 e^{3t} + c_3 e^{5t}$

$\quad 2c_1 - 2c_2 + 2c_3 = 0, \qquad 2c_1 - 2c_3 = 0, \qquad c_1 + c_2 + c_3 = 4$

$\quad x_1(t) = 2e^{t} - 4e^{3t} + 2e^{5t}, \quad x_2(t) = 2e^{t} - 2e^{5t}, \quad x_3(t) = e^{t} + 2e^{3t} + e^{5t}$

29. $\quad x_1(t) = 3c_1 e^{-2t} + c_2 e^{t} + c_3 e^{3t}, \quad x_2(t) = -2c_1 e^{-2t} - c_2 e^{t} - c_3 e^{3t}, \quad x_3(t) = 2c_1 e^{-2t} + c_2 e^{t}$

$\quad 3c_1 + c_2 + c_3 = 1, \qquad -2c_1 - c_2 - c_3 = 2, \qquad 2c_1 + c_2 = 3$

$\quad x_1(t) = 9e^{-2t} - 3e^{t} - 5e^{3t}, \quad x_2(t) = -6e^{-2t} + 3e^{t} + 5e^{3t}, \quad x_3(t) = 6e^{-2t} - 3e^{t}$

31. $\quad x_1(t) = c_1 e^{-t} + c_4 e^{t}, \quad x_2(t) = c_3 e^{t}, \quad x_3(t) = c_2 e^{-t} + 3c_4 e^{t}, \quad x_4(t) = c_1 e^{-t} - 2c_3 e^{t}$

$\quad c_1 + c_4 = 1, \qquad c_3 = 1, \qquad c_2 + 3c_4 = 1, \qquad c_1 - 2c_3 = 1$

$\quad x_1(t) = 3e^{-t} - 2e^{t}, \quad x_2(t) = e^{t}, \quad x_3(t) = 7e^{-t} - 6e^{t}, \quad x_4(t) = 3e^{-t} - 2e^{t}$

33. **(a)** $\quad$ $\mathbf{x}_2 = t\mathbf{x}_1$, so neither is a constant multiple of the other.

$\qquad$ **(b)** $\quad$ $W(\mathbf{x}_1, \mathbf{x}_2) = 0$, whereas Theorem 2 would imply that $W \neq 0$ if $\mathbf{x}_1$ and $\mathbf{x}_2$ were independent solutions of a system of the indicated form.

35. $\quad$ Suppose $W(a) = x_{11}(a)x_{22}(a) - x_{12}(a)x_{21}(a) = 0$. Then the coefficient determinant of the homogeneous linear system $c_1 x_{11}(a) + c_2 x_{12}(a) = 0$, $c_1 x_{21}(a) + c_2 x_{22}(a) = 0$ vanishes. The system therefore has a non-trivial solution $\{c_1, c_2\}$ such that $c_1 \mathbf{x}_1(a) + c_2 \mathbf{x}_2(a) = \mathbf{0}$. Then $\mathbf{x}(t) = c_1 \mathbf{x}_1(t) + c_2 \mathbf{x}_2(t)$ is a solution of $\mathbf{x}' = \mathbf{Px}$ such that $\mathbf{x}(a) = \mathbf{0}$. It therefore follows (by uniqueness of solutions) that $\mathbf{x}(t) \equiv \mathbf{0}$, that is, $c_1 \mathbf{x}_1(t) + c_2 \mathbf{x}_2(t) \equiv \mathbf{0}$ with c_1 and c_2 not both zero. Thus the solution vectors $\mathbf{x}_1$ and $\mathbf{x}_2$ are linearly dependent.

37.   Suppose that $c_1\mathbf{x}_1(t) + c_2\mathbf{x}_2(t) + \cdots + c_n\mathbf{x}_n(t) \equiv \mathbf{0}$. Then the ith scalar component of this vector equation is $c_1 x_{i1}(t) + c_2 x_{i2}(t) + \cdots + c_n x_{in}(t) \equiv 0$. Hence the fact that the scalar functions $x_{i1}(t)$, $x_{i2}(t)$, $\cdots$, $x_{in}(t)$ are linear linearly independent implies that $c_1 = c_2 = \cdots c_n = 0$. Consequently the vector functions $\mathbf{x}_1(t), \mathbf{x}_2(t), \cdots, \mathbf{x}_n(t)$ are linearly independent.

SECTION 7.3

THE EIGENVALUE METHOD
FOR LINEAR SYSTEMS

In each of Problems 1–16 we give the characteristic equation, the eigenvalues λ_1 and λ_2 of the coefficient matrix of the given system, the corresponding equations determining the associated eigenvectors $\mathbf{v}_1 = [a_1 \ \ b_1]^T$ and $\mathbf{v}_2 = [a_2 \ \ b_2]^T$, these eigenvectors, and the resulting scalar components $x_1(t)$ and $x_2(t)$ of a general solution $\mathbf{x}(t) = c_1\mathbf{v}_1 e^{\lambda_1 t} + c_2\mathbf{v}_2 e^{\lambda_2 t}$ of the system.

1.   Characteristic equation $\lambda^2 - 2\lambda - 3 = 0$

Eigenvalues $\lambda_1 = -1$ and $\lambda_2 = 3$

Eigenvector equations $\begin{bmatrix} 2 & 2 \\ 2 & 2 \end{bmatrix}\begin{bmatrix} a_1 \\ b_1 \end{bmatrix} = \begin{bmatrix} 0 \\ 0 \end{bmatrix}$ and $\begin{bmatrix} -2 & 2 \\ 2 & -2 \end{bmatrix}\begin{bmatrix} a_2 \\ b_2 \end{bmatrix} = \begin{bmatrix} 0 \\ 0 \end{bmatrix}$

Eigenvectors $\mathbf{v}_1 = [1 \ \ -1]^T$ and $\mathbf{v}_2 = [1 \ \ 1]^T$

$x_1(t) = c_1 e^{-t} + c_2 e^{3t}, \quad x_2(t) = -c_1 e^{-t} + c_2 e^{3t}$

The left-hand figure below shows a direction field and some typical solution curves for the system in Problem 1.

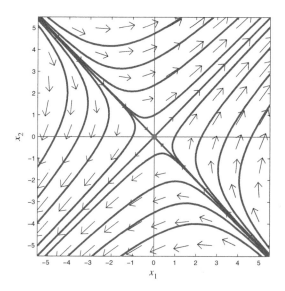 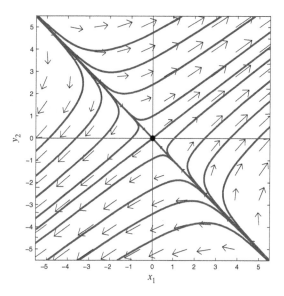

3. Characteristic equation $\quad \lambda^2 - 5\lambda - 6 = 0$

Eigenvalues $\quad \lambda_1 = -1$ and $\lambda_2 = 6$

Eigenvector equations $\begin{bmatrix} 4 & 4 \\ 3 & 3 \end{bmatrix} \begin{bmatrix} a_1 \\ b_1 \end{bmatrix} = \begin{bmatrix} 0 \\ 0 \end{bmatrix}$ and $\begin{bmatrix} -3 & 4 \\ 3 & -4 \end{bmatrix} \begin{bmatrix} a_2 \\ b_2 \end{bmatrix} = \begin{bmatrix} 0 \\ 0 \end{bmatrix}$

Eigenvectors $\mathbf{v}_1 = \begin{bmatrix} 1 & -1 \end{bmatrix}^T$ and $\mathbf{v}_2 = \begin{bmatrix} 4 & 3 \end{bmatrix}^T$

$x_1(t) = c_1 e^{-t} + 4c_2 e^{6t}, \quad x_2(t) = -c_1 e^{-t} + 3c_2 e^{6t}$

The equations

$$x_1(0) = c_1 + 4c_2 = 1$$

$$x_2(0) = -c_1 + 3c_2 = 1$$

yield $c_1 = -1/7$ and $c_2 = 2/7$, so the desired particular solution is given by

$$x_1(t) = \tfrac{1}{7}(-e^{-t} + 8e^{6t}), \quad x_2(t) = \tfrac{1}{7}(e^{-t} + 6e^{6t}).$$

See the right-hand figure at the bottom of the preceding page.

5. Characteristic equation $\quad \lambda^2 - 4\lambda - 5 = 0$

Eigenvalues $\quad \lambda_1 = -1$ and $\lambda_2 = 5$

Eigenvector equations $\begin{bmatrix} 7 & -7 \\ 1 & -1 \end{bmatrix} \begin{bmatrix} a_1 \\ b_1 \end{bmatrix} = \begin{bmatrix} 0 \\ 0 \end{bmatrix}$ and $\begin{bmatrix} 1 & -7 \\ 1 & -7 \end{bmatrix} \begin{bmatrix} a_2 \\ b_2 \end{bmatrix} = \begin{bmatrix} 0 \\ 0 \end{bmatrix}$

Eigenvectors $\mathbf{v}_1 = \begin{bmatrix} 1 & 1 \end{bmatrix}^T$ and $\mathbf{v}_2 = \begin{bmatrix} 7 & 1 \end{bmatrix}^T$

$x_1(t) = c_1 e^{-t} + 7c_2 e^{5t}, \quad x_2(t) = c_1 e^{-t} + c_2 e^{5t}$

The figure below shows a direction field and some typical solution curves.

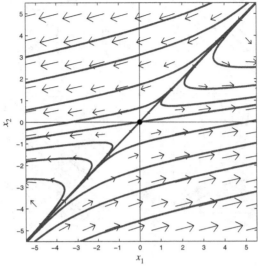

7.   Characteristic equation $\lambda^2 + 8\lambda - 9 = 0$

Eigenvalues $\lambda_1 = 1$ and $\lambda_2 = -9$

Eigenvector equations $\begin{bmatrix} -4 & 4 \\ 6 & -6 \end{bmatrix}\begin{bmatrix} a_1 \\ b_1 \end{bmatrix} = \begin{bmatrix} 0 \\ 0 \end{bmatrix}$ and $\begin{bmatrix} 6 & 4 \\ 6 & 4 \end{bmatrix}\begin{bmatrix} a_2 \\ b_2 \end{bmatrix} = \begin{bmatrix} 0 \\ 0 \end{bmatrix}$

Eigenvectors $\mathbf{v}_1 = \begin{bmatrix} 1 & 1 \end{bmatrix}^T$ and $\mathbf{v}_2 = \begin{bmatrix} 2 & -3 \end{bmatrix}^T$

$x_1(t) = c_1 e^t + 2c_2 e^{-9t}, \quad x_2(t) = c_1 e^t - 3c_2 e^{-9t}$

The figure below shows a direction field and some typical solution curves.

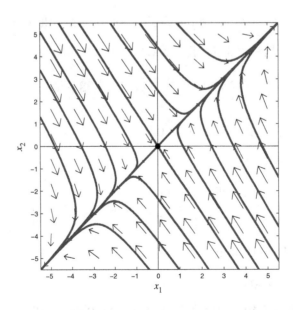

9.   Characteristic equation $\lambda^2 + 16 = 0$

Eigenvalue $\lambda = 4i$

Eigenvector equation $\begin{bmatrix} 2-4i & -5 \\ 4 & -2-4i \end{bmatrix} \begin{bmatrix} a \\ b \end{bmatrix} = \begin{bmatrix} 0 \\ 0 \end{bmatrix}$

Eigenvector $\mathbf{v} = [5 \quad 2-4i]^{\mathrm{T}}$

The real and imaginary parts of

$$\mathbf{x}(t) = \mathbf{v}\, e^{4it} = \begin{bmatrix} 5\cos 4t + 5i\sin 4t \\ (2\cos 4t + 4\sin 4t) + i\,(2\sin 4t - 4\cos 4t) \end{bmatrix}$$

yield the general solution

$$x_1(t) = 5c_1\cos 4t + 5c_2\sin 4t$$

$$x_2(t) = c_1(2\cos 4t + 4\sin 4t) + c_2(2\sin 4t - 4\cos 4t).$$

The initial conditions $x_1(0) = 2$ and $x_2(0) = 3$ give $c_1 = 2/5$ and $c_2 = -11/20$, so the desired particular solution is

$$x_1(t) = 2\cos 4t - \tfrac{11}{4}\sin 4t$$
$$x_2(t) = 3\cos 4t + \tfrac{1}{2}\sin 4t.$$

The figure below shows a direction field and some typical solution curves.

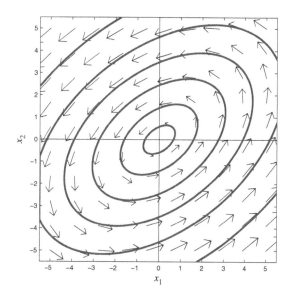

11.   Characteristic equation $\lambda^2 - 2\lambda + 5 = 0$

Eigenvalue $\lambda = 1 - 2i$

Eigenvector equation $\begin{bmatrix} 2i & -2 \\ 2 & 2i \end{bmatrix} \begin{bmatrix} a \\ b \end{bmatrix} = \begin{bmatrix} 0 \\ 0 \end{bmatrix}$

Eigenvector $\mathbf{v} = \begin{bmatrix} 1 & i \end{bmatrix}^T$

The real and imaginary parts of

$$\mathbf{x}(t) = \begin{bmatrix} 1 & i \end{bmatrix}^T e^t (\cos 2t - i \sin 2t)$$
$$= e^t \begin{bmatrix} \cos 2t & \sin 2t \end{bmatrix}^T + i e^t \begin{bmatrix} -\sin 2t & \cos 2t \end{bmatrix}^T$$

yield the general solution

$$x_1(t) = e^t(c_1 \cos 2t - c_2 \sin 2t)$$
$$x_2(t) = e^t(c_1 \sin 2t + c_2 \cos 2t).$$

The particular solution with $x_1(0) = 0$ and $x_2(0) = 4$ is obtained with $c_1 = 0$ and $c_2 = 4$, so

$$x_1(t) = -4e^t \sin 2t, \qquad x_2(t) = 4e^t \cos 2t.$$

The left-hand figure below shows a direction field and some typical solution curves.

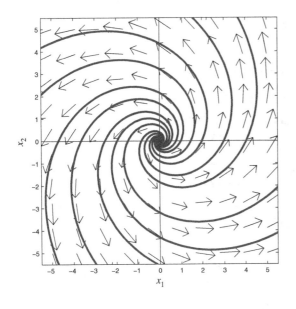 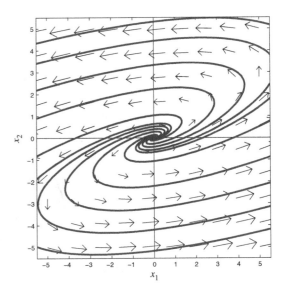

13.   Characteristic equation $\lambda^2 - 4\lambda + 13 = 0$

Eigenvalue $\lambda = 2 - 3i$

Eigenvector equation $\begin{bmatrix} 3+3i & -9 \\ 2 & -3+3i \end{bmatrix} \begin{bmatrix} a \\ b \end{bmatrix} = \begin{bmatrix} 0 \\ 0 \end{bmatrix}$

Eigenvector $\mathbf{v} = [3 \quad 1+i]^T$

$$\mathbf{x}(t) = \mathbf{v}e^{(2-3i)t} = e^{2t}\begin{bmatrix} 3\cos 3t - 3i\sin 3t \\ (\cos 3t + \sin 3t) + i(\cos 3t - \sin 3t) \end{bmatrix}$$

$x_1(t) = 3e^{2t}(c_1\cos 3t - c_2\sin 3t)$

$x_2(t) = e^{2t}[(c_1 + c_2)\cos 3t + (c_1 - c_2)\sin 3t].$

The right-hand figure above shows a direction field and some typical solution curves.

15. Characteristic equation $\lambda^2 - 10\lambda + 41 = 0$
Eigenvalue $\lambda = 5 - 4i$

Eigenvector equation $\begin{bmatrix} 2+4i & -5 \\ 4 & -2+4i \end{bmatrix}\begin{bmatrix} a \\ b \end{bmatrix} = \begin{bmatrix} 0 \\ 0 \end{bmatrix}$

Eigenvector $\mathbf{v} = [5 \quad 2+4i]^T$

$$\mathbf{x}(t) = \mathbf{v}e^{(5-4i)t} = e^{5t}\begin{bmatrix} 5\cos 4t - 5i\sin 4t \\ (2\cos 4t + 4\sin 4t) + i(4\cos 4t - 2\sin 4t) \end{bmatrix}$$

$x_1(t) = 5e^{5t}(c_1\cos 4t - c_2\sin 4t)$

$x_2(t) = e^{5t}[(2c_1 + 4c_2)\cos 4t + (4c_1 - 2c_2)\sin 4t]$

The figure below shows a direction field and some typical solution curves.

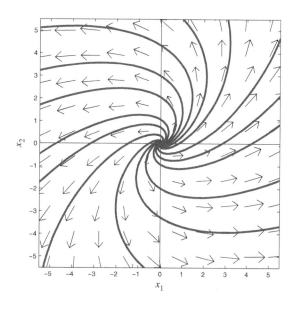

17. Characteristic equation $-\lambda^3 + 15\lambda^2 - 54\lambda = 0$

Eigenvalues $\lambda_1 = 9, \quad \lambda_2 = 6, \quad \lambda_3 = 0$

Eigenvector equations

$$\begin{bmatrix} -5 & 1 & 4 \\ 1 & -2 & 1 \\ 4 & 1 & -5 \end{bmatrix}\begin{bmatrix} a_1 \\ b_1 \\ c_1 \end{bmatrix} = \begin{bmatrix} 0 \\ 0 \\ 0 \end{bmatrix}, \quad \begin{bmatrix} -2 & 1 & 4 \\ 1 & 1 & 1 \\ 4 & 1 & -2 \end{bmatrix}\begin{bmatrix} a_2 \\ b_2 \\ c_2 \end{bmatrix} = \begin{bmatrix} 0 \\ 0 \\ 0 \end{bmatrix}, \quad \begin{bmatrix} 4 & 1 & 4 \\ 1 & 7 & 1 \\ 4 & 1 & 4 \end{bmatrix}\begin{bmatrix} a_3 \\ b_3 \\ c_3 \end{bmatrix} = \begin{bmatrix} 0 \\ 0 \\ 0 \end{bmatrix}$$

Eigenvectors $\mathbf{v}_1 = [1 \ 1 \ 1]^T, \qquad \mathbf{v}_2 = [1 \ -2 \ 1]^T, \qquad \mathbf{v}_3 = [1 \ 0 \ -1]^T$

$x_1(t) = c_1 e^{9t} + c_2 e^{6t} + c_3$

$x_2(t) = c_1 e^{9t} - 2c_2 e^{6t}$

$x_3(t) = c_1 e^{9t} + c_2 e^{6t} - c_3$

19.   Characteristic equation $-\lambda^3 + 12\lambda^2 - 45\lambda + 54 = 0$

Eigenvalues $\lambda_1 = 6, \ \lambda_2 = 3, \ \lambda_3 = 3$

Eigenvector equations

$$\begin{bmatrix} -2 & 1 & 1 \\ 1 & -2 & 1 \\ 1 & 1 & -2 \end{bmatrix}\begin{bmatrix} a_1 \\ b_1 \\ c_1 \end{bmatrix} = \begin{bmatrix} 0 \\ 0 \\ 0 \end{bmatrix}, \quad \begin{bmatrix} 1 & 1 & 1 \\ 1 & 1 & 1 \\ 1 & 1 & 1 \end{bmatrix}\begin{bmatrix} a_2 \\ b_2 \\ c_2 \end{bmatrix} = \begin{bmatrix} 0 \\ 0 \\ 0 \end{bmatrix}, \quad \begin{bmatrix} 1 & 1 & 1 \\ 1 & 1 & 1 \\ 1 & 1 & 1 \end{bmatrix}\begin{bmatrix} a_3 \\ b_3 \\ c_3 \end{bmatrix} = \begin{bmatrix} 0 \\ 0 \\ 0 \end{bmatrix}$$

Eigenvectors $\mathbf{v}_1 = [1 \ 1 \ 1]^T, \qquad \mathbf{v}_2 = [1 \ -2 \ 1]^T, \qquad \mathbf{v}_3 = [1 \ 0 \ -1]^T$

$x_1(t) = c_1 e^{6t} + c_2 e^{3t} + c_3 e^{3t}$

$x_2(t) = c_1 e^{6t} - 2c_2 e^{3t}$

$x_3(t) = c_1 e^{6t} + c_2 e^{3t} - c_3 e^{3t}$

21.   Characteristic equation $-\lambda^3 + \lambda = 0$

Eigenvalues $\lambda_1 = 0, \ \lambda_2 = 1, \ \lambda_3 = -1$

Eigenvector equations

$$\begin{bmatrix} 5 & 0 & -6 \\ 2 & -1 & -2 \\ 4 & -2 & -4 \end{bmatrix}\begin{bmatrix} a_1 \\ b_1 \\ c_1 \end{bmatrix} = \begin{bmatrix} 0 \\ 0 \\ 0 \end{bmatrix}, \quad \begin{bmatrix} 4 & 0 & -6 \\ 2 & -2 & -2 \\ 4 & -2 & -5 \end{bmatrix}\begin{bmatrix} a_2 \\ b_2 \\ c_2 \end{bmatrix} = \begin{bmatrix} 0 \\ 0 \\ 0 \end{bmatrix}, \quad \begin{bmatrix} 6 & 0 & -6 \\ 2 & 0 & -2 \\ 4 & -2 & -3 \end{bmatrix}\begin{bmatrix} a_3 \\ b_3 \\ c_3 \end{bmatrix} = \begin{bmatrix} 0 \\ 0 \\ 0 \end{bmatrix}$$

Eigenvectors $\mathbf{v}_1 = [6 \ 2 \ 5]^T, \qquad \mathbf{v}_2 = [3 \ 1 \ 2]^T, \qquad \mathbf{v}_3 = [2 \ 1 \ 2]^T$

$x_1(t) = 6c_1 + 3c_2 e^t + 2c_3 e^{-t}$

$x_2(t) = 2c_1 + c_2 e^t + c_3 e^{-t}$

$x_3(t) = 5c_1 + 2c_2 e^t + 2c_3 e^{-t}$

23. Characteristic equation $-\lambda^3 + 3\lambda^2 + 4\lambda - 12 = 0$

Eigenvalues $\lambda_1 = 2, \ \lambda_2 = -2, \ \lambda_3 = 3$

Eigenvector equations

$$\begin{bmatrix} 1 & 1 & 1 \\ -5 & -5 & -1 \\ 5 & 5 & 1 \end{bmatrix} \begin{bmatrix} a_1 \\ b_1 \\ c_1 \end{bmatrix} = \begin{bmatrix} 0 \\ 0 \\ 0 \end{bmatrix}, \quad \begin{bmatrix} 5 & 1 & 1 \\ -5 & -1 & -1 \\ 5 & 5 & 5 \end{bmatrix} \begin{bmatrix} a_2 \\ b_2 \\ c_2 \end{bmatrix} = \begin{bmatrix} 0 \\ 0 \\ 0 \end{bmatrix}, \quad \begin{bmatrix} 0 & 1 & 1 \\ -5 & -6 & -1 \\ 5 & 5 & 0 \end{bmatrix} \begin{bmatrix} a_3 \\ b_3 \\ c_3 \end{bmatrix} = \begin{bmatrix} 0 \\ 0 \\ 0 \end{bmatrix}$$

Eigenvectors $\mathbf{v}_1 = [1 \ -1 \ 0]^T, \quad \mathbf{v}_2 = [0 \ 1 \ -1]^T, \quad \mathbf{v}_3 = [1 \ -1 \ 1]^T$

$x_1(t) = c_1 e^{2t} \qquad\qquad + c_3 e^{3t}$

$x_2(t) = -c_1 e^{2t} + c_2 e^{-2t} - c_3 e^{3t}$

$x_3(t) = \qquad\quad - c_2 e^{-2t} + c_3 e^{3t}$

25. Characteristic equation $-\lambda^3 + 4\lambda^2 - 13\lambda = 0$

Eigenvalues $\lambda = 0$ and $2 \pm 3i$

With $\lambda = 1$ the eigenvector equation

$$\begin{bmatrix} 5 & 5 & 2 \\ -6 & -6 & -5 \\ 6 & 6 & 5 \end{bmatrix} \begin{bmatrix} a_1 \\ b_1 \\ c_1 \end{bmatrix} = \begin{bmatrix} 0 \\ 0 \\ 0 \end{bmatrix} \quad \text{gives eigenvector } \mathbf{v}_1 = [1 \ -1 \ 0]^T.$$

With $\lambda = 2 + 3i$ we solve the eigenvector equation

$$\begin{bmatrix} 3-3i & 5 & 2 \\ -6 & -8-3i & -5 \\ 6 & 6 & 3-3i \end{bmatrix} \begin{bmatrix} a \\ b \\ c \end{bmatrix} = \begin{bmatrix} 0 \\ 0 \\ 0 \end{bmatrix}$$

to find the complex-valued eigenvector $\mathbf{v} = [1+i \ -2 \ 2]^T$. The corresponding complex-valued solution is

$$\mathbf{x}(t) = \mathbf{v}\, e^{(2+3i)t} = e^{2t} \begin{bmatrix} (\cos 3t - \sin 3t) + i(\cos 3t + \sin 3t) \\ -2\cos 3t - 2i\sin 3t \\ 2\cos 3t + 2i\sin 3t \end{bmatrix}.$$

The scalar components of the resulting general solution are

$$x_1(t) = c_1 + e^{2t}[(c_2 + c_3)\cos 3t + (-c_2 + c_3)\sin 3t]$$

$$x_2(t) = -c_1 + 2e^{2t}(-c_2\cos 3t - c_3\sin 3t)$$

$$x_3(t) = 2e^{2t}(c_2\cos 3t + c_3\sin 3t).$$

27. The coefficient matrix

$$\mathbf{A} = \begin{bmatrix} -0.2 & 0 \\ 0.2 & -0.4 \end{bmatrix}$$

has characteristic equation $\lambda^2 + 0.6\lambda + 0.08 = 0$ with eigenvalues $\lambda_1 = -0.2$ and $\lambda_2 = -0.4$. We find easily that the associated eigenvectors are $\mathbf{v_1} = \begin{bmatrix} 1 & 1 \end{bmatrix}^T$ and $\mathbf{v_2} = \begin{bmatrix} 0 & 1 \end{bmatrix}^T$, so we get the general solution

$$x_1(t) = c_1 e^{-0.2t}, \quad x_2(t) = c_1 e^{-0.2t} + c_2 e^{-0.4t}.$$

The initial conditions $x_1(0) = 15$, $x_2(0) = 0$ give $c_1 = 15$ and $c_2 = -15$, so we get

$$x_1(t) = 15e^{-0.2t}, \quad x_2(t) = 15e^{-0.2t} - 15e^{-0.4t}.$$

To find the maximum value of $x_2(t)$, we solve the equation $x_2'(t) = 0$ for $t = 5\ln 2$, which gives the maximum value $x_2(5\ln 2) = 3.75$ lb. The following figure shows the graphs of $x_1(t)$ and $x_2(t)$.

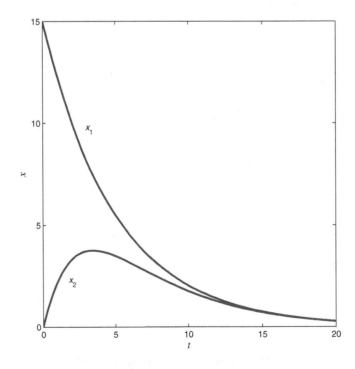

29. The coefficient matrix

$$\mathbf{A} = \begin{bmatrix} -0.2 & 0.4 \\ 0.2 & -0.4 \end{bmatrix}$$

has eigenvalues $\lambda_1 = 0$ and $\lambda_2 = -0.6$, with eigenvectors $\mathbf{v}_1 = [2 \quad 1]^T$ and $\mathbf{v}_2 = [1 \quad -1]^T$ that yield the general solution

$$x_1(t) = 2c_1 + c_2 e^{-0.6t}, \quad x_2(t) = c_1 - c_2 e^{-0.6t}.$$

The initial conditions $x_1(0) = 15$, $x_2(0) = 0$ give $c_1 = c_2 = 5$, so we get

$$x_1(t) = 10 + 5e^{-0.6t}, \quad x_2(t) = 5 - 5e^{-0.6t}.$$

The figure below shows the graphs of $x_1(t)$ and $x_2(t)$.

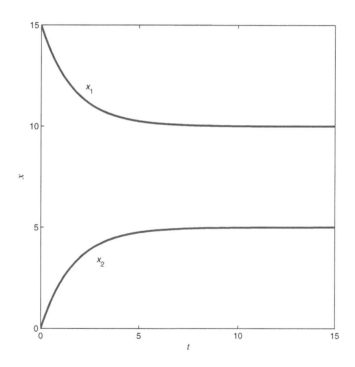

31. The coefficient matrix

$$\mathbf{A} = \begin{bmatrix} -1 & 0 & 0 \\ 1 & -2 & 0 \\ 0 & 2 & -3 \end{bmatrix}$$

has as eigenvalues its diagonal elements $\lambda_1 = -1$, $\lambda_2 = -2$, and $\lambda_3 = -3$. We find readily that the associated eigenvectors are $\mathbf{v}_1 = [1 \quad 1 \quad 1]^T$, $\mathbf{v}_2 = [0 \quad 1 \quad 2]^T$, and $\mathbf{v}_3 = [0 \quad 0 \quad 1]^T$. The resulting general solution is solution is given by

$$x_1(t) = c_1 e^{-t}$$
$$x_2(t) = c_1 e^{-t} + c_2 e^{-2t}$$
$$x_3(t) = c_1 e^{-t} + 2c_2 e^{-2t} + c_3 e^{-3t}.$$

The initial conditions $x_1(0) = 27$, $x_2(0) = x_2(0) = 0$ give $c_1 = c_3 = 27$, $c_2 = -27$, so we get

$$x_1(t) = 27 e^{-t}$$
$$x_2(t) = 27 e^{-t} - 27 e^{-2t}$$
$$x_3(t) = 27 e^{-t} - 54 e^{-2t} + 27 e^{-3t}.$$

The equation $x_3'(t) = 0$ simplifies to the equation

$$3e^{-2t} - 4e^{-t} + 1 = \left(3e^{-t} - 1\right)\left(e^{-t} - 1\right) = 0$$

with positive solution $t_m = \ln 3$. Thus the maximum amount of salt ever in tank 3 is $x_3(\ln 3) = 4$ pounds. The figure below shows the graphs of $x_1(t)$, $x_2(t)$, and $x_3(t)$.

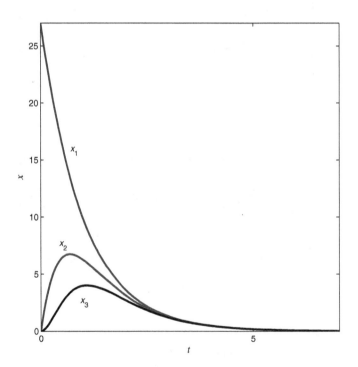

33.   The coefficient matrix

$$\mathbf{A} = \begin{bmatrix} -4 & 0 & 0 \\ 4 & -6 & 0 \\ 0 & 6 & -2 \end{bmatrix}$$

has as eigenvalues its diagonal elements $\lambda_1 = -4$, $\lambda_2 = -6$, and $\lambda_3 = -2$. We find readily that the associated eigenvectors are $\mathbf{v}_1 = [-1 \quad -2 \quad 6]^T$, $\mathbf{v}_2 = [0 \quad -2 \quad 3]^T$, and $\mathbf{v}_3 = [0 \quad 0 \quad 1]^T$. The resulting general solution is solution is given by

$$x_1(t) = -c_1 e^{-4t}$$
$$x_2(t) = -2c_1 e^{-4t} - 2c_2 e^{-6t}$$
$$x_3(t) = 6c_1 e^{-4t} + 3c_2 e^{-6t} + c_3 e^{-2t}.$$

The initial conditions $x_1(0) = 45$, $x_2(0) = x_2(0) = 0$ give $c_1 = -45$, $c_2 = 45$, $c_3 = 135$, so we get

$$x_1(t) = 45 e^{-4t}$$
$$x_2(t) = 90 e^{-4t} - 90 e^{-6t}$$
$$x_3(t) = -270 e^{-4t} + 135 e^{-6t} + 135 e^{-2t}.$$

The equation $x_3'(t) = 0$ simplifies to the equation

$$3e^{-4t} - 4e^{-2t} + 1 = \left(3e^{-2t} - 1\right)\left(e^{-2t} - 1\right) = 0$$

with positive solution $t_m = \frac{1}{2}\ln 3$. Thus the maximum amount of salt ever in tank 3 is $x_3(\frac{1}{2}\ln 3) = 20$ pounds. The figure below shows the graphs of $x_1(t)$, $x_2(t)$, and $x_3(t)$.

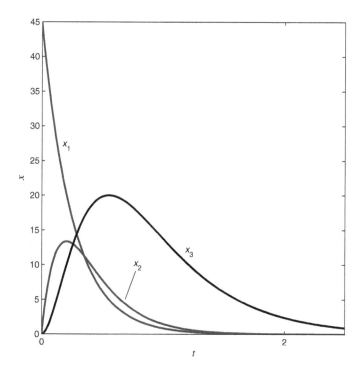

35. The coefficient matrix

$$A = \begin{bmatrix} -6 & 0 & 3 \\ 6 & -20 & 0 \\ 0 & 20 & -3 \end{bmatrix}$$

has characteristic equation $-\lambda^3 - 29\lambda^2 - 198\lambda = -\lambda(\lambda - 18)(\lambda - 11) = 0$ with eigenvalues $\lambda_0 = 0$, $\lambda_1 = -18$, and $\lambda_2 = -11$. We find that associated eigenvectors are $\mathbf{v}_0 = [10 \quad 3 \quad 20]^T$, $\mathbf{v}_1 = [-1 \quad -3 \quad 4]^T$, and $\mathbf{v}_2 = [-3 \quad -2 \quad 5]^T$. The resulting general solution is solution is given by

$$x_1(t) = 10c_0 - c_1 e^{-18t} - 3c_2 e^{-11t}$$
$$x_2(t) = 3c_0 - 3c_1 e^{-18t} - 2c_2 e^{-11t}$$
$$x_3(t) = 20c_0 + 4c_1 e^{-18t} + 5c_2 e^{-11t}.$$

The initial conditions $x_1(0) = 33$, $x_2(0) = x_2(0) = 0$ give $c_1 = 1$, $c_2 = 55/7$, $c_3 = -72/7$, so we get

$$x_1(t) = 10 - \tfrac{1}{7}\left(55e^{-18t} - 216e^{-11t}\right)$$
$$x_2(t) = 3 - \tfrac{1}{7}\left(165e^{-18t} - 144e^{-11t}\right)$$
$$x_3(t) = 20 + \tfrac{1}{7}\left(220e^{-18t} - 360e^{-11t}\right).$$

Thus the limiting amounts of salt in tanks 1, 2, and 3 are 10 lb, 3 lb, and 20 lb. The figure below shows the graphs of $x_1(t)$, $x_2(t)$, and $x_3(t)$.

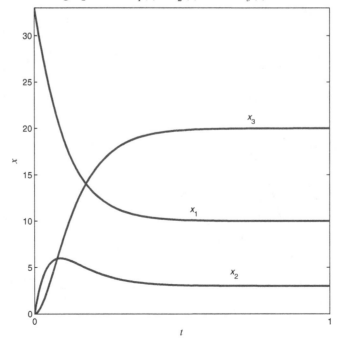

37. The coefficient matrix

$$\mathbf{A} = \begin{bmatrix} -1 & 0 & 2 \\ 1 & -3 & 0 \\ 0 & 3 & -2 \end{bmatrix}$$

has characteristic equation $-\lambda^3 - 6\lambda^2 - 11\lambda = 0$ with eigenvalues $\lambda_0 = 0$, $\lambda_1 = -3 - i\sqrt{2}$, and $\lambda_2 = -3 + i\sqrt{2}$. The eigenvector equation

$$\begin{bmatrix} -1 & 0 & 2 \\ 1 & -3 & 0 \\ 0 & 3 & -2 \end{bmatrix} \begin{bmatrix} a \\ b \\ c \end{bmatrix} = \begin{bmatrix} 0 \\ 0 \\ 0 \end{bmatrix}$$

associated with the eigenvalue $\lambda_0 = 0$ yields the associated eigenvector $\mathbf{v}_0 = \begin{bmatrix} 6 & 2 & 3 \end{bmatrix}^T$ and consequently the constant solution $\mathbf{x}_0(t) \equiv \mathbf{v}_0$. Then the eigenvector equation

$$\begin{bmatrix} 2 + i\sqrt{2} & 0 & 2 \\ 1 & i\sqrt{2} & 0 \\ 0 & 3 & 1 + i\sqrt{2} \end{bmatrix} \begin{bmatrix} a \\ b \\ c \end{bmatrix} = \begin{bmatrix} 0 \\ 0 \\ 0 \end{bmatrix}$$

associated with $\lambda_1 = -3 - i\sqrt{2}$ yields the complex-valued eigenvector $\mathbf{v}_1 = \begin{bmatrix} (-2 + i\sqrt{2})/3 & (-1 - i\sqrt{2})/3 & 1 \end{bmatrix}^T$. The corresponding complex solution is

$$\mathbf{x}_1(t) = \mathbf{v}_1 e^{(-3 - i\sqrt{2})t}$$

$$= \frac{1}{3} e^{-3t} \begin{bmatrix} \left(-2\cos(t\sqrt{2}) + \sqrt{2}\sin(t\sqrt{2})\right) + i\left(\sqrt{2}\cos(t\sqrt{2}) + 2\sin(t\sqrt{2})\right) \\ \left(-\cos(t\sqrt{2}) - \sqrt{2}\sin(t\sqrt{2})\right) + i\left(-\sqrt{2}\cos(t\sqrt{2}) + \sin(t\sqrt{2})\right) \\ 3\cos(t\sqrt{2}) - 3i\sin(t\sqrt{2}) \end{bmatrix}.$$

The scalar components of resulting general solution $\mathbf{x} = c_0\mathbf{x}_0 + c_1\,\mathrm{Re}(\mathbf{x}_1) + c_2\,\mathrm{Im}(\mathbf{x}_1)$ are given by

$$x_1(t) = 6c_0 + \tfrac{1}{3}e^{-3t}\left[\left(-2c_1 + \sqrt{2}\,c_2\right)\cos(t\sqrt{2}) + \left(\sqrt{2}\,c_1 + 2c_2\right)\sin(t\sqrt{2})\right]$$

$$x_2(t) = 2c_0 + \tfrac{1}{3}e^{-3t}\left[\left(-c_1 - \sqrt{2}\,c_2\right)\cos(t\sqrt{2}) + \left(-\sqrt{2}\,c_1 + c_2\right)\sin(t\sqrt{2})\right]$$

$$x_3(t) = 3c_0 + e^{-3t}\left[c_1\cos(t\sqrt{2}) - c_2\sin(t\sqrt{2})\right].$$

When we impose the initial conditions $x_1(0) = 55$, $x_2(0) = x_2(0) = 0$ we find that $c_0 = 5$, $c_1 = -15$, and $c_2 = 45/\sqrt{2}$. This finally gives the particular solution

$$x_1(t) = 30 + e^{-3t}\left[25\cos(t\sqrt{2}) + 10\sqrt{2}\sin(t\sqrt{2})\right]$$

$$x_2(t) = 10 - e^{-3t}\left[10\cos(t\sqrt{2}) - \tfrac{25}{2}\sqrt{2}\sin(t\sqrt{2})\right]$$

$$x_3(t) = 15 - e^{-3t}\left[15\cos(t\sqrt{2}) + \tfrac{45}{2}\sqrt{2}\sin(t\sqrt{2})\right].$$

Thus the limiting amounts of salt in tanks 1, 2, and 3 are 30 lb, 10 lb, and 15 lb. The figure below shows the graphs of $x_1(t)$, $x_2(t)$, and $x_3(t)$.

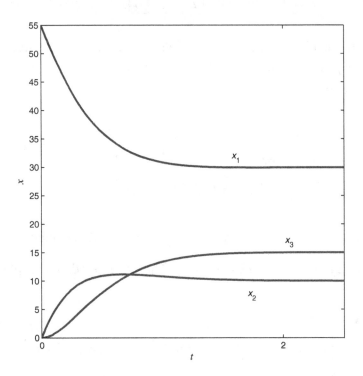

In Problems 38–41 the Maple command **with(linalg):eigenvects(A)**, the Mathematica command **Eigensystem[A]**, or the MATLAB command **[V,D] = eig(A)** can be used to find the eigenvalues and associated eigenvectors of the given coefficient matrix **A**.

39.   Characteristic equation: $(\lambda^2 - 1)(\lambda^2 - 4) = 0$

Eigenvalues and associated eigenvectors:

$$\lambda = 1, \qquad \mathbf{v} = [3 \quad -2 \quad 4 \quad 1]^\mathrm{T}$$
$$\lambda = -1, \qquad \mathbf{v} = [0 \quad 0 \quad 1 \quad 0]^\mathrm{T}$$
$$\lambda = 2, \qquad \mathbf{v} = [0 \quad 1 \quad 0 \quad 0]^\mathrm{T}$$
$$\lambda = -2, \qquad \mathbf{v} = [1 \quad -1 \quad 0 \quad 0]^\mathrm{T}$$

Scalar solution equations:

$$x_1(t) = 3c_1e^t \qquad\qquad + c_4e^{-2t}$$

$$x_2(t) = -2c_1e^t \qquad + c_3e^{2t} - c_4e^{-2t}$$

$$x_3(t) = 4c_1e^t + c_2e^{-t}$$

$$x_4(t) = c_1e^t$$

41. The eigenvectors associated with the respective eigenvalues $\lambda_1 = -3$, $\lambda_2 = -6$, $\lambda_3 = 10$, and $\lambda_4 = 15$ are

$$\mathbf{v}_1 = \begin{bmatrix} 1 & 0 & 0 & -1 \end{bmatrix}^T$$
$$\mathbf{v}_2 = \begin{bmatrix} 0 & 1 & -1 & 0 \end{bmatrix}^T$$
$$\mathbf{v}_3 = \begin{bmatrix} -2 & 1 & 1 & -2 \end{bmatrix}^T$$
$$\mathbf{v}_4 = \begin{bmatrix} 1 & 2 & 2 & 1 \end{bmatrix}^T.$$

Hence the general solution has scalar component functions

$$x_1(t) = c_1e^{-3t} \qquad - 2c_3e^{10t} + c_4e^{15t}$$

$$x_2(t) = \qquad c_2e^{-6t} + c_3e^{10t} + 2c_4e^{15t}$$

$$x_3(t) = \qquad - c_2e^{-6t} + c_3e^{10t} + 2c_4e^{15t}$$

$$x_4(t) = -c_1e^{-3t} \qquad - 2c_3e^{10t} + c_4e^{15t}.$$

The given initial conditions are satisfied by choosing $c_1 = c_2 = 0$, $c_3 = -1$, and $c_4 = 1$, so the desired particular solution is given by

$$x_1(t) = 2e^{10t} + e^{15t} = x_4(t)$$
$$x_2(t) = -e^{10t} + 2e^{15t} = x_3(t) .$$

In Problems 42–50 we give a general solution in the form $\mathbf{x}(t) = c_1\mathbf{v}_1e^{\lambda_1 t} + c_2\mathbf{v}_2e^{\lambda_2 t} + \cdots$ that exhibits explicitly the eigenvalues $\lambda_1, \lambda_2, \ldots$ and corresponding eigenvectors $\mathbf{v}_1, \mathbf{v}_2, \ldots$ of the given coefficient matrix $\mathbf{A}$.

43. $\mathbf{x}(t) = c_1 \begin{bmatrix} 3 \\ -1 \\ 5 \end{bmatrix} e^{-2t} + c_2 \begin{bmatrix} 1 \\ 1 \\ 1 \end{bmatrix} e^{4t} + c_3 \begin{bmatrix} 1 \\ -1 \\ 3 \end{bmatrix} e^{8t}$

45. $\mathbf{x}(t) = c_1 \begin{bmatrix} 1 \\ 1 \\ 1 \\ -1 \end{bmatrix} e^{-3t} + c_2 \begin{bmatrix} 1 \\ 2 \\ -1 \\ 1 \end{bmatrix} + c_3 \begin{bmatrix} 2 \\ 1 \\ 1 \\ 1 \end{bmatrix} e^{3t} + c_4 \begin{bmatrix} 1 \\ -1 \\ 2 \\ -1 \end{bmatrix} e^{6t}$

47. $\mathbf{x}(t) = c_1 \begin{bmatrix} 2 \\ 2 \\ 1 \\ -1 \end{bmatrix} e^{-3t} + c_2 \begin{bmatrix} 1 \\ 2 \\ -1 \\ 1 \end{bmatrix} e^{3t} + c_3 \begin{bmatrix} 2 \\ 1 \\ 1 \\ 1 \end{bmatrix} e^{6t} + c_4 \begin{bmatrix} 1 \\ -1 \\ 2 \\ -1 \end{bmatrix} e^{9t}$

49. $\mathbf{x}(t) = c_1 \begin{bmatrix} 1 \\ 0 \\ 3 \\ 1 \\ 1 \end{bmatrix} e^{-3t} + c_2 \begin{bmatrix} 0 \\ 3 \\ 0 \\ -1 \\ 1 \end{bmatrix} + c_3 \begin{bmatrix} 1 \\ 7 \\ 1 \\ 1 \\ 1 \end{bmatrix} e^{3t} + c_4 \begin{bmatrix} 0 \\ 1 \\ 0 \\ 1 \\ 1 \end{bmatrix} e^{6t} + c_5 \begin{bmatrix} 2 \\ 0 \\ 5 \\ 2 \\ 1 \end{bmatrix} e^{9t}$

SECTION 7.4

A GALLERY OF SOLUTION CURVES OF LINEAR SYSTEMS

This section emphasizes the connection between the algebraic properties of the matrix **A**—specifically, its eigenvalues and eigenvectors—and the characteristic pattern of the phase diagram of the system $\mathbf{x}' = \mathbf{Ax}$.

In Problems 1-16 the eigenvalues, eigenvectors and phase portraits appear in the solutions to Section 7.3. Thus here we simply categorize each phase portrait according to the gallery in Fig. 7.4.16.

1. Saddle point (real eigenvalues of opposite sign)

3. Saddle point (real eigenvalues of opposite sign)

5. Saddle point (real eigenvalues of opposite sign)

7. Saddle point (real eigenvalues of opposite sign)

9. Center (pure imaginary eigenvalues)

11. Spiral source (complex conjugate eigenvalues with positive real part)

13. Spiral source (complex conjugate eigenvalues with positive real part)

15. Spiral source (complex conjugate eigenvalues with positive real part)

In Problems 17-28 we "pigeonhole" each phase portrait according to the gallery in Fig. 7.4.16 and give the nature of the eigenvalues of the matrix **A**. Where appropriate we further give approximate values of the corresponding eigenvectors.

17. Center; pure imaginary eigenvalues

19. Saddle point; real eigenvalues of opposite sign; $\mathbf{v}_1 \approx \begin{bmatrix} 0 & 1 \end{bmatrix}^{\mathrm{T}}$ corresponds to the negative eigenvalue and $\mathbf{v}_2 \approx \begin{bmatrix} -1 & 1 \end{bmatrix}^{\mathrm{T}}$ to the positive one.

21. Proper nodal source; repeated positive real eigenvalue with linearly independent eigenvectors

23. Spiral sink; complex conjugate eigenvalues with negative real part

25. Saddle point; real eigenvalues of opposite sign; $\mathbf{v}_1 \approx \begin{bmatrix} 1 & 1 \end{bmatrix}^{\mathrm{T}}$ corresponds to the positive eigenvalue and $\mathbf{v}_2 \approx \begin{bmatrix} 4 & -1 \end{bmatrix}^{\mathrm{T}}$ to the negative one.

27. Improper nodal source; distinct positive real eigenvalues; $\mathbf{v}_1 \approx \begin{bmatrix} 2 & 3 \end{bmatrix}^{\mathrm{T}}$, $\mathbf{v}_2 \approx \begin{bmatrix} 2 & -1 \end{bmatrix}^{\mathrm{T}}$

29. **a)** If $v_0 = 0$, then $v(t) \equiv 0$ for all t, so that the point $(u,v) = (u_0 e^{-2t}, 0)$ (in oblique coordinates) lies on the u-axis. The reverse argument applies if $u_0 = 0$.

b) If both u_0 and v_0 are nonzero, then solving $u = u_0 e^{-2t}$ for t gives $t = -\dfrac{1}{2} \ln \dfrac{u}{u_0}$,

whereas solving $v = v_0 e^{5t}$ for t gives $t = \dfrac{1}{5} \ln \dfrac{v}{v_0}$. We can thus eliminate t to conclude that

$-\dfrac{1}{2} \ln \dfrac{u}{u_0} = \dfrac{1}{5} \ln \dfrac{v}{v_0}$. Then solving for v gives $\ln \dfrac{v}{v_0} = -\dfrac{5}{2} \ln \dfrac{u}{u_0}$, or

$\dfrac{v}{v_0} = \exp\left(-\dfrac{5}{2} \ln \dfrac{u}{u_0}\right) = \left(\dfrac{u}{u_0}\right)^{-5/2}$, or finally $v = v_0 \left(\dfrac{u}{u_0}\right)^{-5/2} = v_0 u_0^{5/2} u^{-5/2} = C u^{-5/2}$, where

$C = v_0 u_0^{5/2}$.

31. If λ is an eigenvalue of $\mathbf{A}$ with associated eigenvector $\mathbf{v}$, then $(\mathbf{A} - \lambda\mathbf{I})\mathbf{v} = \mathbf{0}$. Taking the negative of both sides of this equation then gives $-(\mathbf{A} - \lambda\mathbf{I})\mathbf{v} = -\mathbf{0} = \mathbf{0}$. However, $-(\mathbf{A} - \lambda\mathbf{I})$ can be written as $(-\mathbf{A}) - (-\lambda)\mathbf{I}$, and so $[(-\mathbf{A}) - (-\lambda)\mathbf{I}]\mathbf{v} = \mathbf{0}$ as well. It follows that $-\lambda$ is an eigenvalue of the matrix $-\mathbf{A}$ with associated eigenvector $\mathbf{v}$. This means that if $\mathbf{A}$ has positive eigenvalues $0 < \lambda_2 < \lambda_1$ with associated eigenvectors $\mathbf{v}_1$ and $\mathbf{v}_2$, then $-\mathbf{A}$ has negative eigenvalues $-\lambda_1 < -\lambda_2 < 0$ associated to these same eigenvectors $\mathbf{v}_1$ and $\mathbf{v}_2$.

33. **a)** Let $\mathbf{v}_1$ and $\mathbf{v}_2$ denote the two linearly independent eigenvectors of $\mathbf{A}$ associated with the eigenvalue λ, so that $\mathbf{A}\mathbf{v}_1 = \lambda\mathbf{v}_1$ and $\mathbf{A}\mathbf{v}_2 = \lambda\mathbf{v}_2$. If $\mathbf{v}$ is any two-dimensional vector, then the fact that $\mathbf{v}_1$ and $\mathbf{v}_2$ are linearly independent implies that $\mathbf{v}$ can be written as a linear combination of $\mathbf{v}_1$ and $\mathbf{v}_2$, so that $\mathbf{v} = c_1\mathbf{v}_1 + c_2\mathbf{v}_2$ for some scalars c_1 and c_2. But then the linearity of matrix multiplication gives

$$\mathbf{A}\mathbf{v} = \mathbf{A}(c_1\mathbf{v}_1 + c_2\mathbf{v}_2) = c_1\mathbf{A}\mathbf{v}_1 + c_2\mathbf{A}\mathbf{v}_2 = c_1\lambda\mathbf{v}_1 + c_2\lambda\mathbf{v}_2 = \lambda(c_1\mathbf{v}_1 + c_2\mathbf{v}_2) = \lambda\mathbf{v},$$

proving that $\mathbf{v}$ is an eigenvector of $\mathbf{A}$.

b) Let $\mathbf{A} = \begin{bmatrix} a & b \\ c & d \end{bmatrix}$. Part a) implies that $\mathbf{A}\mathbf{v} = \lambda\mathbf{v}$ for all vectors $\mathbf{v}$, so in particular, if we take $\mathbf{v} = \begin{bmatrix} 1 & 0 \end{bmatrix}^T$, then $\mathbf{A}\mathbf{v} = \begin{bmatrix} a & c \end{bmatrix}^T = \lambda\mathbf{v} = \begin{bmatrix} \lambda & 0 \end{bmatrix}^T$, proving that $a = \lambda$ and $c = 0$. Similarly, if we take $\mathbf{v} = \begin{bmatrix} 0 & 1 \end{bmatrix}^T$, then $\mathbf{A}\mathbf{v} = \begin{bmatrix} b & d \end{bmatrix}^T = \lambda\mathbf{v} = \begin{bmatrix} 0 & \lambda \end{bmatrix}^T$, proving that $b = 0$ and $d = \lambda$. Thus A is given by Eq. (22).

35. Write the given equation as $M(x_1, x_2)\,dx_1 + N(x_1, x_2)\,dx_2 = 0$, where $M(x_1, x_2) = 6x_2 - 8x_1$ and $N(x_1, x_2) = 6x_1 - 17x_2$. The equation is exact because $\dfrac{\partial M}{\partial x_2} = 6 = \dfrac{\partial N}{\partial x_1}$. Its general solution is therefore given by $F(x_1, x_2) = k$, where $F(x_1, x_2)$ (as discussed in Section 1.6) satisfies the conditions $\dfrac{\partial F}{\partial x_1} = M$ and $\dfrac{\partial F}{\partial x_2} = N$ and k is a constant. The first condition implies that

$$F(x_1, x_2) = \int M\,dx_1 = 6x_1x_2 - 4x_1^2 + h(x_2),$$

which specifies F up to the unknown function $h(x_2)$. Then the second condition gives

$$\frac{\partial}{\partial x_2}\left[6x_1x_2 - 4x_1^2 + h(x_2)\right] = 6x_1 + h'(x_2) = 6x_1 - 17x_2,$$

or simply $h'(x_2) = -17x_2$, which means that up to a constant, $h(x_2) = -\dfrac{17}{2}x_2^2$.

Altogether then, the general solution of the given equation is

$$F(x_1, x_2) = 6x_1 x_2 - 4x_1^2 - \frac{17}{2}x_2^2 = k.$$

37. With the same values of A, B, and C we find that $\dfrac{B}{A-C} = \dfrac{6}{-4+\dfrac{17}{2}} = \dfrac{6}{9/2} = \dfrac{4}{3}$. Thus it

suffices to confirm that $\theta = \arctan\dfrac{2}{4}$ satisfies $\tan 2\theta = \dfrac{4}{3}$. However, this is readily verified using the double-angle formula for the tangent function:

$$\tan 2\theta = \frac{2\tan\theta}{1 - \tan^2\theta} = \frac{2\cdot\dfrac{2}{4}}{1 - \left(\dfrac{2}{4}\right)^2} = \frac{4}{3}.$$

39. **a)** The characteristic equation of $\mathbf{A}$ is given by $\det(\mathbf{A} - \lambda\mathbf{I}) = 0$, that is

$$\begin{vmatrix} a - \lambda & b \\ c & d - \lambda \end{vmatrix} = (a - \lambda)(d - \lambda) - bc = \lambda^2 - (a + d)\lambda + (ad - bc) = 0.$$

b) The quadratic formula shows that if the solutions to the characteristic equation are pure imaginary, then the coefficient $-(a + d)$ of λ must vanish. Hence the trace $T(\mathbf{a}) = a + d = 0$, which means that $d = -a$. For the same reason, the constant term $ad - bc$ must be positive. Substituting $d = -a$ then gives $ad - bc = -a^2 - bc > 0$, which is impossible if $c = 0$.

SECTION 7.5

SECOND-ORDER SYSTEMS AND MECHANICAL APPLICATIONS

This section uses the eigenvalue method to exhibit realistic applications of linear systems. If a computer system like Maple, *Mathematica*, MATLAB, or even a TI calculator is available, then a system of more than three railway cars, or a multistory building with four or more floors (as in the project), can be investigated. However, the problems in the text are intended for manual solution.

Problems 1–7 involve the system

$$m_1 x_1'' = -(k_1 + k_2) x_1 + k_2 x_2$$
$$m_2 x_2'' = k_2 x_1 - (k_2 + k_3) x_2$$

with various values of m_1, m_2 and k_1, k_2, k_3. In each problem we divide the first equation by m_1 and the second one by m_2 to obtain a second-order linear system $\mathbf{x}'' = \mathbf{Ax}$ in the standard form of Theorem 1 in this section. If the eigenvalues λ_1 and λ_2 are both negative, then the natural (circular) frequencies of the system are $\omega_1 = \sqrt{-\lambda_1}$ and $\omega_2 = \sqrt{-\lambda_2}$, and—according to Eq. (11) in Theorem 1 of this section—the eigenvectors $\mathbf{v}_1$ and $\mathbf{v}_2$ associated with λ_1 and λ_2 determine the natural modes of oscillations at these frequencies.

1. The matrix $\mathbf{A} = \begin{bmatrix} -2 & 2 \\ 2 & -2 \end{bmatrix}$ has eigenvalues $\lambda_0 = 0$ and $\lambda_1 = -4$ with associated

    eigenvectors $\mathbf{v}_0 = \begin{bmatrix} 1 & 1 \end{bmatrix}^T$ and $\mathbf{v}_1 = \begin{bmatrix} 1 & -1 \end{bmatrix}^T$. Thus we have the special case described in Eq. (12) of Theorem 1, and a general solution is given by

    $$x_1(t) = a_1 + a_2 t + b_1 \cos 2t + b_2 \sin 2t,$$
    $$x_2(t) = a_1 + a_2 t - b_1 \cos 2t - b_2 \sin 2t.$$

    The natural frequencies are $\omega_1 = 0$ and $\omega_2 = 2$. In the degenerate natural mode with "frequency" $\omega_1 = 0$ the two masses move by translation without oscillating. At frequency $\omega_2 = 2$ they oscillate in opposite directions with equal amplitudes.

3. The matrix $\mathbf{A} = \begin{bmatrix} -3 & 2 \\ 1 & -2 \end{bmatrix}$ has eigenvalues $\lambda_1 = -1$ and $\lambda_2 = -4$, with associated

    eigenvectors $\mathbf{v}_1 = \begin{bmatrix} 1 & 1 \end{bmatrix}^T$ and $\mathbf{v}_2 = \begin{bmatrix} 2 & -1 \end{bmatrix}^T$. Hence a general solution is given by

    $$x_1(t) = a_1 \cos t + a_2 \sin t + 2b_1 \cos 2t + 2b_2 \sin 2t,$$
    $$x_2(t) = a_1 \cos t + a_2 \sin t - b_1 \cos 2t - b_2 \sin 2t.$$

    The natural frequencies are $\omega_1 = 1$ and $\omega_2 = 2$. In the natural mode with frequency ω_1, the two masses m_1 and m_2 move in the same direction with equal amplitudes of oscillation. In the natural mode with frequency ω_2 they move in opposite directions with the amplitude of oscillation of m_1 twice that of m_2.

5. The matrix $\mathbf{A} = \begin{bmatrix} -3 & 1 \\ 1 & -3 \end{bmatrix}$ has eigenvalues $\lambda_1 = -2$ and $\lambda_2 = -4$ with associated eigenvectors $\mathbf{v}_1 = \begin{bmatrix} 1 & 1 \end{bmatrix}^T$ and $\mathbf{v}_2 = \begin{bmatrix} 1 & -1 \end{bmatrix}^T$. Hence a general solution is given by

$$x_1(t) = a_1 \cos\left(\sqrt{2}t\right) + a_2 \sin\left(\sqrt{2}t\right) + b_1 \cos 2t + b_2 \sin 2t,$$
$$x_2(t) = a_1 \cos\left(\sqrt{2}t\right) + a_2 \sin\left(\sqrt{2}t\right) - b_1 \cos 2t - b_2 \sin 2t.$$

The natural frequencies are $\omega_1 = \sqrt{2}$ and $\omega_2 = 2$. In the natural mode with frequency ω_1, the two masses m_1 and m_2 move in the same direction with equal amplitudes of oscillation. At frequency ω_2 they move in opposite directions with equal amplitudes.

7. The matrix $\mathbf{A} = \begin{bmatrix} -10 & 6 \\ 6 & -10 \end{bmatrix}$ has eigenvalues $\lambda_1 = -4$ and $\lambda_2 = -16$ with associated eigenvectors $\mathbf{v}_1 = \begin{bmatrix} 1 & 1 \end{bmatrix}^T$ and $\mathbf{v}_2 = \begin{bmatrix} 1 & -1 \end{bmatrix}^T$. Hence a general solution is given by

$$x_1(t) = a_1 \cos 2t + a_2 \sin 2t + b_1 \cos 4t + b_2 \sin 4t,$$
$$x_2(t) = a_1 \cos 2t + a_2 \sin 2t - b_1 \cos 4t - b_2 \sin 4t.$$

The natural frequencies are $\omega_1 = 2$ and $\omega_2 = 4$. In the natural mode with frequency ω_1, the two masses m_1 and m_2 move in the same direction with equal amplitudes of oscillation. At frequency ω_2 they move in opposite directions with equal amplitudes.

9. Substitution of the trial solution $x_1 = c_1 \cos 3t$, $x_2 = c_2 \cos 3t$ in the system

$$x_1'' = -3x_1 + 2x_2 \quad 2x_2'' = 2x_1 - 4x_2 + 120\cos 3t$$

yields $c_1 = 3$ and $c_2 = -9$, so a general solution is given by

$$x_1(t) = a_1 \cos t + a_2 \sin t + 2b_1 \cos 2t + 2b_2 \sin 2t + 3\cos 3t,$$
$$x_2(t) = a_1 \cos t + a_2 \sin t - b_1 \cos 2t - b_2 \sin 2t - 9\cos 3t.$$

Imposition of the initial conditions $x_1(0) = x_2(0) = x_1'(0) = x_2'(0) = 0$ now yields $a_1 = 5$, $a_2 = 0$, $b_1 = -4$, and $b_2 = 0$. The resulting particular solution is

$$x_1(t) = 5\cos t - 8\cos 2t + 3\cos 3t,$$
$$x_2(t) = 5\cos t + 4\cos 2t - 9\cos 3t.$$

We have a superposition of three oscillations, in which the two masses move

- in the same direction with frequency $\omega_1 = 1$ and equal amplitudes;
- in opposite directions with frequency $\omega_2 = 2$ and with the amplitude of motion of m_1 being twice that of m_2;

- in opposite directions with frequency $\omega_3 = 3$ and with the amplitude of motion of m_2 being 3 times that of m_1.

11.   (a) The matrix $\mathbf{A} = \begin{bmatrix} -40 & 8 \\ 12 & -60 \end{bmatrix}$ has eigenvalues $\lambda_1 = -36$ and $\lambda_2 = -64$ with associated eigenvalues $\mathbf{v}_1 = \begin{bmatrix} 2 & 1 \end{bmatrix}^T$ and $\mathbf{v}_2 = \begin{bmatrix} 1 & -3 \end{bmatrix}^T$. Hence a general solution is given by

$$x(t) = 2a_1 \cos 6t + 2a_2 \sin 6t + b_1 \cos 8t + b_2 \sin 8t,$$
$$y(t) = a_1 \cos 6t + a_2 \sin 6t - 3b_1 \cos 8t - 3b_2 \sin 8t.$$

The natural frequencies are $\omega_1 = 6$ and $\omega_2 = 8$. In mode 1 the two masses oscillate in the same direction with frequency $\omega_1 = 6$ and with the amplitude of motion of m_1 being twice that of m_2. In mode 2 the two masses oscillate in opposite directions with frequency $\omega_2 = 8$ and with the amplitude of motion of m_2 being 3 times that of m_1.

(b)   Substitution of the trial solution $x = c_1 \cos 7t$, $y = c_2 \cos 7t$ in the system

$$x'' = -40x + 8y - 195\cos 7t, \quad y'' = 12x - 60y - 195\cos 7t$$

yields $c_1 = 19$ and $c_2 = 3$, so a general solution is given by

$$x(t) = 2a_1 \cos 6t + 2a_2 \sin 6t + b_1 \cos 8t + b_2 \sin 8t + 19\cos 7t,$$
$$y(t) = a_1 \cos 6t + a_2 \sin 6t - 3b_1 \cos 8t - 3b_2 \sin 8t + 3\cos 7t.$$

Imposition of the initial conditions $x(0) = 19$, $x'(0) = 12$, $y(0) = 3$, and $y'(0) = 6$ now yields $a_1 = 0$, $a_2 = 1$, $b_1 = 0$, and $b_2 = 0$. The resulting particular solution is

$$x(t) = 2\sin 6t + 19\cos 7t,$$
$$y(t) = \sin 6t + 3\cos 7t.$$

Thus the expected oscillation with frequency $\omega_2 = 8$ is missing, and we have a superposition of (only two) oscillations, in which the two masses move

- in the same direction with frequency $\omega_1 = 6$ and with the amplitude of motion of m_1 being twice that of m_2;

- in the same direction with frequency $\omega_3 = 7$ and with the amplitude of motion of m_1 being $\dfrac{19}{3}$ times that of m_2.

13. The coefficient matrix $\mathbf{A} = \begin{bmatrix} -4 & 2 & 0 \\ 2 & -4 & 2 \\ 0 & 2 & -4 \end{bmatrix}$ has characteristic polynomial

$$-\lambda^3 - 12\lambda^2 - 40\lambda - 32 = -(\lambda + 4)(\lambda^2 + 8\lambda + 8).$$

Its eigenvalues $\lambda_1 = -4$, $\lambda_2 = -4 - 2\sqrt{2}$, and $\lambda_3 = -4 + 2\sqrt{2}$ have associated eigenvectors $\mathbf{v}_1 = \begin{bmatrix} 1 & 0 & -1 \end{bmatrix}^T$, $\mathbf{v}_2 = \begin{bmatrix} 1 & -\sqrt{2} & 1 \end{bmatrix}^T$, and $\mathbf{v}_3 = \begin{bmatrix} 1 & \sqrt{2} & 1 \end{bmatrix}^T$. Hence the system's three natural modes of oscillation have

- Natural frequency $\omega_1 = 2$ with amplitude ratios $1 : 0 : -1$;

- Natural frequency $\omega_2 = \sqrt{4 + 2\sqrt{2}}$ with amplitude ratios $1 : -\sqrt{2} : 1$.

- Natural frequency $\omega_3 = \sqrt{4 - 2\sqrt{2}}$ with amplitude ratios $1 : \sqrt{2} : 1$.

15. First we need the general solution of the homogeneous system $\mathbf{x}'' = \mathbf{A}\mathbf{x}$ with

$$\mathbf{A} = \begin{bmatrix} -50 & 25/2 \\ 50 & -50 \end{bmatrix}.$$

The eigenvalues of A are $\lambda_1 = -25$ and $\lambda_2 = -75$, so the natural frequencies of the system are $\omega_1 = 5$ and $\omega_2 = 5\sqrt{3}$. The associated eigenvectors are $\mathbf{v}_1 = \begin{bmatrix} 1 & 2 \end{bmatrix}^T$ and $\mathbf{v}_2 = \begin{bmatrix} 1 & -2 \end{bmatrix}^T$, so the complementary solution $\mathbf{x}_c(t)$ is given by

$$x_1(t) = a_1 \cos 5t + a_2 \sin 5t + b_1 \cos\left(5\sqrt{3}t\right) + b_2 \sin\left(5\sqrt{3}t\right),$$

$$x_2(t) = 2a_1 \cos 5t + 2a_2 \sin 5t - 2b_1 \cos\left(5\sqrt{3}t\right) - 2b_2 \sin\left(5\sqrt{3}t\right).$$

When we substitute the trial solution $\mathbf{x}_p(t) = \begin{bmatrix} c_1 & c_2 \end{bmatrix}^T \cos 10t$ in the nonhomogeneous system, we find that $c_1 = \dfrac{4}{3}$ and $c_2 = -\dfrac{16}{3}$, so a particular solution $\mathbf{x}_p(t)$ is described by

$$x_1(t) = \frac{4}{3}\cos 10t, \quad x_2(t) = -\frac{16}{3}\cos 10t.$$

Finally, when we impose the zero initial conditions on the solution $\mathbf{x}(t) = \mathbf{x}_c(t) + \mathbf{x}_p(t)$ we find that $a_1 = \dfrac{2}{3}$, $a_2 = 0$, $b_1 = -2$, and $b_2 = 0$. Thus the solution we seek is described by

$$x_1(t) = \frac{2}{3}\cos 5t - 2\cos\left(5\sqrt{3}t\right) + \frac{4}{3}\cos 10t,$$

$$x_2(t) = \frac{4}{3}\cos 5t + 4\cos\left(5\sqrt{3}t\right) + \frac{16}{3}\cos 10t$$

We have a superposition of two oscillations with the natural frequencies $\omega_1 = 5$ and $\omega_2 = 5\sqrt{3}$ and a forced oscillation with frequency $\omega = 10$. In each of the two natural oscillations the amplitude of motion of m_2 is twice that of m_1, while in the forced oscillation the amplitude of motion of m_2 is four times that of m_1.

17.   With $c_1 = c_2 = 2$, it follows from Problem 16 that the natural frequencies and associated eigenvectors are $\omega_1 = 0$, $\mathbf{v}_1 = \begin{bmatrix} 1 & 1 \end{bmatrix}^T$ and $\omega_2 = 2$, $\mathbf{v}_2 = \begin{bmatrix} 1 & -1 \end{bmatrix}^T$. Hence Theorem 1 gives the general solution

$$x_1(t) = a_1 + b_1 t + a_2 \cos 2t + b_2 \sin 2t,$$

$$x_2(t) = a_1 + b_1 t - a_2 \cos 2t - b_2 \sin 2t.$$

The initial conditions $x_1'(0) = v_0$, $x_1(0) = x_2(0) = x_2'(0) = 0$ yield $a_1 = a_2 = 0$, $b_1 = \dfrac{v_0}{2}$, and $b_2 = \dfrac{v_0}{4}$, so

$$x_1(t) = \frac{v_0}{4}\left(2t + \sin 2t\right), \quad x_2(t) = \frac{v_0}{4}\left(2t - \sin 2t\right),$$

while $x_2 - x_1 = \dfrac{v_0}{4}\left(-2\sin 2t\right) < 0$; that is, until $t = \dfrac{\pi}{2}$. Finally, $x_1'\left(\dfrac{\pi}{2}\right) = 0$ and

$$x_2'\left(\frac{\pi}{2}\right) = v_0.$$

19.   With $c_1 = 1$ and $c_2 = 3$, it follows from Problem 16 that the natural frequencies and associated eigenvectors are $\omega_1 = 0$, $\mathbf{v}_1 = \begin{bmatrix} 1 & 1 \end{bmatrix}^T$ and $\omega_2 = 2$, $\mathbf{v}_2 = \begin{bmatrix} 1 & -3 \end{bmatrix}^T$. Hence Theorem 1 gives the general solution

$$x_1(t) = a_1 + b_1 t + a_2 \cos 2t + b_2 \sin 2t,$$

$$x_2(t) = a_1 + b_1 t - 3a_2 \cos 2t - 3b_2 \sin 2t.$$

The initial conditions $x_1'(0) = v_0$, $x_1(0) = x_2(0) = x_2'(0) = 0$ yield $a_1 = a_2 = 0$, $b_1 = \dfrac{3v_0}{4}$, and $b_2 = \dfrac{v_0}{8}$, so

$$x_1(t) = \frac{v_0}{8}\left(6t + \sin 2t\right), \quad x_2(t) = \frac{v_0}{8}\left(6t - 3\sin 2t\right),$$

while $x_2 - x_1 = \dfrac{v_0}{8}\left(-4\sin 2t\right) < 0$; that is, until $t = \dfrac{\pi}{2}$. Finally, $x_1'\left(\dfrac{\pi}{2}\right) = \dfrac{v_0}{2}$ and

$x_2'\left(\dfrac{\pi}{2}\right) = \dfrac{3v_0}{2}$.

The method of solution in each of Problems 20–23 is the same as that in Example 2 in this section. Thus, looking at the equations in (26), we need to solve the equations

$$
\begin{array}{rcrcrcl}
b_1 & + & 2b_2 & + & 4b_3 & = & x_1'(0) \\
b_1 & & & - & 12b_3 & = & x_2'(0) \\
b_1 & - & 2b_2 & + & 4b_3 & = & x_3'(0)
\end{array}
$$

for the coefficients b_1, b_2, b_3 after inserting given initial values $x_1'(0)$, $x_2'(0)$, $x_3'(0)$ of the three railway cars.

21.   With $x_1'(0) = 2v_0$, $x_2'(0) = 0$, and $x_3'(0) = -v_0$, substitution of the resulting coefficient values b_1, b_2, b_3 in (25) gives the railway car displacement functions

$$x_1(t) = \frac{1}{32}v_0\left(12t + 24\sin 2t + \sin 4t\right),$$

$$x_2(t) = \frac{1}{32}v_0\left(12t \qquad\qquad - 3\sin 4t\right),$$

$$x_3(t) = \frac{1}{32}v_0\left(12t - 24\sin 2t + \sin 4t\right).$$

We then see (substituting $\sin 4t = 2\sin 2t\cos 2t$) that

$$x_2(t) - x_1(t) = -\frac{1}{8}v_0\left(\sin 4t + 6\sin 2t\right) = -\frac{1}{4}v_0\left(\sin 2t\right)\left(\cos 2t + 3\right)$$

remains negative until $t = \dfrac{\pi}{2}$ (as does $x_3(t) - x_2(t)$, similarly) at which time the cars separate with velocities

$$x_1'\left(\frac{\pi}{2}\right) = -v_0, \quad x_2'\left(\frac{\pi}{2}\right) = 0, \quad x_3'\left(\frac{\pi}{2}\right) = 2v_0\,.$$

Thus the car in the center remains fixed thereafter, whereas the first and third cars rebound in opposite directions, having exchanged their original velocities.

23.   With $x_1'(0) = 3v_0$, $x_2'(0) = 2v_0$, and $x_3'(0) = 2v_0$, substitution of the resulting coefficient values b_1, b_2, b_3 in (25) gives the railway car displacement functions

$$x_1(t) = \frac{1}{32}v_0\left(76t + 8\sin 2t + \sin 4t\right),$$

$$x_2(t) = \frac{1}{32}v_0\left(76t \qquad\quad - 3\sin 4t\right),$$

$$x_3(t) = \frac{1}{32}v_0\left(76t - 8\sin 2t + \sin 4t\right).$$

We then see (substituting $\sin 4t = 2\sin 2t \cos 2t$) that

$$x_2(t) - x_1(t) = -\frac{1}{8}v_0\left(2\sin 2t + \sin 4t\right) = -\frac{1}{4}v_0\left(\sin 2t\right)\left(1 + \cos 2t\right)$$

remains negative until $t = \dfrac{\pi}{2}$ (as does $x_3(t) - x_2(t)$, similarly) at which time the cars separate with velocities

$$x_1'\left(\frac{\pi}{2}\right) = 2v_0, \quad x_2'\left(\frac{\pi}{2}\right) = 2v_0, \quad x_3'\left(\frac{\pi}{2}\right) = 3v_0.$$

Thus the car in the center proceeds thereafter with the same velocity it had originally, whereas the first and third cars rebound in opposite directions, having exchanged their original velocities.

25.   **(a)** The matrix

$$\mathbf{A} = \begin{bmatrix} -160/3 & 320/3 \\ 8 & -116 \end{bmatrix}$$

has eigenvalues $\lambda_1 = -41.8285$ and $\lambda_2 = -127.5049$, so the natural frequencies are

$$\omega_1 \approx 6.4675\,\text{rad/sec} \approx 1.0293\,\text{Hz}, \quad \omega_2 \approx 11.2918\,\text{rad/sec} \approx 1.7971\,\text{Hz}.$$

(b) Resonance occurs at the two critical speeds

$$v_1 = \frac{20\omega_1}{\pi} \approx 41\,\text{ft/sec} \approx 28\,\text{mi/h}, \quad v_2 = \frac{20\omega_2}{\pi} \approx 72\,\text{ft/sec} \approx 49\,\text{mi/h}.$$

27. $100x'' = -4000x$, $800\theta'' = 100000\theta$.

Obviously the matrix $\mathbf{A} = \begin{bmatrix} -40 & 0 \\ 0 & -125 \end{bmatrix}$ has eigenvalues $\lambda_1 = -40$ and $\lambda_2 = -125$.

Up-and-down: $\omega_1 = \sqrt{40}$, $v_1 \approx 40.26\,\text{ft/sec} \approx 27\,\text{mi/h}$;

Angular: $\omega_2 = \sqrt{125}$, $v_2 \approx 71.18\,\text{ft/sec} \approx 49\,\text{mi/h}$.

29. $100x'' = -3000x - 5000\theta$, $800\theta'' = -5000x - 75000\theta$.

The matrix $\mathbf{A} = \begin{bmatrix} -30 & -50 \\ -25/4 & -375/4 \end{bmatrix}$ has eigenvalues $\lambda_1, \lambda_2 = \dfrac{5}{8}\left(-99 \pm \sqrt{3401}\right)$.

$\omega_1 \approx 5.0424$, $v_1 \approx 32.10\,\text{ft/sec} \approx 22\,\text{mi/h}$;

$\omega_2 \approx 9.9158$, $v_2 \approx 63.13\,\text{ft/sec} \approx 43\,\text{mi/h}$.

SECTION 7.6

MULTIPLE EIGENVALUE SOLUTIONS

In each of Problems 1–6 we give first the characteristic equation with repeated (multiplicity 2) eigenvalue λ. In each case we find that $(\mathbf{A} - \lambda\mathbf{I})^2 = \mathbf{0}$. Then $\mathbf{w} = \begin{bmatrix} 1 & 0 \end{bmatrix}^{\mathsf{T}}$ is a generalized eigenvector and $\mathbf{v} = (\mathbf{A} - \lambda\mathbf{I})\mathbf{w} \neq \mathbf{0}$ is an ordinary eigenvector associated with λ. We give finally the scalar component functions $x_1(t)$, $x_2(t)$ of the general solution

$$\mathbf{x}(t) = c_1\mathbf{v}e^{\lambda t} + c_2(\mathbf{v}t + \mathbf{w})e^{\lambda t}$$

of the given system $\mathbf{x}' = \mathbf{A}\mathbf{x}$.

1.   Characteristic equation $\lambda^2 + 6\lambda + 9 = 0$
          Repeated eigenvalue $\lambda = -3$
          Generalized eigenvector $\mathbf{w} = \begin{bmatrix} 1 & 0 \end{bmatrix}^{\mathsf{T}}$
          $\mathbf{v} = (\mathbf{A} - \lambda\mathbf{I})\mathbf{w} = \begin{bmatrix} 1 & 1 \\ -1 & -1 \end{bmatrix}\begin{bmatrix} 1 \\ 0 \end{bmatrix} = \begin{bmatrix} 1 \\ -1 \end{bmatrix}$

          $x_1(t) = (c_1 + c_2 + c_2 t)e^{-3t}$

          $x_2(t) = (-c_1 \quad - c_2 t)e^{-3t}$.

The left-hand figure below shows a direction field and typical solution curves.

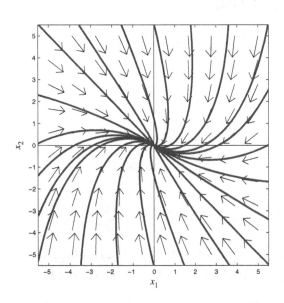

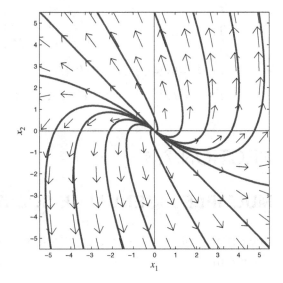

3.   Characteristic equation $\lambda^2 - 6\lambda + 9 = 0$
Repeated eigenvalue $\lambda = 3$
Generalized eigenvector $\mathbf{w} = \begin{bmatrix} 1 & 0 \end{bmatrix}^T$

$$\mathbf{v} = (\mathbf{A} - \lambda\mathbf{I})\mathbf{w} = \begin{bmatrix} -2 & -2 \\ 2 & 2 \end{bmatrix}\begin{bmatrix} 1 \\ 0 \end{bmatrix} = \begin{bmatrix} -2 \\ 2 \end{bmatrix}$$

$$x_1(t) = (-2c_1 + c_2 - 2c_2t)e^{3t}$$

$$x_2(t) = (\ 2c_1 + \qquad 2c_2t)e^{3t}.$$

The right-hand figure above shows a direction field and typical solution curves.

5.   Characteristic equation $\lambda^2 - 10\lambda + 25 = 0$
Repeated eigenvalue $\lambda = 5$
Generalized eigenvector $\mathbf{w} = \begin{bmatrix} 1 & 0 \end{bmatrix}^T$

$$\mathbf{v} = (\mathbf{A} - \lambda\mathbf{I})\mathbf{w} = \begin{bmatrix} 2 & 1 \\ -4 & -2 \end{bmatrix}\begin{bmatrix} 1 \\ 0 \end{bmatrix} = \begin{bmatrix} 2 \\ -4 \end{bmatrix}$$

$$x_1(t) = (\ 2c_1 + c_2 + 2c_2t)e^{5t}$$

$$x_2(t) = (-4c_1 \qquad - 4c_2t)e^{5t}.$$

The figure below shows a direction field and typical solution curves.

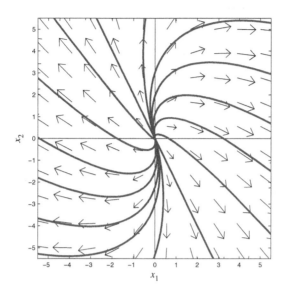

In each of Problems 7–10 the characteristic polynomial is easily calculated by expansion along the row or column of **A** that contains two zeros. The matrix **A** has only two distinct eigenvalues, so we write $\lambda_1, \lambda_2, \lambda_3$ with either $\lambda_1 = \lambda_2$ or $\lambda_2 = \lambda_3$. Nevertheless, we find that it has 3 linearly independent eigenvectors $\mathbf{v}_1, \mathbf{v}_2$, and $\mathbf{v}_3$. We list also the scalar components $x_1(t), x_2(t), x_3(t)$ of the general solution $\mathbf{x}(t) = c_1\mathbf{v}_1e^{\lambda_1 t} + c_2\mathbf{v}_2e^{\lambda_2 t} + c_3\mathbf{v}_3e^{\lambda_3 t}$ of the system.

7.   Characteristic equation $-\lambda^3 + 13\lambda^2 - 40\lambda + 36 = -(\lambda - 2)^2(\lambda - 9)$

Eigenvalues $\lambda = 2, 2, 9$

Eigenvectors $[1 \quad 1 \quad 0]^{\mathrm{T}}, [1 \quad 0 \quad 1]^{\mathrm{T}}, [0 \quad 1 \quad 0]^{\mathrm{T}}$

$x_1(t) = c_1e^{2t} + c_2e^{2t}$

$x_2(t) = c_1e^{2t} \qquad + c_3e^{9t}$

$x_3(t) = \qquad c_2e^{2t}$

9.   Characteristic equation $-\lambda^3 + 19\lambda^2 - 115\lambda + 225 = -(\lambda - 5)^2(\lambda - 9)$

Eigenvalues $\lambda = 5, 5, 9$

Eigenvectors $[1 \quad 2 \quad 0]^{\mathrm{T}}, [7 \quad 0 \quad 2]^{\mathrm{T}}, [3 \quad 0 \quad 1]^{\mathrm{T}}$

$x_1(t) = c_1e^{5t} + 7c_2e^{5t} + 3c_3e^{9t}$

$x_2(t) = 2c_1e^{5t}$

$x_3(t) = \qquad 2c_2e^{5t} + c_3e^{9t}$

In each of Problems 11–14, the characteristic equation is $-\lambda^3 - 3\lambda^2 - 3\lambda - 1 = -(\lambda+1)^3$.
Hence $\lambda = -1$ is a triple eigenvalue of defect 2, and we find that $(\mathbf{A} - \lambda\mathbf{I})^3 = \mathbf{0}$. In each
problem we start with $\mathbf{v}_3 = [1 \ \ 0 \ \ 0]^T$ and then calculate $\mathbf{v}_2 = (\mathbf{A} - \lambda\mathbf{I})\mathbf{v}_3$ and
$\mathbf{v}_1 = (\mathbf{A} - \lambda\mathbf{I})\mathbf{v}_2 \neq \mathbf{0}$. It follows that $(\mathbf{A} - \lambda\mathbf{I})\mathbf{v}_1 = (\mathbf{A} - \lambda\mathbf{I})^2\mathbf{v}_2 = (\mathbf{A} - \lambda\mathbf{I})^3\mathbf{v}_3 = \mathbf{0}$, so the vector
$\mathbf{v}_1$ (if nonzero) is an ordinary eigenvector associated with the triple eigenvalue λ. Hence
$\{\mathbf{v}_1, \mathbf{v}_2, \mathbf{v}_3\}$ is a length 3 chain of generalized eigenvectors, and the corresponding general
solution is described by

$$\mathbf{x}(t) = e^{-t}[c_1\mathbf{v}_1 + c_2(\mathbf{v}_1 t + \mathbf{v}_2) + c_3(\mathbf{v}_1 t^2/2 + \mathbf{v}_2 t + \mathbf{v}_3)].$$

We give the scalar components $x_1(t)$, $x_2(t)$, $x_3(t)$ of $\mathbf{x}(t)$.

11. $\mathbf{v}_1 = [0 \ \ 1 \ \ 0]^T$, $\mathbf{v}_2 = [-2 \ \ -1 \ \ 1]^T$, $\mathbf{v}_3 = [1 \ \ 0 \ \ 0]^T$

$x_1(t) = e^{-t}(-2c_2 + c_3 - 2c_3 t)$

$x_2(t) = e^{-t}(c_1 - c_2 + c_2 t - c_3 t + c_3 t^2/2)$

$x_3(t) = e^{-t}(c_2 + c_3 t)$

13.   Here we are stymied initially, because if $\mathbf{v}_3 = [1 \ \ 0 \ \ 0]^T$ then $(\mathbf{A} - \lambda\mathbf{I})\mathbf{v}_3 = \mathbf{0}$ does not
qualify as a (nonzero) generalized eigenvector. We there make a fresh start with
$\mathbf{v}_3 = [0 \ \ 1 \ \ 0]^T$, and now we get the desired nonzero generalized eigenvectors upon
successive multiplication by $\mathbf{A} - \lambda\mathbf{I}$.

$\mathbf{v}_1 = [1 \ \ 0 \ \ 0]^T$, $\mathbf{v}_2 = [0 \ \ 2 \ \ 1]^T$, $\mathbf{v}_3 = [0 \ \ 1 \ \ 0]^T$

$x_1(t) = e^{-t}(c_1 + c_2 t + c_3 t^2/2)$

$x_2(t) = e^{-t}(2c_2 + c_3 + 2c_3 t)$

$x_3(t) = e^{-t}(c_2 + c_3 t)$

In each of Problems 15–18, the characteristic equation is $-\lambda^3 + 3\lambda^2 - 3\lambda + 1 = -(\lambda-1)^3$.
Hence $\lambda = 1$ is a triple eigenvalue of defect 1, and we find that $(\mathbf{A} - \lambda\mathbf{I})^2 = \mathbf{0}$. First we find
the two linearly independent (ordinary) eigenvectors $\mathbf{u}_1$ and $\mathbf{u}_2$ associated with λ. Then we
start with $\mathbf{v}_2 = [1 \ \ 0 \ \ 0]^T$ and calculate $\mathbf{v}_1 = (\mathbf{A} - \lambda\mathbf{I})\mathbf{v}_2 \neq \mathbf{0}$. It follows that
$(\mathbf{A} - \lambda\mathbf{I})\mathbf{v}_1 = (\mathbf{A} - \lambda\mathbf{I})^2\mathbf{v}_2 = \mathbf{0}$, so $\mathbf{v}_1$ is an ordinary eigenvector associated with λ. However,
$\mathbf{v}_1$ is a linear combination of $\mathbf{u}_1$ and $\mathbf{u}_2$, so $\mathbf{v}_1 e^t$ is a linear combination of the independent
solutions $\mathbf{u}_1 e^t$ and $\mathbf{u}_2 e^t$. But $\{\mathbf{v}_1, \mathbf{v}_2\}$ is a length 2 chain of generalized eigenvectors
associated with λ, so $(\mathbf{v}_1 t + \mathbf{v}_2)e^t$ is the desired third independent solution. The corresponding
general solution is described by

$$\mathbf{x}(t) = e^t [c_1\mathbf{u}_1 + c_2\mathbf{u}_2 + c_3(\mathbf{v}_1\, t + \mathbf{v}_2)]$$

We give the scalar components $x_1(t), x_2(t), x_3(t)$ of $\mathbf{x}(t)$.

15. $\mathbf{u}_1 = [3 \ \ -1 \ \ 0]^T$ $\mathbf{u}_2 = [0 \ \ 0 \ \ 1]^T$

$\mathbf{v}_1 = [-3 \ \ 1 \ \ 1]^T$ $\mathbf{v}_2 = [1 \ \ 0 \ \ 0]^T$

$x_1(t) = e^t(3c_1 + c_3 - 3c_3\, t)$

$x_2(t) = e^t(-c_1 + c_3\, t)$

$x_3(t) = e^t(c_2 + c_3\, t)$

17. $\mathbf{u}_1 = [2 \ \ 0 \ \ -9]^T$ $\mathbf{u}_2 = [1 \ \ -3 \ \ 0]^T$

$\mathbf{v}_1 = [0 \ \ 6 \ \ -9]^T$ $\mathbf{v}_2 = [0 \ \ 1 \ \ 0]^T$

(Either $\mathbf{v}_2 = [1 \ \ 0 \ \ 0]^T$ or $\mathbf{v}_2 = [0 \ \ 0 \ \ 1]^T$ can be used also, but they yield different forms of the solution than given in the book's answer section.)

$x_1(t) = e^t(2c_1 + c_2)$

$x_2(t) = e^t(-3c_2 + c_3 + 6c_3\, t)$

$x_3(t) = e^t(-9c_1 - 9c_3\, t)$

19.   Characteristic equation $\lambda^4 - 2\lambda^2 + 1 = 0$

Double eigenvalue $\lambda = -1$ with eigenvectors

$$\mathbf{v}_1 = [1 \ \ 0 \ \ 0 \ \ 1]^T \ \ \text{and} \ \ \mathbf{v}_2 = [0 \ \ 0 \ \ 1 \ \ 0]^T.$$

Double eigenvalue $\lambda = +1$ with eigenvectors

$$\mathbf{v}_3 = [0 \ \ 1 \ \ 0 \ \ -2]^T \ \ \text{and} \ \ \mathbf{v}_4 = [1 \ \ 0 \ \ 3 \ \ 0]^T.$$

General solution

$$\mathbf{x}(t) = e^{-t}(c_1\mathbf{v}_1 + c_2\mathbf{v}_2) + e^t(c_3\mathbf{v}_3 + c_4\mathbf{v}_4)$$

Scalar components

$$x_1(t) = c_1 e^{-t} + c_4 e^t$$

$$x_2(t) = c_3 e^t$$

$$x_3(t) = c_2 e^{-t} + 3c_4 e^t$$

$$x_4(t) = c_1 e^{-t} - 2c_3 e^t$$

21.   Characteristic equation $\lambda^4 - 4\lambda^3 + 6\lambda^2 - 4\lambda + 1 = (\lambda - 1)^4 = 0$
Eigenvalue $\lambda = 1$ with multiplicity 4 and defect 2.

We find that $(\mathbf{A} - \lambda\mathbf{I})^2 \neq 0$ but $(\mathbf{A} - \lambda\mathbf{I})^3 = 0$. We therefore start with
$\mathbf{v}_3 = [1 \ \ 0 \ \ 0 \ \ 0]^T$ and define $\mathbf{v}_2 = (\mathbf{A} - \lambda\mathbf{I})\mathbf{v}_3$ and $\mathbf{v}_1 = (\mathbf{A} - \lambda\mathbf{I})\mathbf{v}_2 \neq 0$, thereby
obtaining the length 3 chain $\{\mathbf{v}_1, \ \mathbf{v}_2, \ \mathbf{v}_3\}$ with

$$\mathbf{v}_1 = [0 \ \ 0 \ \ 0 \ \ 1]^T, \quad \mathbf{v}_2 = [-2 \ \ 1 \ \ 1 \ \ 0]^T, \quad \mathbf{v}_3 = [1 \ \ 0 \ \ 0 \ \ 0]^T.$$

Then we find the second ordinary eigenvector $\mathbf{v}_4 = [0 \ \ \ 0 \ \ 1 \ \ 0]^T$. The corresponding
general solution
$$\mathbf{x}(t) = e^t \left[c_1\mathbf{v}_1 + c_2(\mathbf{v}_1 t + \mathbf{v}_2) + c_3(\mathbf{v}_1 t^2/2 + \mathbf{v}_2 t + \mathbf{v}_3) + c_4\mathbf{v}_4 \right]$$

has scalar components

$$x_1(t) = e^t(-2c_2 + c_3 - 2c_3 t)$$

$$x_2(t) = e^t(c_2 + c_3 t)$$

$$x_3(t) = e^t(c_2 + c_4 + c_3 t).$$

$$x_4(t) = e^t(c_1 + c_2 t + c_3 t^2/2.)$$

23. $\lambda_1 = -1:$ $\{\mathbf{v}_1\}$ with $\mathbf{v}_1 = [1 \ \ -1 \ \ 2]^T$
    $\lambda_2 = 3:$ $\{\mathbf{v}_2\}$ with $\mathbf{v}_2 = [4 \ \ \ 0 \ \ 9]^T$ and
        $\{\mathbf{v}_3\}$ with $\mathbf{v}_3 = [0 \ \ \ 2 \ \ 1]^T$

Scalar components

$$x_1(t) = \quad c_1 e^{-t} + 4c_2 e^{3t}$$

$$x_2(t) = -c_1 e^{-t} \qquad\quad + 2c_3 e^{3t}$$

$$x_3(t) = 2c_1 e^{-t} + 9c_2 e^{3t} + \quad c_3 e^{3t}$$

In Problems 25, 26, and 28 there is given a single eigenvalue λ of multiplicity 3. We find that
$(\mathbf{A} - \lambda\mathbf{I})^2 \neq 0$ but $(\mathbf{A} - \lambda\mathbf{I})^3 = 0$. We therefore start with $\mathbf{v}_3 = [1 \ \ 0 \ \ 0]^T$ and define
$\mathbf{v}_2 = (\mathbf{A} - \lambda\mathbf{I})\mathbf{v}_3$ and $\mathbf{v}_1 = (\mathbf{A} - \lambda\mathbf{I})\mathbf{v}_2 \neq 0$, thereby obtaining the length 3 chain $\{\mathbf{v}_1, \ \mathbf{v}_2, \ \mathbf{v}_3\}$ of
generalized eigenvectors based on the ordinary eigenvector $\mathbf{v}_1$. We list the scalar components of
the corresponding general solution

$$\mathbf{x}(t) = c_1\mathbf{v}_1 e^{\lambda t} + c_2(\mathbf{v}_1 t + \mathbf{v}_2)e^{\lambda t} + c_3(\mathbf{v}_1 t^2/2 + \mathbf{v}_2 t + \mathbf{v}_3)e^{\lambda t}.$$

25. $\{\mathbf{v}_1, \mathbf{v}_2, \mathbf{v}_3\}$ with

$$\mathbf{v}_1 = [-1 \quad 0 \quad -1]^{\mathrm{T}}, \quad \mathbf{v}_2 = [-4 \quad -1 \quad 0]^{\mathrm{T}}, \quad \mathbf{v}_3 = [1 \quad 0 \quad 0]^{\mathrm{T}}$$

Scalar components

$$x_1(t) = e^{2t}(-c_1 - 4c_2 + c_3 - c_2 t - 4c_3 t - c_3 t^2 / 2)$$
$$x_2(t) = e^{2t}(-c_2 - c_3 t)$$
$$x_3(t) = e^{2t}(-c_1 - c_2 t - c_3 t^2 / 2)$$

27.   We find that the triple eigenvalue $\lambda = 2$ has the two linearly independent eigenvectors $[1 \quad 1 \quad 0]^{\mathrm{T}}$ and $[-1 \quad 0 \quad 1]^{\mathrm{T}}$. Next we find that $(\mathbf{A} - \lambda \mathbf{I}) \neq \mathbf{0}$ but $(\mathbf{A} - \lambda \mathbf{I})^2 = \mathbf{0}$. We therefore start with $\mathbf{v}_2 = [1 \quad 0 \quad 0]^{\mathrm{T}}$ and define

$$\mathbf{v}_1 = (\mathbf{A} - \lambda \mathbf{I})\mathbf{v}_2 = [-5 \quad 3 \quad 8]^{\mathrm{T}} \neq \mathbf{0},$$

thereby obtaining the length 2 chain $\{\mathbf{v}_1, \mathbf{v}_2\}$ of generalized eigenvectors based on the ordinary eigenvector $\mathbf{v}_1$. If we take $\mathbf{v}_3 = [1 \quad 1 \quad 0]^{\mathrm{T}}$, then the general solution $\mathbf{x}(t) = e^{2t}[c_1 \mathbf{v}_1 + c_2(\mathbf{v}_1 t + \mathbf{v}_2) + c_3 \mathbf{v}_3]$ has scalar components

$$x_1(t) = e^{2t}(-5c_1 + c_2 + c_3 - 5c_2 t)$$
$$x_2(t) = e^{2t}(3c_1 + 3c_2 t)$$
$$x_3(t) = e^{2t}(8c_1 + 8c_2 t).$$

In Problems 29 and 30 the matrix $\mathbf{A}$ has two distinct eigenvalues λ_1 and λ_2 each having multiplicity 2 and defect 1. First, we select $\mathbf{v}_2$ so that $\mathbf{v}_1 = (\mathbf{A} - \lambda_1 \mathbf{I})\mathbf{v}_2 \neq \mathbf{0}$ but $(\mathbf{A} - \lambda_1 \mathbf{I})\mathbf{v}_1 = \mathbf{0}$, so $\{\mathbf{v}_1, \mathbf{v}_2\}$ is a length 2 chain based on $\mathbf{v}_1$. Next, we select $\mathbf{u}_2$ so that $\mathbf{u}_1 = (\mathbf{A} - \lambda_1 \mathbf{I})\mathbf{u}_2 \neq \mathbf{0}$ but $(\mathbf{A} - \lambda_1 \mathbf{I})\mathbf{u}_1 = \mathbf{0}$, so $\{\mathbf{u}_1, \mathbf{u}_2\}$ is a length 2 chain based on $\mathbf{u}_1$. We give the scalar components of the corresponding general solution

$$\mathbf{x}(t) = e^{\lambda_1 t}[c_1 \mathbf{v}_1 + c_2(\mathbf{v}_1 t + \mathbf{v}_2)] + e^{\lambda_2 t}[c_3 \mathbf{u}_1 + c_4(\mathbf{u}_1 t + \mathbf{u}_2)].$$

29. $\lambda = -1$: $\{\mathbf{v}_1, \mathbf{v}_2\}$ with $\mathbf{v}_1 = [1 \quad -3 \quad -1 \quad -2]^{\mathrm{T}}$ and $\mathbf{v}_2 = [0 \quad 1 \quad 0 \quad 0]^{\mathrm{T}}$,
        $\lambda = 2$: $\{\mathbf{u}_1, \mathbf{u}_2\}$ with $\mathbf{u}_1 = [0 \quad -1 \quad 1 \quad 0]^{\mathrm{T}}$ and $\mathbf{u}_2 = [0 \quad 0 \quad 2 \quad 1]^{\mathrm{T}}$

Scalar components

$$x_1(t) = e^{-t}(c_1 + c_2 t)$$
$$x_2(t) = e^{-t}(-3c_1 + c_2 - 3c_2 t) + e^{2t}(-c_3 - c_4 t)$$
$$x_3(t) = e^{-t}(-c_1 - c_2 t) + e^{2t}(c_3 + 2c_4 + c_4 t)$$
$$x_4(t) = e^{-t}(-2c_1 - 2c_2 t) + e^{2t}(c_4)$$

31. We have the single eigenvalue $\lambda = 1$ of multiplicity 4. Starting with $\mathbf{v}_3 = [1\ \ 0\ \ 0\ \ 0]^T$, we calculate $\mathbf{v}_2 = (\mathbf{A} - \lambda \mathbf{I})\mathbf{v}_3$ and $\mathbf{v}_1 = (\mathbf{A} - \lambda \mathbf{I})\mathbf{v}_2 \neq \mathbf{0}$, and find that $(\mathbf{A} - \lambda \mathbf{I})\mathbf{v}_1 = \mathbf{0}$. Therefore $\{\mathbf{v}_1, \mathbf{v}_2, \mathbf{v}_3\}$ is a length 3 chain based on the ordinary eigenvector $\mathbf{v}_1$. Next, the eigenvector equation $(\mathbf{A} - \lambda \mathbf{I})\mathbf{v} = \mathbf{0}$ yields the second linearly independent eigenvector $\mathbf{v}_4 = [0\ \ 1\ \ 3\ \ 0]^T$. With

$$\mathbf{v}_1 = [42\ \ 7\ \ {-21}\ \ {-42}]^T, \qquad \mathbf{v}_2 = [34\ \ 22\ \ {-10}\ \ {-27}]^T,$$

$$\mathbf{v}_3 = [1\ \ 0\ \ 0\ \ 0]^T \quad \text{and} \quad \mathbf{v}_4 = [0\ \ 1\ \ 3\ \ 0]$$

the general solution

$$\mathbf{x}(t) = e^t \left[c_1\mathbf{v}_1 + c_2(\mathbf{v}_1 t + \mathbf{v}_2) + c_3(\mathbf{v}_1 t^2/2 + \mathbf{v}_2 t + \mathbf{v}_3) + c_4\mathbf{v}_4 \right]$$

has scalar components

$$x_1(t) = e^t(42c_1 + 34c_2 + c_3 + 42c_2 t + 34c_3 t + 21c_3 t^2)$$

$$x_2(t) = e^t(7c_1 + 22c_2 + c_4 + 7c_2 t + 22c_3 t + 7c_3 t^2/2)$$

$$x_3(t) = e^t(-21c_1 - 10c_2 + 3c_4 - 21c_2 t - 10c_3 t - 21c_3 t^2/2)$$

$$x_4(t) = e^t(-42c_1 - 27c_2 - 42c_2 t - 27c_3 t - 21c_3 t^2).$$

33. The chain $\{\mathbf{v}_1, \mathbf{v}_2\}$ was found using the matrices

$$\mathbf{A} - \lambda \mathbf{I} = \begin{bmatrix} 4i & -4 & 1 & 0 \\ 4 & 4i & 0 & 1 \\ 0 & 0 & 4i & -4 \\ 0 & 0 & 4 & 4i \end{bmatrix} \rightarrow \begin{bmatrix} 1 & i & 0 & 0 \\ 0 & 0 & 1 & 0 \\ 0 & 0 & 0 & 1 \\ 0 & 0 & 0 & 0 \end{bmatrix}$$

and

$$(\mathbf{A} - \lambda \mathbf{I})^2 = \begin{bmatrix} -32 & -32i & 8i & -8 \\ 32i & -32 & 8 & 8i \\ 0 & 0 & -32 & -32i \\ 0 & 0 & 32i & -32 \end{bmatrix} \rightarrow \begin{bmatrix} 1 & i & 0 & 0 \\ 0 & 0 & 1 & i \\ 0 & 0 & 0 & 0 \\ 0 & 0 & 0 & 0 \end{bmatrix}$$

where $\rightarrow$ signifies reduction to row-echelon form. The resulting real-valued solution vectors are

$$\mathbf{x}_1(t) = e^{3t} \begin{bmatrix} \cos 4t & \sin 4t & 0 & 0 \end{bmatrix}^T$$

$$\mathbf{x}_2(t) = e^{3t} \begin{bmatrix} -\sin 4t & \cos 4t & 0 & 0 \end{bmatrix}^T$$

$$\mathbf{x}_3(t) = e^{3t} \begin{bmatrix} t\cos 4t & t\sin 4t & \cos 4t & \sin 4t \end{bmatrix}^T$$

$$\mathbf{x}_4(t) = e^{3t} \begin{bmatrix} -t\sin 4t & t\cos 4t & -\sin 4t & \cos 4t \end{bmatrix}^T.$$

35. The coefficient matrix

$$\mathbf{A} = \begin{bmatrix} 0 & 0 & 1 & 0 \\ 0 & 0 & 0 & 1 \\ -1 & 1 & -2 & 1 \\ 1 & -1 & 1 & -2 \end{bmatrix}$$

has eigenvalues

$\lambda = 0$ with eigenvector $\mathbf{v}_1 = \begin{bmatrix} 1 & 1 & 0 & 0 \end{bmatrix}^T$
$\lambda = -1$ with eigenvectors $\mathbf{v}_2 = \begin{bmatrix} 1 & 0 & -1 & 0 \end{bmatrix}^T$ and $\mathbf{v}_3 = \begin{bmatrix} 0 & 1 & 0 & -1 \end{bmatrix}^T$,
$\lambda = -2$ with eigenvector $\mathbf{v}_4 = \begin{bmatrix} 1 & -1 & -2 & 2 \end{bmatrix}^T.$

When we impose the given initial conditions on the general solution

$$\mathbf{x}(t) = c_1\mathbf{v}_1 + c_2\mathbf{v}_2 e^{-t} + c_3\mathbf{v}_3 e^{-t} + c_4\mathbf{v}_4 e^{-2t}$$

we find that $c_1 = v_0$, $c_2 = c_3 = -v_0$, $c_4 = 0$. Hence the position functions of the two masses are given by

$$x_1(t) = x_2(t) = v_0(1 - e^{-t}).$$

Each mass travels a distance v_0 before stopping.

In Problems 37–46 we use the eigenvectors and generalized eigenvectors found in Problems 23–32 to construct a matrix $\mathbf{Q}$ such that $\mathbf{J} = \mathbf{Q}^{-1}\mathbf{A}\mathbf{Q}$ is a Jordan normal form of the given matrix $\mathbf{A}$.

37. $\mathbf{v}_1 = \begin{bmatrix} 1 & -1 & 2 \end{bmatrix}^T,$ $\mathbf{v}_2 = \begin{bmatrix} 4 & 0 & 9 \end{bmatrix}^T,$ $\mathbf{v}_3 = \begin{bmatrix} 0 & 2 & 1 \end{bmatrix}^T$

$$\mathbf{Q} = \begin{bmatrix} \mathbf{v}_1 & \mathbf{v}_2 & \mathbf{v}_3 \end{bmatrix} = \begin{bmatrix} 1 & 4 & 0 \\ -1 & 0 & 2 \\ 2 & 9 & 1 \end{bmatrix}$$

$$\mathbf{J} = \mathbf{Q}^{-1}\mathbf{A}\mathbf{Q} = \frac{1}{2}\begin{bmatrix} -18 & -4 & 8 \\ 5 & 1 & -2 \\ -9 & -1 & 4 \end{bmatrix}\begin{bmatrix} 39 & 8 & -16 \\ -36 & -5 & 16 \\ 72 & 16 & -29 \end{bmatrix}\begin{bmatrix} 1 & 4 & 0 \\ -1 & 0 & 2 \\ 2 & 9 & 1 \end{bmatrix} = \begin{bmatrix} -1 & 0 & 0 \\ 0 & 3 & 0 \\ 0 & 0 & 3 \end{bmatrix}$$

39. $\mathbf{v}_1 = [-1 \quad 0 \quad -1]^T, \quad \mathbf{v}_2 = [-4 \ -1 \ 0]^T, \quad \mathbf{v}_3 = [1 \quad 0 \quad 0]^T$

$$\mathbf{Q} = [\mathbf{v}_1 \quad \mathbf{v}_2 \quad \mathbf{v}_3] = \begin{bmatrix} -1 & -4 & 1 \\ 0 & -1 & 0 \\ -1 & 0 & 0 \end{bmatrix}$$

$$\mathbf{J} = \mathbf{Q}^{-1}\mathbf{AQ} = \begin{bmatrix} 0 & 0 & -1 \\ 0 & -1 & 0 \\ 1 & -4 & -1 \end{bmatrix}\begin{bmatrix} -2 & 17 & 4 \\ -1 & 6 & 1 \\ 0 & 1 & 2 \end{bmatrix}\begin{bmatrix} -1 & -4 & 1 \\ 0 & -1 & 0 \\ -1 & 0 & 0 \end{bmatrix} = \begin{bmatrix} 2 & 1 & 0 \\ 0 & 2 & 1 \\ 0 & 0 & 2 \end{bmatrix}$$

41. $\mathbf{v}_1 = [-5 \quad 3 \quad 8]^T, \quad \mathbf{v}_2 = [1 \quad 0 \quad 0]^T, \quad \mathbf{v}_3 = [1 \quad 1 \quad 0]^T$

$$\mathbf{Q} = [\mathbf{v}_1 \quad \mathbf{v}_2 \quad \mathbf{v}_3] = \begin{bmatrix} -5 & 1 & 1 \\ 3 & 0 & 1 \\ 8 & 0 & 0 \end{bmatrix}$$

$$\mathbf{J} = \mathbf{Q}^{-1}\mathbf{AQ} = \frac{1}{8}\begin{bmatrix} 0 & 0 & 1 \\ 8 & -8 & 8 \\ 0 & 8 & -3 \end{bmatrix}\begin{bmatrix} -3 & 5 & -5 \\ 3 & -1 & 3 \\ 8 & -8 & 10 \end{bmatrix}\begin{bmatrix} -5 & 1 & 1 \\ 3 & 0 & 1 \\ 8 & 0 & 0 \end{bmatrix} = \begin{bmatrix} 2 & 1 & 0 \\ 0 & 2 & 0 \\ 0 & 0 & 2 \end{bmatrix}$$

43. $\mathbf{v}_1 = [1 \ -3 \ -1 \ -2]^T, \quad \mathbf{v}_2 = [0 \quad 1 \quad 0 \quad 0]^T,$
$\mathbf{u}_1 = [0 \ -1 \quad 1 \quad 0]T, \quad \mathbf{u}_2 = [0 \quad 0 \quad 2 \quad 1]^T$

$$\mathbf{Q} = [\mathbf{v}_1 \quad \mathbf{v}_2 \quad \mathbf{u}_1 \quad \mathbf{u}_2] = \begin{bmatrix} 1 & 0 & 0 & 0 \\ -3 & 1 & -1 & 0 \\ -1 & 0 & 1 & 2 \\ -2 & 0 & 0 & 1 \end{bmatrix}$$

$\mathbf{J} = \mathbf{Q}^{-1}\mathbf{AQ}$

$$= \begin{bmatrix} 1 & 0 & 0 & 0 \\ 0 & 1 & 1 & -2 \\ -3 & 0 & 1 & -2 \\ 2 & 0 & 0 & 1 \end{bmatrix}\begin{bmatrix} -1 & 1 & 1 & -2 \\ 7 & -4 & -6 & 11 \\ 5 & -1 & 1 & 3 \\ 6 & -2 & -2 & 6 \end{bmatrix}\begin{bmatrix} 1 & 0 & 0 & 0 \\ -3 & 1 & -1 & 0 \\ -1 & 0 & 1 & 2 \\ -2 & 0 & 0 & 1 \end{bmatrix} = \begin{bmatrix} -1 & 1 & 0 & 0 \\ 0 & -1 & 0 & 0 \\ 0 & 0 & 2 & 1 \\ 0 & 0 & 0 & 2 \end{bmatrix}$$

45. $\mathbf{v}_1 = [42 \quad 7 \quad -21 \quad -42]^T,$ $\mathbf{v}_2 = [34 \quad 22 \quad -10 \quad -27]^T,$

  $\mathbf{v}_3 = [1 \quad 0 \quad 0 \quad 0]^T,$ $\mathbf{v}_4 = [0 \quad 1 \quad 3 \quad 0]^T$

$$\mathbf{Q} = [\mathbf{v}_1 \quad \mathbf{v}_2 \quad \mathbf{v}_3 \quad \mathbf{v}_4] = \begin{bmatrix} 42 & 34 & 1 & 0 \\ 7 & 22 & 0 & 1 \\ -21 & -10 & 0 & 3 \\ -42 & -27 & 0 & 0 \end{bmatrix}$$

$$\mathbf{J} = \mathbf{Q}^{-1}\mathbf{AQ}$$

$$= \frac{1}{2058}\begin{bmatrix} 0 & -81 & 27 & -76 \\ 0 & 126 & -42 & 42 \\ 2058 & -882 & 294 & 1764 \\ 0 & -147 & 735 & -392 \end{bmatrix}\begin{bmatrix} 2 & 1 & -2 & 1 \\ 0 & 3 & -5 & 3 \\ 0 & -13 & 22 & -12 \\ 0 & -27 & 45 & -25 \end{bmatrix}\begin{bmatrix} 42 & 34 & 1 & 0 \\ 7 & 22 & 0 & 1 \\ -21 & -10 & 0 & 3 \\ -42 & -27 & 0 & 0 \end{bmatrix}$$

$$= \begin{bmatrix} 1 & 1 & 0 & 0 \\ 0 & 1 & 1 & 0 \\ 0 & 0 & 1 & 0 \\ 0 & 0 & 0 & 1 \end{bmatrix}$$

SECTION 7.7

NUMERICAL METHODS FOR SYSTEMS

In Problems 1-8 we first write the given system in the form $x' = f(t,x,y) \quad y' = g(t,x,y)$. Then we use the template

$$h = 0.1; \qquad\qquad t_1 = t_0 + h$$
$$x_1 = x_0 + hf(t_0,x_0,y_0); \quad y_1 = y_0 + hg(t_0,x_0,y_0)$$
$$x_2 = x_1 + hf(t_1,x_1,y_1); \quad y_2 = y_1 + hg(t_1,x_1,y_1)$$

(with the given values of t_0, x_0, and y_0) to calculate the Euler approximations $x_1 \approx x(0.1)$ and $y_1 \approx y(0.1)$, and $x_2 \approx x(0.2)$ and $y_2 \approx y(0.2)$, in part (a). We give these approximations and the actual values $x_{act} = x(0.2)$, $y_{act} = y(0.2)$ in tabular form. We use the template

$$h = 0.2; \qquad\qquad t_1 = t_0 + h$$
$$u_1 = x_0 + hf(t_0,x_0,y_0); \qquad v_1 = y_0 + hg(t_0,x_0,y_0)$$
$$x_1 = x_0 + \frac{1}{2}h\big[f(t_0,x_0,y_0) + f(t_1,u_1,v_1)\big] \quad y_1 = y_0 + \frac{1}{2}h\big[g(t_0,x_0,y_0) + g(t_1,u_1,v_1)\big]$$

to calculate the improved Euler approximations $u_1 \approx x(0.2)$ and $u_1 \approx y(0.2)$, and $x_1 \approx x(0.2)$ and $y_1 \approx y(0.2)$, in part (b). We give these approximations and the actual values $x_{act} = x(0.2)$, $y_{act} = y(0.2)$ in tabular form. We use the template

$$h = 0.2;$$

$$F_1 = f(t_0, x_0, y_0); \qquad\qquad G_1 = g(t_0, x_0, y_0);$$

$$F_2 = f\left(t_0 + \frac{h}{2}, x_0 + \frac{h}{2}F_1, y_0 + \frac{1}{2}hG_1\right); \quad G_2 = g\left(t_0 + \frac{h}{2}, x_0 + \frac{h}{2}F_1, y_0 + \frac{h}{2}G_1\right);$$

$$F_3 = f\left(t_0 + \frac{h}{2}, x_0 + \frac{h}{2}F_2, y_0 + \frac{h}{2}G_2\right); \quad G_3 = g\left(t_0 + \frac{h}{2}, x_0 + \frac{h}{2}F_2, y_0 + \frac{h}{2}G_2\right);$$

$$F_4 = f(t_0 + h, x_0 + hF_3, y_0 + hG_3); \qquad G_4 = g(t_0 + h, x_0 + hF_3, y_0 + hG_3);$$

$$x_1 = x_0 + \frac{h}{6}(F_1 + 2F_2 + 2F_3 + F_4); \qquad y_1 = y_0 + \frac{h}{6}(G_1 + 2G_2 + 2G_3 + G_4)$$

to calculate the intermediate slopes and Runge-Kutta approximations $x_1 \approx x(0.2)$ and $y_1 \approx y(0.2)$ for part (c). Again, we give the results in tabular form.

1. **(a)**

x_1	y_1	x_2	y_2	x_{act}	y_{act}
0.4	2.2	0.88	2.5	1.0034	2.6408

(b)

u_1	v_1	x_1	y_1	x_{act}	y_{act}
0.8	2.4	0.96	2.6	1.0034	2.6408

(c)

F_1	G_1	F_2	G_2	F_3	G_3	F_4	G_4
4	2	4.8	3	5.08	3.26	6.32	4.684
x_1	y_1	x_{act}	y_{act}				
1.0027	2.6401	1.0034	2.6408				

3.　**(a)**

x_1	y_1	x_2	y_2	x_{act}	y_{act}
1.7	1.5	2.81	2.31	3.6775	2.9628

(b)

u_1	v_1	x_1	y_1	x_{act}	y_{act}
2.4	2	3.22	2.62	3.6775	2.9628

(c)

F_1	G_1	F_2	G_2	F_3	G_3	F_4	G_4
7	5	11.1	8.1	13.57	9.95	23.102	17.122
x_1	y_1	x_{act}	y_{act}				
3.6481	2.9407	3.6775	2.9628				

5.　**(a)**

x_1	y_1	x_2	y_2	x_{act}	y_{act}
0.9	3.2	–0.52	2.92	-0.5793	2.4488

(b)

u_1	v_1	x_1	y_1	x_{act}	y_{act}
–0.2	3.4	–0.84	2.44	-0.5793	2.4488

(c)

F_1	G_1	F_2	G_2	F_3	G_3	F_4	G_4
–11	2	–14.2	–2.8	–12.44	–3.12	–12.856	–6.704
x_1	y_1	x_{act}	y_{act}				
–0.5712	2.4485	-0.5793	2.4488				

7.   (a)

x_1	y_1	x_2	y_2	x_{act}	y_{act}
2.5	1.3	3.12	1.68	3.2820	1.7902

(b)

u_1	v_1	x_1	y_1	x_{act}	y_{act}
3	1.6	3.24	1.76	3.2820	1.7902

(c)

F_1	G_1	F_2	G_2	F_3	G_3	F_4	G_4
5	3	6.2	3.8	6.48	4	8.088	5.096

x_1	y_1	x_{act}	y_{act}				
3.2816	1.7899	3.2820	1.7902				

In Problems 9-11 we use the same Runge-Kutta template as in part (c) of Problems 1–8 above, and give both the Runge-Kutta approximate values with step sizes $h = 0.1$ and $h = 0.05$, and also the actual values.

9.

With $h = 0.1$:	$x(1) \approx 3.99261$	$y(1) \approx 6.21770$
With $h = 0.05$:	$x(1) \approx 3.99234$	$y(1) \approx 6.21768$
Actual values:	$x(1) \approx 3.99232$	$y(1) \approx 6.21768$

11.

With $h = 0.1$:	$x(1) \approx -0.05832$	$y(1) \approx 0.56664$
With $h = 0.05$:	$x(1) \approx -0.05832$	$y(1) \approx 0.56665$
Actual values:	$x(1) \approx -0.05832$	$y(1) \approx 0.56665$

13. With $y = x'$ we want to solve numerically the initial value problem

$$x' = y, \qquad x(0) = 0,$$
$$y' = -32 - 0.04y, \quad y(0) = 288.$$

When we run Program RK2DIM with step size $h = 0.1$ we find that the change of sign in the velocity v occurs as follows:

t	x	v
7.6	1050.2	+2.8
7.7	1050.3	-0.4

Thus the bolt attains a maximum height of about 1050 feet in about 7.7 seconds.

15. With $y = x'$ and with x in miles and in seconds, we want to solve numerically the initial value problem

$$x' = y, \qquad x(0) = 0,$$
$$y' = \frac{-95485.5}{x^2 + 7920x + 15681600}, \quad y(0) = 1.$$

We find (running RK2DIM with $h = 1$) that the projectile reaches a maximum height of about 83.83 miles in about $168 \text{ sec} = 2 \text{ min } 48 \text{ sec}$.

17. The data in Problem 16 indicate that the range increases when the initial angle is decreased below $45°$. The further data

Angle	Range
41.0	352.1
40.5	352.6
40.0	352.9
39.5	352.8
39.0	352.7
35.0	350.8

indicate that a maximum range of about 353 ft is attained with $\alpha \approx 40°$.

19. First we run program **rkn** (with $h = 0.1$) with $v_0 = 250 \text{ ft/sec}$ and obtain the following results:

t	x	y
5.0	457.43	103.90
6.0	503.73	36.36

Interpolation gives $x = 494.4$ when $y = 50$. Then a run with $v_0 = 255$ ft/sec gives the following results:

t	x	y
5.5	486.75	77.46
6.0	508.86	41.62

Finally, a run with $v_0 = 253$ ft/sec gives these results:

t	x	y
5.5	484.77	75.44
6.0	506.82	39.53

Now $x \approx 500$ ft when $y = 50$ ft. Thus Babe Ruth's home run ball had an initial velocity of 253 ft/sec.

21.   A run with $h = 0.1$ indicates that the projectile has a range of about $21,400$ ft ≈ 4.05 mi and a flight time of about 46 sec. It attains a maximum height of about 8970 ft in about 17.5 sec. At time $t \approx 23$ sec it has its minimum velocity of about 368 ft/sec. It hits the ground ($t \approx 23$ sec) at an angle of about $77°$ with a velocity of about 518 ft/sec.

CHAPTER 8

MATRIX EXPONENTIAL METHODS

SECTION 8.1

MATRIX EXPONENTIALS AND LINEAR SYSTEMS

In Problems 1–8 we first use the eigenvalues and eigenvectors of the coefficient matrix $\mathbf{A}$ to find first a fundamental matrix $\Phi(t)$ for the homogeneous system $\mathbf{x}' = \mathbf{A}\mathbf{x}$. Then we apply the formula $\mathbf{x}(t) = \Phi(t)\Phi(0)^{-1}\mathbf{x}_0$ to find the solution vector $\mathbf{x}(t)$ that satisfies the initial condition $\mathbf{x}(0) = \mathbf{x}_0$. Formulas (11) and (12) in the text provide inverses of 2×2 and 3×3 matrices.

1. Eigensystem: $\lambda_1 = 1$, $\mathbf{v}_1 = \begin{bmatrix} 1 & -1 \end{bmatrix}^{\mathrm{T}}$; $\lambda_2 = 3$, $\mathbf{v}_1 = \begin{bmatrix} 1 & 1 \end{bmatrix}^{\mathrm{T}}$

$$\Phi(t) = \begin{bmatrix} e^{\lambda_1 t}\mathbf{v}_1 & e^{\lambda_2 t}\mathbf{v}_2 \end{bmatrix} = \begin{bmatrix} e^t & e^{3t} \\ -e^t & e^{3t} \end{bmatrix}$$

$$\mathbf{x}(t) = \begin{bmatrix} e^t & e^{3t} \\ -e^t & e^{3t} \end{bmatrix} \cdot \frac{1}{2}\begin{bmatrix} 1 & -1 \\ 1 & 1 \end{bmatrix} \cdot \begin{bmatrix} 3 \\ -2 \end{bmatrix} = \frac{1}{2}\begin{bmatrix} 5e^t + e^{3t} \\ -5e^t + e^{3t} \end{bmatrix}$$

3. Eigensystem: $\lambda = 4i$, $\mathbf{v} = \begin{bmatrix} 1+2i & 2 \end{bmatrix}^{\mathrm{T}}$;

$$\Phi(t) = \begin{bmatrix} \mathrm{Re}\left(\mathbf{v}e^{\lambda t}\right) & \mathrm{Im}\left(\mathbf{v}e^{\lambda t}\right) \end{bmatrix} = \begin{bmatrix} \cos 4t - 2\sin 4t & 2\cos 4t + \sin 4t \\ 2\cos 4t & 2\sin 4t \end{bmatrix}$$

$$\mathbf{x}(t) = \begin{bmatrix} \cos 4t - 2\sin 4t & 2\cos 4t + \sin 4t \\ 2\cos 4t & 2\sin 4t \end{bmatrix} \cdot \frac{1}{4}\begin{bmatrix} 0 & 2 \\ 2 & -1 \end{bmatrix} \cdot \begin{bmatrix} 0 \\ 1 \end{bmatrix} = \frac{1}{4}\begin{bmatrix} -5\sin 4t \\ 4\cos 4t - 2\sin 4t \end{bmatrix}$$

5. Eigensystem: $\lambda = 3i$, $\mathbf{v} = \begin{bmatrix} -1+i & 3 \end{bmatrix}^{\mathrm{T}}$;

$$\Phi(t) = \begin{bmatrix} \mathrm{Re}\left(\mathbf{v}e^{\lambda t}\right) & \mathrm{Im}\left(\mathbf{v}e^{\lambda t}\right) \end{bmatrix} = \begin{bmatrix} -\cos 3t - \sin 3t & \cos 3t - \sin 3t \\ 3\cos 3t & 3\sin 3t \end{bmatrix}$$

$$\mathbf{x}(t) = \begin{bmatrix} -\cos 3t - \sin 3t & \cos 3t - \sin 3t \\ 3\cos 3t & 3\sin 3t \end{bmatrix} \cdot \frac{1}{3}\begin{bmatrix} 0 & 1 \\ 3 & 1 \end{bmatrix} \cdot \begin{bmatrix} 1 \\ -1 \end{bmatrix} = \frac{1}{3}\begin{bmatrix} 3\cos 3t - \sin 3t \\ -3\cos 3t + 6\sin 3t \end{bmatrix}$$

7. Eigensystem:

$$\lambda_1 = 0, \quad \mathbf{v}_1 = \begin{bmatrix} 6 & 2 & 5 \end{bmatrix}^T; \quad \lambda_2 = 1, \quad \mathbf{v}_2 = \begin{bmatrix} 3 & 1 & 2 \end{bmatrix}^T; \quad \lambda_3 = -1, \quad \mathbf{v}_3 = \begin{bmatrix} 2 & 1 & 2 \end{bmatrix}^T;$$

$$\Phi(t) = \begin{bmatrix} e^{\lambda_1 t}\mathbf{v}_1 & e^{\lambda_2 t}\mathbf{v}_2 & e^{\lambda_3 t}\mathbf{v}_3 \end{bmatrix} = \begin{bmatrix} 6 & 3e^t & 2e^{-t} \\ 2 & e^t & e^{-t} \\ 5 & 2e^t & 2e^{-t} \end{bmatrix}$$

$$\mathbf{x}(t) = \begin{bmatrix} 6 & 3e^t & 2e^{-t} \\ 2 & e^t & e^{-t} \\ 5 & 2e^t & 2e^{-t} \end{bmatrix} \cdot \begin{bmatrix} 0 & -2 & 1 \\ 1 & 2 & -2 \\ -1 & 3 & 0 \end{bmatrix} \cdot \begin{bmatrix} 2 \\ 1 \\ 0 \end{bmatrix} = \begin{bmatrix} -12 + 12e^t + 2e^{-t} \\ -4 + 4e^t + e^{-t} \\ -10 + 8e^t + 2e^{-t} \end{bmatrix}$$

In each of Problems 9-20 we first solve the given linear system to find two linearly independent solutions $\mathbf{x}_1$ and $\mathbf{x}_2$, then set up the fundamental matrix $\Phi(t) = \begin{bmatrix} \mathbf{x}_1(t) & \mathbf{x}_2(t) \end{bmatrix}$, and finally calculate the matrix exponential $e^{At} = \Phi(t)\Phi(0)^{-1}$.

9. Eigensystem: $\lambda_1 = 1, \quad \mathbf{v}_1 = \begin{bmatrix} 1 & 1 \end{bmatrix}^T; \quad \lambda_2 = 3, \quad \mathbf{v}_2 = \begin{bmatrix} 2 & 1 \end{bmatrix}^T;$

$$\Phi(t) = \begin{bmatrix} e^{\lambda_1 t}\mathbf{v}_1 & e^{\lambda_2 t}\mathbf{v}_2 \end{bmatrix} = \begin{bmatrix} e^t & 2e^{3t} \\ e^t & e^{3t} \end{bmatrix}$$

$$e^{At} = \begin{bmatrix} e^t & 2e^{3t} \\ e^t & e^{3t} \end{bmatrix} \begin{bmatrix} -1 & 2 \\ 1 & -1 \end{bmatrix} = \begin{bmatrix} -e^t + 2e^{3t} & 2e^t - 2e^{3t} \\ -e^t + e^{3t} & 2e^t - e^{3t} \end{bmatrix}$$

11. Eigensystem: $\lambda_1 = 2, \quad \mathbf{v}_1 = \begin{bmatrix} 1 & 1 \end{bmatrix}^T; \quad \lambda_2 = 3, \quad \mathbf{v}_2 = \begin{bmatrix} 3 & 2 \end{bmatrix}^T;$

$$\Phi(t) = \begin{bmatrix} e^{\lambda_1 t}\mathbf{v}_1 & e^{\lambda_2 t}\mathbf{v}_2 \end{bmatrix} = \begin{bmatrix} e^{2t} & 3e^{3t} \\ e^{2t} & 2e^{3t} \end{bmatrix}$$

$$e^{At} = \begin{bmatrix} e^{2t} & 3e^{3t} \\ e^{2t} & 2e^{3t} \end{bmatrix} \begin{bmatrix} -2 & 3 \\ 1 & -1 \end{bmatrix} = \begin{bmatrix} -2e^{2t} + 3e^{3t} & 3e^{2t} - 3e^{3t} \\ -2e^{2t} + 2e^{3t} & 3e^{2t} - 2e^{3t} \end{bmatrix}$$

13. Eigensystem: $\lambda_1 = 1, \quad \mathbf{v}_1 = \begin{bmatrix} 1 & 1 \end{bmatrix}^T; \quad \lambda_2 = 3, \quad \mathbf{v}_2 = \begin{bmatrix} 4 & 3 \end{bmatrix}^T$

$$\Phi(t) = \begin{bmatrix} e^{\lambda_1 t}\mathbf{v}_1 & e^{\lambda_2 t}\mathbf{v}_2 \end{bmatrix} = \begin{bmatrix} e^t & 4e^{3t} \\ e^t & 3e^{3t} \end{bmatrix}$$

$$e^{At} = \begin{bmatrix} e^t & 4e^{3t} \\ e^t & 3e^{3t} \end{bmatrix} \begin{bmatrix} -3 & 4 \\ 1 & -1 \end{bmatrix} = \begin{bmatrix} -3e^t + 4e^{3t} & 4e^t - 4e^{3t} \\ -3e^t + 3e^{3t} & 4e^t - 3e^{3t} \end{bmatrix}$$

15. Eigensystem: $\lambda_1 = 1$, $\mathbf{v}_1 = \begin{bmatrix} 2 & 1 \end{bmatrix}^T$; $\lambda_2 = 2$, $\mathbf{v}_2 = \begin{bmatrix} 5 & 2 \end{bmatrix}^T$;

$$\Phi(t) = \begin{bmatrix} e^{\lambda_1 t}\mathbf{v}_1 & e^{\lambda_2 t}\mathbf{v}_2 \end{bmatrix} = \begin{bmatrix} 2e^t & 5e^{2t} \\ e^t & 2e^{2t} \end{bmatrix}$$

$$e^{\mathbf{A}t} = \begin{bmatrix} 2e^t & 5e^{2t} \\ e^t & 2e^{2t} \end{bmatrix}\begin{bmatrix} -2 & 5 \\ 1 & -2 \end{bmatrix} = \begin{bmatrix} -4e^t + 5e^{2t} & 10e^t - 10e^{2t} \\ -2e^t + 2e^{2t} & 5e^t - 4e^{2t} \end{bmatrix}$$

17. Eigensystem: $\lambda_1 = 2$, $\mathbf{v}_1 = \begin{bmatrix} 1 & -1 \end{bmatrix}^T$; $\lambda_2 = 4$, $\mathbf{v}_2 = \begin{bmatrix} 1 & 1 \end{bmatrix}^T$;

$$\Phi(t) = \begin{bmatrix} e^{\lambda_1 t}\mathbf{v}_1 & e^{\lambda_2 t}\mathbf{v}_2 \end{bmatrix} = \begin{bmatrix} e^{2t} & e^{4t} \\ -e^{2t} & e^{4t} \end{bmatrix}$$

$$e^{\mathbf{A}t} = \begin{bmatrix} e^{2t} & e^{4t} \\ -e^{2t} & e^{4t} \end{bmatrix} \cdot \frac{1}{2}\begin{bmatrix} 1 & -1 \\ 1 & 1 \end{bmatrix} = \frac{1}{2}\begin{bmatrix} e^{2t} + e^{4t} & -e^{2t} + e^{4t} \\ -e^{2t} + e^{4t} & e^{2t} + e^{4t} \end{bmatrix}$$

19. Eigensystem: $\lambda_1 = 5$, $\mathbf{v}_1 = \begin{bmatrix} 1 & -2 \end{bmatrix}^T$; $\lambda_2 = 10$, $\mathbf{v}_2 = \begin{bmatrix} 2 & 1 \end{bmatrix}^T$;

$$\Phi(t) = \begin{bmatrix} e^{\lambda_1 t}\mathbf{v}_1 & e^{\lambda_2 t}\mathbf{v}_2 \end{bmatrix} = \begin{bmatrix} e^{5t} & 2e^{10t} \\ -2e^{5t} & e^{10t} \end{bmatrix}$$

$$e^{\mathbf{A}t} = \begin{bmatrix} e^{5t} & 2e^{10t} \\ -2e^{5t} & e^{10t} \end{bmatrix} \cdot \frac{1}{5}\begin{bmatrix} 1 & -2 \\ 2 & 1 \end{bmatrix} = \frac{1}{5}\begin{bmatrix} e^{5t} + 4e^{10t} & -2e^{5t} + 2e^{10t} \\ -2e^{5t} + 2e^{10t} & 4e^{5t} + e^{10t} \end{bmatrix}$$

21. $\mathbf{A}^2 = \mathbf{0}$, so $e^{\mathbf{A}t} = \mathbf{I} + \mathbf{A}t = \begin{bmatrix} 1+t & -t \\ t & 1-t \end{bmatrix}$.

23. $\mathbf{A}^3 = \mathbf{0}$, so $e^{\mathbf{A}t} = \mathbf{I} + \mathbf{A}t + \frac{1}{2}\mathbf{A}^2 t^2 = \begin{bmatrix} 1+t & -t & -t-t^2 \\ t & 1-t & t-t^2 \\ 0 & 0 & 1 \end{bmatrix}$.

25. $\mathbf{A} = 2\mathbf{I} + \mathbf{B}$, where $\mathbf{B}^2 = \mathbf{0}$, so $e^{\mathbf{A}t} = e^{2\mathbf{I}t}e^{\mathbf{B}t} = (e^{2t}\mathbf{I})(\mathbf{I} + \mathbf{B}t)$. Hence

$$e^{\mathbf{A}t} = \begin{bmatrix} e^{2t} & 5te^{2t} \\ 0 & e^{2t} \end{bmatrix}, \quad \mathbf{x}(t) = e^{\mathbf{A}t}\begin{bmatrix} 4 \\ 7 \end{bmatrix} = e^{2t}\begin{bmatrix} 4+35t \\ 7 \end{bmatrix}.$$

27. $\mathbf{A} = \mathbf{I} + \mathbf{B}$, where $\mathbf{B}^3 = \mathbf{0}$, so $e^{\mathbf{A}t} = e^{\mathbf{I}t}e^{\mathbf{B}t} = (e^t\mathbf{I})\left(\mathbf{I} + t + \frac{1}{2}\mathbf{B}^2 t^2\right)$. Hence

$$e^{\mathbf{A}t} = \begin{bmatrix} e^t & 2te^t & \left(3t+2t^2\right)e^t \\ 0 & e^t & 2te^t \\ 0 & 0 & e^t \end{bmatrix}, \quad \mathbf{x}(t) = e^{\mathbf{A}t}\begin{bmatrix} 4 \\ 5 \\ 6 \end{bmatrix} = e^t\begin{bmatrix} 4+28t+12t^2 \\ 5+12t \\ 6 \end{bmatrix}.$$

29. $\mathbf{A} = \mathbf{I} + \mathbf{B}$, where $\mathbf{B}^4 = \mathbf{0}$, so $e^{\mathbf{A}t} = e^{\mathbf{I}t}e^{\mathbf{B}t} = \left(e^t\mathbf{I}\right)\left(\mathbf{I} + \mathbf{B}t + \dfrac{1}{2}\mathbf{B}^2t^2 + \dfrac{1}{6}\mathbf{B}^3t^3\right)$. Hence

$$e^{\mathbf{A}t} = e^t\begin{bmatrix} 1 & 2t & 3t+6t^2 & 4t+6t^2+4t^3 \\ 0 & 1 & 6t & 3t+6t^2 \\ 0 & 0 & 1 & 2t \\ 0 & 0 & 0 & 1 \end{bmatrix}, \quad \mathbf{x}(t) = e^{\mathbf{A}t}\begin{bmatrix} 1 \\ 1 \\ 1 \\ 1 \end{bmatrix} = e^t\begin{bmatrix} 1+9t+12t^2+4t^3 \\ 1+9t+6t^2 \\ 1+2t \\ 1 \end{bmatrix}.$$

33. $e^{\mathbf{A}t} = \mathbf{I}\cosh t + \mathbf{A}\sinh t = \begin{bmatrix} \cosh t & \sinh t \\ \sinh t & \cosh t \end{bmatrix}$, so the general solution of $\mathbf{x}' = \mathbf{A}\mathbf{x}$ is

$$\mathbf{x}(t) = e^{\mathbf{A}t}\mathbf{c} = \begin{bmatrix} c_1\cosh t + c_2\sinh t \\ c_1\sinh t + c_2\cosh t \end{bmatrix}.$$

In Problems 35–40 we give first the linearly independent generalized eigenvectors $\mathbf{u}_1, \mathbf{u}_2, \ldots, \mathbf{u}_n$ of the matrix $\mathbf{A}$ and the corresponding solution vectors $\mathbf{x}_1(t), \mathbf{x}_2(t), \ldots, \mathbf{x}_n(t)$ defined by Eq. (34) in the text, then the fundamental matrix $\Phi(t) = \begin{bmatrix} \mathbf{x}_1(t) & \mathbf{x}_2(t) & \cdots & \mathbf{x}_n(t) \end{bmatrix}$. Finally we calculate the exponential matrix $e^{\mathbf{A}t} = \Phi(t)\Phi(0)^{-1}$.

35. $\lambda = 3$: $\mathbf{u}_1 = \begin{bmatrix} 4 & 0 \end{bmatrix}^{\mathrm{T}}$, $\mathbf{u}_2 = \begin{bmatrix} 0 & 1 \end{bmatrix}^{\mathrm{T}}$;

$\{\mathbf{u}_1, \mathbf{u}_2\}$ is a length 2 chain based on the ordinary (rank 1) eigenvector $\mathbf{u}_1$, so $\mathbf{u}_2$ is a generalized eigenvector of rank 2.

$$\mathbf{x}_1(t) = e^{\lambda t}\mathbf{u}_1, \quad \mathbf{x}_2(t) = e^{\lambda t}\left[\mathbf{u}_2 + (\mathbf{A}-\lambda\mathbf{I})\mathbf{u}_2 t\right]$$

$$\Phi(t) = \begin{bmatrix} \mathbf{x}_1(t) & \mathbf{x}_2(t) \end{bmatrix} = e^{3t}\begin{bmatrix} 4 & 4t \\ 0 & 1 \end{bmatrix}$$

$$e^{\mathbf{A}t} = e^{3t}\begin{bmatrix} 4 & 4t \\ 0 & 1 \end{bmatrix} \cdot \frac{1}{4}\begin{bmatrix} 1 & 0 \\ 0 & 4 \end{bmatrix} = e^{3t}\begin{bmatrix} 1 & 4t \\ 0 & 1 \end{bmatrix}$$

37. $\lambda_1 = 2$: $\mathbf{u}_1 = \begin{bmatrix} 1 & 0 & 0 \end{bmatrix}^{\mathrm{T}}$, $\mathbf{x}_1(t) = e^{\lambda_1 t}\mathbf{u}_1$;

$\lambda_2 = 1$: $\mathbf{u}_2 = \begin{bmatrix} 9 & -3 & 0 \end{bmatrix}^{\mathrm{T}}$, $\mathbf{u}_3 = \begin{bmatrix} 10 & 1 & -1 \end{bmatrix}^{\mathrm{T}}$;

$\{\mathbf{u}_2, \mathbf{u}_3\}$ is a length 2 chain based on the ordinary (rank 1) eigenvector $\mathbf{u}_2$, so $\mathbf{u}_3$ is a generalized eigenvector of rank 2.

$$\mathbf{x}_2(t) = e^{\lambda_2 t}\mathbf{u}_2, \quad \mathbf{x}_3(t) = e^{\lambda_2 t}\left[\mathbf{u}_3 + (\mathbf{A} - \lambda_2\mathbf{I})\mathbf{u}_3 t\right]$$

$$\Phi(t) = \begin{bmatrix} \mathbf{x}_1(t) & \mathbf{x}_2(t) & \mathbf{x}_3(t) \end{bmatrix} = \begin{bmatrix} e^{2t} & 9e^t & (10+9t)e^t \\ 0 & -3e^t & (1-3t)e^t \\ 0 & 0 & -e^t \end{bmatrix}$$

$$e^{\mathbf{A}t} = \begin{bmatrix} e^{2t} & 9e^t & (10+9t)e^t \\ 0 & -3e^t & (1-3t)e^t \\ 0 & 0 & -e^t \end{bmatrix} \cdot \frac{1}{3}\begin{bmatrix} 3 & 9 & 13 \\ 0 & -1 & -1 \\ 0 & 0 & -3 \end{bmatrix}$$

$$= \begin{bmatrix} e^{2t} & -3e^t + 3e^{2t} & (-13-9t)e^t + 13e^{2t} \\ 0 & e^t & 3te^t \\ 0 & 0 & e^t \end{bmatrix}$$

$$= \begin{bmatrix} e^{2t} & -3e^t + 3e^{2t} & (-13-9t)e^t + 13e^{2t} \\ 0 & e^t & 3te^t \\ 0 & 0 & e^t \end{bmatrix}$$

39. $\lambda_2 = 1: \mathbf{u}_1 = \begin{bmatrix} 3 & 0 & 0 & 0 \end{bmatrix}^T, \mathbf{u}_2 = \begin{bmatrix} 0 & 1 & 0 & 0 \end{bmatrix}^T;$

$\{\mathbf{u}_1, \mathbf{u}_2\}$ is a length 2 chain based on the ordinary (rank 1) eigenvector $\mathbf{u}_1$, so $\mathbf{u}_2$ is a generalized eigenvector of rank 2.

$$\mathbf{x}_1(t) = e^{\lambda_1 t}\mathbf{u}_1, \quad \mathbf{x}_2(t) = e^{\lambda_1 t}\left[\mathbf{u}_2 + (\mathbf{A} - \lambda_1\mathbf{I})\mathbf{u}_2 t\right]$$

$\lambda_2 = 2: \mathbf{u}_3 = \begin{bmatrix} 144 & 36 & 12 & 0 \end{bmatrix}^T, \mathbf{u}_4 = \begin{bmatrix} 0 & 27 & 17 & 4 \end{bmatrix}^T;$

$\{\mathbf{u}_3, \mathbf{u}_4\}$ is a length 2 chain based on the ordinary (rank 1) eigenvector $\mathbf{u}_3$, so $\mathbf{u}_4$ is a generalized eigenvector of rank 2.

$$\mathbf{x}_3(t) = e^{\lambda_2 t}\mathbf{u}_3, \quad \mathbf{x}_4(t) = e^{\lambda_2 t}\left[\mathbf{u}_4 + (\mathbf{A} - \lambda_2\mathbf{I})\mathbf{u}_4 t\right]$$

$$\Phi(t) = \begin{bmatrix} \mathbf{x}_1(t) & \mathbf{x}_2(t) & \mathbf{x}_3(t) & \mathbf{x}_4(t) \end{bmatrix} = \begin{bmatrix} 3e^t & 3te^t & 144e^{2t} & 144te^{2t} \\ 0 & e^t & 36e^{2t} & (27+36t)e^{2t} \\ 0 & 0 & 12e^{2t} & (17+12t)e^{2t} \\ 0 & 0 & 0 & 4e^{2t} \end{bmatrix}$$

$$e^{At} = \begin{bmatrix} 3e^t & 3te^t & 144e^{2t} & 144te^{2t} \\ 0 & e^t & 36e^{2t} & (27+36t)e^{2t} \\ 0 & 0 & 12e^{2t} & (17+12t)e^{2t} \\ 0 & 0 & 0 & 4e^{2t} \end{bmatrix} \cdot \frac{1}{48} \begin{bmatrix} 16 & 0 & -192 & 816 \\ 0 & 48 & -144 & 288 \\ 0 & 0 & 4 & -17 \\ 0 & 0 & 0 & 12 \end{bmatrix}$$

$$= \begin{bmatrix} e^t & 3te^t & (-12-9t)e^t +12te^{2t} & (51+18t)e^t +(-51+36t)e^{2t} \\ 0 & e^t & -3e^t +3e^{2t} & 6e^t +(-6+9t)e^{2t} \\ 0 & 0 & e^{2t} & 3te^{2t} \\ 0 & 0 & 0 & e^{2t} \end{bmatrix}$$

SECTION 8.2

NONHOMOGENEOUS LINEAR SYSTEMS

1.   Substitution of the trial solution $x_p(t) = a$, $y_p(t) = b$ yields the equations

$$a + 2b + 3 = 0, \quad 2a + b - 2 = 0$$

with solution $a = \dfrac{7}{3}$, $b = -\dfrac{8}{3}$. Thus we obtain the particular solution

$$x(t) = \frac{7}{3}, \quad y(t) = -\frac{8}{3}.$$

3.   When we substitute the trial solution

$$x_p = a_1 + b_1 t + c_1 t^2, \quad y_p = a_2 + b_2 t + c_2 t^2$$

and collect coefficients, we get the equations

$$3a_1 + 4a_2 = b_1 \quad 3b_1 + 4b_2 = 2c_1 \quad 3c_1 + 4c_2 = 0$$
$$3a_1 + 2a_2 = b_2 \quad 3b_1 + 2b_2 = 2c_2 \quad 3c_1 + 2c_2 + 1 = 0$$

Working backwards, we solve first for $c_1 = -\dfrac{2}{3}$ and $c_2 = \dfrac{1}{2}$, then for $b_1 = \dfrac{10}{9}$ and

$b_2 = -\dfrac{7}{6}$, and finally for $a_1 = -\dfrac{31}{27}$ and $a_2 = \dfrac{41}{36}$. This determines the particular solution

$x_p(t)$, $y_p(t)$. Next, the coefficient matrix of the associated homogeneous system has

eigenvalues $\lambda_1 = -1$ and $\lambda_2 = 6$, with eigenvectors $\mathbf{v}_1 = \begin{bmatrix} 1 & -1 \end{bmatrix}^T$ and $\mathbf{v}_2 = \begin{bmatrix} 4 & 3 \end{bmatrix}^T$,

respectively, so the complementary solution is given by

$$x_c(t) = c_1 e^{-t} + 4c_2 e^{6t}, \quad y_c(t) = -c_1 e^{-t} + 3c_2 e^{6t}.$$

When we impose the initial conditions $x(0) = 0$, $y(0) = 0$ on the general solution

$x(t) = x_c(t) + x_p(t)$, $y(t) = y_c(t) + y_p(t)$ we find that $c_1 = \dfrac{8}{7}$ and $c_2 = \dfrac{1}{756}$. This finally gives the desired particular solution

$$x(t) = \frac{1}{756}\left(864e^{-t} + 4e^{6t} - 868 + 840t - 504t^2\right),$$

$$y(t) = \frac{1}{756}\left(-864e^{-t} + 3e^{6t} + 861 - 882t + 378t^2\right).$$

5. The coefficient matrix of the associated homogeneous system has eigenvalues $\lambda_1 = -1$ and $\lambda_2 = 5$, so the nonhomogeneous term e^{-t} duplicates part of the complementary solution. We therefore try the particular solution

$$x_p(t) = a_1 + b_1 e^{-t} + c_1 t e^{-t}, \quad y_p(t) = a_2 + b_2 e^{-t} + c_2 t e^{-t}.$$

Upon solving the six linear equations we get by collecting coefficients after substitution of this trial solution into the given nonhomogeneous system, we obtain the particular solution

$$x(t) = \frac{1}{3}\left(-12 - e^{-t} - 7t e^{-t}\right), \quad y(t) = \frac{1}{3}\left(-6 - 7t e^{-t}\right).$$

7. First we try the particular solution

$$x_p(t) = a_1 \sin t + b_1 \cos t, \quad y_p(t) = a_2 \sin t + b_2 \cos t.$$

Upon solving the four linear equations we get by collecting coefficients after substitution of this trial solution into the given nonhomogeneous system, we find that $a_1 = -\dfrac{21}{82}$,

$b_1 = -\dfrac{25}{82}$, $a_2 = -\dfrac{15}{41}$, and $b_2 = -\dfrac{12}{41}$. The coefficient matrix of the associated

homogeneous system has eigenvalues $\lambda_1 = 1$ and $\lambda_2 = -9$, with eigenvectors $\mathbf{v}_1 = \begin{bmatrix} 1 & 1 \end{bmatrix}^T$ and $\mathbf{v}_2 = \begin{bmatrix} 2 & -3 \end{bmatrix}^T$, respectively, so the complementary solution is given by

$$x_c(t) = c_1 e^t + 2c_2 e^{-9t}, \quad y_c(t) = c_1 e^t - 3c_2 e^{-9t}.$$

When we impose the initial conditions $x(0) = 1$, $y(0) = 0$ we find that $c_1 = \dfrac{9}{10}$ and

$c_2 = \dfrac{83}{410}$. It follows that the desired particular solution $x = x_c + x_p$, $y = y_c + y_p$ is given by

$$x(t) = \frac{1}{410}\left(369e^{t} + 166e^{-9t} - 125\cos t - 105\sin t\right),$$

$$y(t) = \frac{1}{410}\left(369e^{t} - 249e^{-9t} - 120\cos t - 150\sin t\right).$$

9. Here the associated homogeneous system is the same as in Problem 8, so the nonhomogeneous term $\cos 2t$ duplicates the complementary function. We therefore substitute the trial solution

$$x_p(t) = a_1 \sin 2t + b_1 \cos 2t + c_1 t \sin 2t + d_1 t \cos 2t$$
$$y_p(t) = a_2 \sin 2t + b_2 \cos 2t + c_2 t \sin 2t + d_2 t \cos 2t$$

and use a computer algebra system to solve the system of 8 linear equations that results when we collect coefficients in the usual way. This gives the particular solution

$$x(t) = \frac{1}{4}\left(\sin 2t + 2t\cos 2t + t\sin 2t\right), \quad y(t) = \frac{1}{4}t\sin 2t .$$

11. The coefficient matrix of the associated homogeneous system has eigenvalues $\lambda_1 = 0$ and $\lambda_2 = 4$, so there is duplication of constant terms. We therefore substitute the particular solution

$$x_p(t) = a_1 + b_1 t, \quad y_p(t) = a_2 + b_2 t$$

and solve the resulting equations for $a_1 = -2$, $a_2 = 0$, $b_1 = -2$, and $b_2 = 1$. The eigenvectors of the coefficient matrix associated with the eigenvalues $\lambda_1 = 0$ and $\lambda_2 = 4$ are $\mathbf{v}_1 = \begin{bmatrix} 2 & -1 \end{bmatrix}^{T}$ and $\mathbf{v}_2 = \begin{bmatrix} 2 & 1 \end{bmatrix}^{T}$, respectively, so the general solution of the given nonhomogeneous system is given by

$$x(t) = 2c_1 + 2c_2 e^{4t} - 2 - 2t, \quad y(t) = -c_1 + c_2 e^{4t} + t .$$

When we impose the initial conditions $x(0) = 1$, $y(0) = -1$ we find readily that $c_1 = \frac{5}{4}$, $c_2 = \frac{1}{4}$. This gives the desired particular solution

$$x(t) = \frac{1}{2}\left(1 - 4t + e^{4t}\right), \quad y(t) = \frac{1}{4}\left(-5 + 4t + e^{4t}\right).$$

13. The coefficient matrix of the associated homogeneous system has eigenvalues $\lambda_1 = 1$ and $\lambda_2 = 3$, so there is duplication of e^t terms. We therefore substitute the trial solution

$$x_p(t) = (a_1 + b_1 t)e^{t}, \quad y_p(t) = (a_2 + b_2 t)e^{t}$$

This leads readily to the particular solution

$$x(t) = \frac{1}{2}(1 + 5t)e^t, \quad y(t) = -\frac{5}{2}te^t.$$

In Problems 15 and 16 the amounts $x_1(t)$ and $x_2(t)$ in the two tanks satisfy the equations

$$x_1' = rc_0 - k_1 x_1, \quad x_2' = k_1 x_1 - k_2 x_2,$$

where $k_i = \dfrac{r}{V_i}$, in terms of the flow rate r, the inflowing concentration c_0, and the volumes V_1 and V_2 of the two tanks.

15.   (a) We solve the initial value problem

$$x_1' = 20 - \frac{x_1}{10}, \quad x_1(0) = 0$$

$$x_2' = \frac{x_1}{10} - \frac{x_2}{20}, \quad x_2(0) = 0$$

for $x_1(t) = 200(1 - e^{-t/10})$, $x_2(t) = 400(1 + e^{-t/10} - 2e^{-t/20})$.

(b) Evidently $x_1(t) \to 200\,\text{gal}$ and $x_2(t) \to 400\,\text{gal}$ as $t \to \infty$.

(c) It takes about 6 min 56 sec for tank 1 to reach a salt concentration of 1 lb/gal, and about 24 min 34 sec for tank 2 to reach this concentration.

In Problems 17–34 we apply the variation of parameters formula in Eq. (28) of Section 8.2. The answers shown below were actually calculated using the *Mathematica* code listed in the application for Section 8.2. For instance, for Problem 17 we first enter the coefficient matrix

```
A = {{6, -7}, {1, -2}};
```

the initial vector

```
x0 = {{0}, {0}};
```

and the vector

```
f[t_] := {{60}, {90}};
```

of nonhomogeneous terms. It simplifies the notation to rename *Mathematica*'s exponential matrix function by defining

```
exp[A_] := MatrixExp[A]
```

Then the integral in the variation of parameters formula is given by

```
integral = Integrate[exp[-A*s] . f[s], {s, 0, t}] // Simplify
```

and yields the output

$$\begin{bmatrix} -102 + 7e^{-5t} + 95e^{t} \\ -96 + e^{-5t} + 95e^{t} \end{bmatrix}.$$

Finally the desired particular solution is given by

```
solution = exp[A*t] . (x0 + integral) // Simplify
```

which yields

$$\begin{bmatrix} 102 - 7e^{-5t} - 95e^{t} \\ 96 - e^{-5t} - 95e^{t} \end{bmatrix}.$$

(Maple and MATLAB versions of this computation are provided in the online Expanded Applications that accompany the textbook.)

In each succeeding problem, we need only substitute the given coefficient matrix **A**, initial vector **x0**, and the vector **f** of nonhomogeneous terms in the above commands, and then re-execute them in turn. We give below only the component functions of the final results.

17. $x_1(t) = 102 - 95e^{-t} - 7e^{5t}$, $x_2(t) = 96 - 95e^{-t} - e^{5t}$

19. $x_1(t) = -70 - 60t + 16e^{-3t} + 54e^{2t}$, $x_2(t) = 5 - 60t - 32e^{-3t} + 27e^{2t}$

21. $x_1(t) = -e^{-t} - 14e^{2t} + 15e^{3t}$, $x_2(t) = -5e^{-t} - 10e^{2t} + 15e^{3t}$

23. $x_1(t) = 3 + 11t + 8t^2$, $x_2(t) = 5 + 17t + 24t^2$

25. $x_1(t) = -1 + 8t + \cos t - 8\sin t$, $x_2(t) = -2 + 4t + 2\cos t - 3\sin t$

27. $x_1(t) = 8t^3 + 6t^4$, $x_2(t) = 3t^2 - 2t^3 + 3t^4$

29. $x_1(t) = t\cos t - \ln(\cos t)\sin t$, $x_2(t) = t\sin t - \ln(\cos t)\cos t$

31. $x_1(t) = (9t^2 + 4t^3)e^{t}$, $x_2(t) = 6t^2 e^{t}$, $x_3(t) = 6te^{t}$

33. $x_1(t) = 15t^2 + 60t^3 + 95t^4 + 12t^5$, $x_2(t) = 15t^2 + 55t^3 + 15t^4$, $x_3(t) = 15t^2 + 20t^3$,
   $x_4(t) = 15t^2$

SECTION 8.3

SPECTRAL DECOMPOSITION METHODS

In Problems 1–20 here we want to use projection matrices to find fundamental matrix solutions of the linear systems given in Problems 1–20 of Section 7.3. In each of Problems 1–16, the coefficient matrix $\mathbf{A}$ is 2×2 with distinct eigenvalues λ_1 and λ_2. We can therefore use the method of Example 1 in Section 8.3. That is, if we define the projection matrices

$$\mathbf{P}_1 = \frac{\mathbf{A}-\lambda_2\mathbf{I}}{\lambda_1-\lambda_2} \quad \text{and} \quad \mathbf{P}_2 = \frac{\mathbf{A}-\lambda_1\mathbf{I}}{\lambda_2-\lambda_1}, \tag{1}$$

then the desired fundamental matrix solution of the system $\mathbf{x}' = \mathbf{A}\mathbf{x}$ is the exponential matrix

$$e^{\mathbf{A}t} = e^{\lambda_1 t}\mathbf{P}_1 + e^{\lambda_2 t}\mathbf{P}_2. \tag{2}$$

We use the eigenvalues λ_1 and λ_2 given in the Section 7.3 solutions for these problems.

1. $\quad \mathbf{A} = \begin{bmatrix} 1 & 2 \\ 2 & 1 \end{bmatrix}; \quad \lambda_1 = -1, \quad \lambda_2 = 3$

$\quad \mathbf{P}_1 = \dfrac{1}{-4}(\mathbf{A}-3\mathbf{I}) = \dfrac{1}{2}\begin{bmatrix} 1 & -1 \\ -1 & 1 \end{bmatrix}, \quad \mathbf{P}_2 = \dfrac{1}{4}(\mathbf{A}+\mathbf{I}) = \dfrac{1}{2}\begin{bmatrix} 1 & 1 \\ 1 & 1 \end{bmatrix}$

$\quad e^{\mathbf{A}t} = e^{-t}\mathbf{P}_1 + e^{3t}\mathbf{P}_2 = \dfrac{1}{2}\begin{bmatrix} e^{-t}+e^{3t} & -e^{-t}+e^{3t} \\ -e^{-t}+e^{3t} & e^{-t}+e^{3t} \end{bmatrix}$

3. $\quad \mathbf{A} = \begin{bmatrix} 3 & 4 \\ 3 & 2 \end{bmatrix}; \quad \lambda_1 = -1, \quad \lambda_2 = 6$

$\quad \mathbf{P}_1 = \dfrac{1}{-7}(\mathbf{A}-6\mathbf{I}) = \dfrac{1}{7}\begin{bmatrix} 3 & -4 \\ -3 & 4 \end{bmatrix}, \quad \mathbf{P}_2 = \dfrac{1}{7}(\mathbf{A}+\mathbf{I}) = \dfrac{1}{7}\begin{bmatrix} 4 & 4 \\ 3 & 3 \end{bmatrix}$

$\quad e^{\mathbf{A}t} = e^{-t}\mathbf{P}_1 + e^{6t}\mathbf{P}_2 = \dfrac{1}{7}\begin{bmatrix} 3e^{-t}+4e^{6t} & -4e^{-t}+4e^{6t} \\ -3e^{-t}+3e^{6t} & 4e^{-t}+3e^{6t} \end{bmatrix}$

5. $\quad \mathbf{A} = \begin{bmatrix} 6 & -7 \\ 1 & -2 \end{bmatrix}; \quad \lambda_1 = -1, \quad \lambda_2 = 5$

$\quad \mathbf{P}_1 = \dfrac{1}{-6}(\mathbf{A}-5\mathbf{I}) = \dfrac{1}{6}\begin{bmatrix} -1 & 7 \\ -1 & 7 \end{bmatrix}, \quad \mathbf{P}_2 = \dfrac{1}{6}(\mathbf{A}+\mathbf{I}) = \dfrac{1}{6}\begin{bmatrix} 7 & -7 \\ 1 & -1 \end{bmatrix}$

$$e^{\mathbf{A}t} = e^{-t}\mathbf{P}_1 + e^{5t}\mathbf{P}_2 = \frac{1}{6}\begin{bmatrix} -e^{-t} + 7e^{5t} & 7e^{-t} - 7e^{5t} \\ -e^{-t} + e^{5t} & 7e^{-t} - e^{5t} \end{bmatrix}$$

7. $\mathbf{A} = \begin{bmatrix} -3 & 4 \\ 6 & -5 \end{bmatrix};\quad \lambda_1 = -9,\quad \lambda_2 = 1$

$$\mathbf{P}_1 = \frac{1}{-10}(\mathbf{A} - \mathbf{I}) = \frac{1}{5}\begin{bmatrix} 2 & -2 \\ -3 & 3 \end{bmatrix},\quad \mathbf{P}_2 = \frac{1}{10}(\mathbf{A} + 9\mathbf{I}) = \frac{1}{10}\begin{bmatrix} 3 & 2 \\ 3 & 2 \end{bmatrix}$$

$$e^{\mathbf{A}t} = e^{-9t}\mathbf{P}_1 + e^{t}\mathbf{P}_2 = \frac{1}{5}\begin{bmatrix} 2e^{-9t} + 3e^{t} & -2e^{-9t} + 2e^{t} \\ -3e^{-9t} + 3e^{t} & 3e^{-9t} + 2e^{t} \end{bmatrix}$$

9. $\mathbf{A} = \begin{bmatrix} 2 & -5 \\ 4 & 2 \end{bmatrix};\quad \lambda_1 = -4i,\quad \lambda_2 = 4i$

$$\mathbf{P}_1 = \frac{1}{-8i}(\mathbf{A} - 4i\mathbf{I}) = \frac{1}{8}\begin{bmatrix} 4+2i & -5i \\ 4i & 4-2i \end{bmatrix},\quad \mathbf{P}_2 = \frac{1}{8i}(\mathbf{A} + 4i\mathbf{I}) = \frac{1}{8}\begin{bmatrix} 4-2i & 5i \\ -4i & 4+2i \end{bmatrix}$$

$$e^{\mathbf{A}t} = e^{-4it}\mathbf{P}_1 + e^{4it}\mathbf{P}_2 = (\cos 4t - i\sin 4t)\mathbf{P}_1 + (\cos 4t + i\sin 4t)\mathbf{P}_2$$

$$= \frac{1}{4}\begin{bmatrix} 4\cos 4t + 2\sin 4t & -5\sin 4t \\ 4\sin 4t & 4\cos 4t - 2\sin 4t \end{bmatrix}$$

11. $\mathbf{A} = \begin{bmatrix} 1 & -2 \\ 2 & 1 \end{bmatrix};\quad \lambda_1 = 1 - 2i,\quad \lambda_2 = 1 + 2i$

$$\mathbf{P}_1 = \frac{1}{-4i}(\mathbf{A} - (1+2i)\mathbf{I}) = \frac{1}{2}\begin{bmatrix} 1 & -i \\ i & 1 \end{bmatrix},\quad \mathbf{P}_2 = \frac{1}{4i}(\mathbf{A} - (1-2i)\mathbf{I}) = \frac{1}{2}\begin{bmatrix} 1 & i \\ -i & 1 \end{bmatrix}$$

$$e^{\mathbf{A}t} = e^{(1-2i)t}\mathbf{P}_1 + e^{(1+2i)t}\mathbf{P}_2 = e^{t}(\cos 2t - i\sin 2t)\mathbf{P}_1 + e^{t}(\cos 2t + i\sin 2t)\mathbf{P}_2$$

$$= e^{t}\begin{bmatrix} \cos 2t & -\sin 2t \\ \sin 2t & \cos 2t \end{bmatrix}$$

13. $\mathbf{A} = \begin{bmatrix} 5 & -9 \\ 2 & -1 \end{bmatrix};\quad \lambda_1 = 2 - 3i,\quad \lambda_2 = 2 + 3i$

$$\mathbf{P}_1 = \frac{1}{-6i}(\mathbf{A} - (2+3i)\mathbf{I}) = \frac{1}{6}\begin{bmatrix} 3+3i & -9i \\ 2i & 3-3i \end{bmatrix}$$

$$\mathbf{P}_2 = \frac{1}{6i}(\mathbf{A} - (2-3i)\mathbf{I}) = \frac{1}{6}\begin{bmatrix} 3-3i & 9i \\ -2i & 3+3i \end{bmatrix}$$

$$e^{\mathbf{A}t} = e^{(2-3i)t}\mathbf{P}_1 + e^{(2+3i)t}\mathbf{P}_2 = e^{2t}(\cos 3t - i\sin 3t)\mathbf{P}_1 + e^{2t}(\cos 3t + i\sin 3t)\mathbf{P}_2$$

$$= \frac{e^{2t}}{3}\begin{bmatrix} 3\cos 3t + \sin 3t & -9\sin 3t \\ 2\sin 3t & 3\cos 3t - \sin 3t \end{bmatrix}$$

15. $\mathbf{A} = \begin{bmatrix} 7 & -5 \\ 4 & 3 \end{bmatrix}$; $\lambda_1 = 5 - 4i$, $\lambda_2 = 5 + 4i$

$$\mathbf{P}_1 = \frac{1}{-8i}\left(\mathbf{A} - (5+4i)\mathbf{I}\right) = \frac{1}{8}\begin{bmatrix} 4+2i & -5i \\ 4i & 4-2i \end{bmatrix}$$

$$\mathbf{P}_2 = \frac{1}{8i}\left(\mathbf{A} - (5-4i)\mathbf{I}\right) = \frac{1}{8}\begin{bmatrix} 4-2i & 5i \\ -4i & 4+2i \end{bmatrix}$$

$$e^{\mathbf{A}t} = e^{(5-4i)t}\mathbf{P}_1 + e^{(5+4i)t}\mathbf{P}_2 = e^{5t}(\cos 4t - i\sin 4t)\mathbf{P}_1 + e^{5t}(\cos 4t + i\sin 4t)\mathbf{P}_2$$

$$= \frac{e^{5t}}{4}\begin{bmatrix} 4\cos 4t + 2\sin 2t & -5\sin 4t \\ 4\sin 4t & 4\cos 4t - 2\sin 2t \end{bmatrix}$$

In Problems 17, 18, and 20 the coefficient matrix $\mathbf{A}$ is 3×3 with distinct eigenvalues $\lambda_1, \lambda_2, \lambda_3$. Looking at Equations (7) and (3) in Section 8.3, we see that

$$e^{\mathbf{A}t} = e^{\lambda_1 t}\mathbf{P}_1 + e^{\lambda_2 t}\mathbf{P}_2 + e^{\lambda_3 t}\mathbf{P}_3 \tag{3}$$

where the projection matrices are defined by

$$\mathbf{P}_1 = \frac{(\mathbf{A} - \lambda_2\mathbf{I})(\mathbf{A} - \lambda_3\mathbf{I})}{(\lambda_1 - \lambda_2)(\lambda_1 - \lambda_3)}, \quad \mathbf{P}_2 = \frac{(\mathbf{A} - \lambda_1\mathbf{I})(\mathbf{A} - \lambda_3\mathbf{I})}{(\lambda_2 - \lambda_1)(\lambda_2 - \lambda_3)}, \quad \mathbf{P}_3 = \frac{(\mathbf{A} - \lambda_1\mathbf{I})(\mathbf{A} - \lambda_2\mathbf{I})}{(\lambda_3 - \lambda_1)(\lambda_3 - \lambda_2)}. \tag{4}$$

17. $\mathbf{A} = \begin{bmatrix} 4 & 1 & 4 \\ 1 & 7 & 1 \\ 4 & 1 & 4 \end{bmatrix}$; $\lambda_1 = 0$, $\lambda_2 = 6$, $\lambda_3 = 9$

$$\mathbf{P}_1 = \frac{1}{54}(\mathbf{A}-6\mathbf{I})(\mathbf{A}-9\mathbf{I}) = \frac{1}{2}\begin{bmatrix} 1 & 0 & -1 \\ 0 & 0 & 0 \\ -1 & 0 & 1 \end{bmatrix}$$

$$\mathbf{P}_2 = \frac{1}{-18}(\mathbf{A}-0\mathbf{I})(\mathbf{A}-9\mathbf{I}) = \frac{1}{6}\begin{bmatrix} 1 & -2 & 1 \\ -2 & 4 & -2 \\ 1 & -2 & 1 \end{bmatrix}$$

$$\mathbf{P}_3 = \frac{1}{27}(\mathbf{A}-0\mathbf{I})(\mathbf{A}-6\mathbf{I}) = \frac{1}{3}\begin{bmatrix} 1 & 1 & 1 \\ 1 & 1 & 1 \\ 1 & 1 & 1 \end{bmatrix}$$

$$e^{\mathbf{A}t} = \mathbf{P}_1 + e^{6t}\mathbf{P}_2 + e^{9t}\mathbf{P}_3 = \frac{1}{6}\begin{bmatrix} 3+e^{6t}+2e^{9t} & -2e^{6t}+2e^{9t} & -3+e^{6t}+2e^{9t} \\ -2e^{6t}+2e^{9t} & 4e^{6t}+2e^{9t} & -2e^{6t}+2e^{9t} \\ -3+e^{6t}+2e^{9t} & -2e^{6t}+2e^{9t} & 3+e^{6t}+2e^{9t} \end{bmatrix}$$

19. $\mathbf{A} = \begin{bmatrix} 4 & 1 & 1 \\ 1 & 4 & 1 \\ 1 & 1 & 4 \end{bmatrix}$; $\lambda_1 = 6, \quad \lambda_2 = 3, \quad \lambda_3 = 3$

Here we have the eigenvalue $\lambda_1 = 6$ of multiplicity 1 and the eigenvalue $\lambda_2 = 3$ of multiplicity 2. By Example 2 in Section 8.3, the desired matrix exponential is given by

$$e^{\mathbf{A}t} = e^{\lambda_1 t}\mathbf{P}_1 + e^{\lambda_2 t}\mathbf{P}_2\left[\mathbf{I}+(\mathbf{A}-\lambda_2\mathbf{I})t\right] \tag{5}$$

where $\mathbf{P}_1$ and $\mathbf{P}_2$ are the projection matrices of $\mathbf{A}$ corresponding to the eigenvalues λ_1 and λ_2. The reciprocal of the characteristic polynomial $p(\lambda) = (\lambda-6)(\lambda-3)^2$ has the partial fractions decomposition

$$\frac{1}{p(\lambda)} = \frac{1}{9(\lambda-6)} - \frac{\lambda}{9(\lambda-3)^2},$$

so $a_1(\lambda)=1/9$ and $a_2(\lambda)=-\lambda/9$ in the notation of Equation (25) in the text. Therefore Equation (26) there gives

$$\mathbf{P}_1 = a_1(\mathbf{A})\cdot(\mathbf{A}-\lambda_2\mathbf{I})^2 = \frac{1}{9}(\mathbf{A}-3\mathbf{I})^2 = \frac{1}{3}\begin{bmatrix} 1 & 1 & 1 \\ 1 & 1 & 1 \\ 1 & 1 & 1 \end{bmatrix},$$

$$\mathbf{P}_2 = a_2(\mathbf{A}) \cdot (\mathbf{A} - \lambda_1 \mathbf{I}) = -\frac{1}{9} \mathbf{A}(\mathbf{A} - 6\mathbf{I}) = \frac{1}{3}\begin{bmatrix} 2 & -1 & -1 \\ -1 & 2 & -1 \\ -1 & -1 & 2 \end{bmatrix}.$$

Finally, Equation (5) above gives

$$e^{\mathbf{A}t} = e^{6t}\mathbf{P}_1 + e^{3t}\mathbf{P}_2\left(\mathbf{I} + (\mathbf{A} - 3\mathbf{I})t\right) = \frac{1}{3}\begin{bmatrix} 2e^{3t} + e^{6t} & -e^{3t} + e^{6t} & -e^{3t} + e^{6t} \\ -e^{3t} + e^{6t} & 2e^{3t} + e^{6t} & -e^{3t} + e^{6t} \\ -e^{3t} + e^{6t} & -e^{3t} + e^{6t} & 2e^{3t} + e^{6t} \end{bmatrix}.$$

In Problems 21–30 here we want to use projection matrices to find fundamental matrix solutions of the linear systems given in Problems 1–10 of Section 7.6. In each of Problems 21–26, the 2×2 coefficient matrix $\mathbf{A}$ has characteristic polynomial of the form $p(\lambda) = (\lambda - \lambda_1)^2$ and thus a single eigenvalue λ_1 of multiplicity 2. Consequently, Example 5 in Section 8.3 gives the desired fundamental matrix

$$e^{\mathbf{A}t} = e^{\lambda_1 t}\left[\mathbf{I} + (\mathbf{A} - \lambda_1\mathbf{I})t\right]. \tag{6}$$

21. $\quad \mathbf{A} = \begin{bmatrix} -2 & 1 \\ -1 & -4 \end{bmatrix}; \qquad \lambda_1 = -3$

$$e^{\mathbf{A}t} = e^{-3t}\left[\mathbf{I} + (\mathbf{A} + 3\mathbf{I})t\right] = e^{-3t}\begin{bmatrix} 1+t & t \\ -t & 1-t \end{bmatrix}$$

23. $\quad \mathbf{A} = \begin{bmatrix} 1 & -2 \\ 2 & 5 \end{bmatrix}; \qquad \lambda_1 = 3$

$$e^{\mathbf{A}t} = e^{3t}\left[\mathbf{I} + (\mathbf{A} - 3\mathbf{I})t\right] = e^{3t}\begin{bmatrix} 1-2t & -2t \\ 2t & 1+2t \end{bmatrix}$$

25. **25.** $\quad \mathbf{A} = \begin{bmatrix} 7 & 1 \\ -4 & 3 \end{bmatrix}; \qquad \lambda_1 = 5$

$$e^{\mathbf{A}t} = e^{5t}\left[\mathbf{I} + (\mathbf{A} - 5\mathbf{I})t\right] = e^{5t}\begin{bmatrix} 1+2t & t \\ -4t & 1-2t \end{bmatrix}$$

Each of the 3×3 coefficient matrices in Problems 27–30 has a characteristic polynomial of the form $p(\lambda) = (\lambda - \lambda_1)((\lambda - \lambda_2)^2$ yielding an eigenvalue λ_1 of multiplicity 1 and an eigenvalue λ_2 of multiplicity 2. We therefore use the method explained in Problem 19 above, and list here the results of the principal steps in the calculation of the fundamental matrix e^{At}.

27. $\quad \mathbf{A} = \begin{bmatrix} 2 & 0 & 0 \\ -7 & 9 & 7 \\ 0 & 0 & 2 \end{bmatrix}; \qquad p(\lambda) = (\lambda-9)(\lambda-2)^2; \qquad \lambda_1 = 9, \quad \lambda_2 = 2$

$$\frac{1}{p(\lambda)} = \frac{1}{49(\lambda-9)} - \frac{\lambda+5}{49(\lambda-3)^2}; \qquad a_1(\lambda) = \frac{1}{49}, \qquad a_2(\lambda) = -\frac{\lambda+5}{49}$$

$$\mathbf{P}_1 = a_1(\mathbf{A})\cdot(\mathbf{A}-\lambda_2\mathbf{I})^2 = \frac{1}{49}(\mathbf{A}-2\mathbf{I})^2 = \begin{bmatrix} 0 & 0 & 0 \\ -1 & 1 & 1 \\ 0 & 0 & 0 \end{bmatrix}$$

$$\mathbf{P}_2 = a_2(\mathbf{A})\cdot(\mathbf{A}-\lambda_1\mathbf{I}) = -\frac{1}{49}(\mathbf{A}+5\mathbf{I})(\mathbf{A}-9\mathbf{I}) = \begin{bmatrix} 1 & 0 & 0 \\ 1 & 0 & -1 \\ 0 & 0 & 1 \end{bmatrix}$$

$$e^{At} = e^{9t}\mathbf{P}_1 + e^{2t}\mathbf{P}_2\left(\mathbf{I}+(\mathbf{A}-2\mathbf{I})t\right) = \begin{bmatrix} e^{2t} & 0 & 0 \\ e^{2t}-e^{9t} & e^{9t} & -e^{2t}+e^{9t} \\ 0 & 0 & e^{2t} \end{bmatrix}$$

29. $\quad \mathbf{A} = \begin{bmatrix} -19 & 12 & 84 \\ 0 & 5 & 0 \\ -8 & 4 & 33 \end{bmatrix}; \qquad p(\lambda) = (\lambda-9)(\lambda-5)^2; \qquad \lambda_1 = 9, \quad \lambda_2 = 5$

$$\frac{1}{p(\lambda)} = \frac{1}{16(\lambda-9)} + \frac{1-\lambda}{16(\lambda-5)^2}; \qquad a_1(\lambda) = \frac{1}{16}, \qquad a_2(\lambda) = \frac{1-\lambda}{16}$$

$$\mathbf{P}_1 = a_1(\mathbf{A})\cdot(\mathbf{A}-\lambda_2\mathbf{I})^2 = \frac{1}{16}(\mathbf{A}-5\mathbf{I})^2 = \begin{bmatrix} -6 & 3 & 21 \\ 0 & 0 & 0 \\ -2 & 1 & 7 \end{bmatrix}$$

$$\mathbf{P}_2 = a_2(\mathbf{A})\cdot(\mathbf{A}-\lambda_1\mathbf{I}) = \frac{1}{16}(\mathbf{I}-\mathbf{A})(\mathbf{A}-9\mathbf{I}) = \begin{bmatrix} 7 & -3 & -21 \\ 0 & 1 & 0 \\ 2 & -1 & -6 \end{bmatrix}$$

$$e^{At} = e^{9t}\mathbf{P}_1 + e^{5t}\mathbf{P}_2\left(\mathbf{I}+(\mathbf{A}-5\mathbf{I})t\right) = \begin{bmatrix} 7e^{5t}-6e^{9t} & -3e^{5t}+3e^{9t} & -21e^{5t}+21e^{9t} \\ 0 & e^{5t} & 0 \\ 2e^{5t}-2e^{9t} & -e^{5t}+e^{9t} & -6e^{5t}+7e^{9t} \end{bmatrix}$$

In Problems 31–40 we use the methods of this section to find the matrix exponentials that were *given* in the statements of Problems 21–30 of Section 8.2. Once e^{At} is known, the desired particular solution $x(t)$ is provided by the variation of parameters formula

$$x(t) = e^{At}\left(x(0) + \int_0^t e^{-As} f(s)\,ds\right) \tag{7}$$

of Section 8.2. We give first the calculation of the matrix exponential using projection matrices and then the final result, which we obtained in each case using a computer algebra system to evaluate the right-hand side in Equation (7) — as described in the remarks preceding Problems 21–30 of Section 8.2. We illustrate this highly formal process by giving intermediate results in Problems 31, 34, and 39.

31. $A = \begin{bmatrix} 4 & -1 \\ 5 & 2 \end{bmatrix}, \quad \lambda_1 = -1, \quad \lambda_2 = 3$

$$P_1 = \frac{1}{-4}(A - 3I) = \frac{1}{4}\begin{bmatrix} -1 & 1 \\ -5 & 5 \end{bmatrix}, \quad P_2 = \frac{1}{4}(A + I) = \frac{1}{4}\begin{bmatrix} 5 & -1 \\ 5 & -1 \end{bmatrix}$$

$$e^{At} = e^{-t}P_1 + e^{3t}P_2 = \frac{1}{4}\begin{bmatrix} -e^{-t} + 5e^{3t} & e^{-t} - e^{3t} \\ -5e^{-t} + 5e^{3t} & 5e^{-t} - e^{3t} \end{bmatrix}$$

$$x(0) = \begin{bmatrix} 0 \\ 0 \end{bmatrix}, \quad f(t) = \begin{bmatrix} 18e^{2t} \\ 30e^{2t} \end{bmatrix}$$

$$e^{-As}f(s) = \frac{1}{4}\begin{bmatrix} -e^{s} + 5e^{-3s} & e^{s} - e^{-3s} \\ -5e^{s} + 5e^{-3s} & 5e^{s} - e^{-3s} \end{bmatrix}\begin{bmatrix} 18e^{2s} \\ 30e^{2s} \end{bmatrix} = \begin{bmatrix} 15e^{-s} + 3e^{3s} \\ 15e^{-s} + 15e^{3s} \end{bmatrix}$$

$$x(0) + \int_0^t e^{-As}f(s)\,ds = \begin{bmatrix} 0 \\ 0 \end{bmatrix} + \int_0^t \begin{bmatrix} 15e^{-s} + 3e^{3s} \\ 15e^{-s} + 15e^{3s} \end{bmatrix} ds = \begin{bmatrix} 14 - 15e^{-t} + e^{3t} \\ 10 - 15e^{-t} + 5e^{3t} \end{bmatrix}$$

$$x(t) = e^{At}\left(x(0) + \int_0^t e^{-As}f(s)\,ds\right) = \frac{1}{4}\begin{bmatrix} -e^{-t} + 5e^{3t} & e^{-t} - e^{3t} \\ -5e^{-t} + 5e^{3t} & 5e^{-t} - e^{3t} \end{bmatrix}\begin{bmatrix} 14 - 15e^{-t} + e^{3t} \\ 10 - 15e^{-t} + 5e^{3t} \end{bmatrix}$$

$$x(t) = \begin{bmatrix} -e^{-t} - 14e^{2t} + 15e^{3t} \\ -5e^{-t} - 10e^{2t} + 15e^{3t} \end{bmatrix}$$

33. $A = \begin{bmatrix} 3 & -1 \\ 9 & -3 \end{bmatrix}, \quad \lambda_1 = 0, \quad \lambda_2 = 0$

$$e^{At} = e^{-0t}\left[I + (A + 0I)t\right] = I + At = \begin{bmatrix} 1 + 3t & -t \\ 9t & 1 - 3t \end{bmatrix} \qquad \text{(as in Problems 21-26)}$$

$$\mathbf{x}(0) = \begin{bmatrix} 3 \\ 5 \end{bmatrix}, \qquad \mathbf{f}(t) = \begin{bmatrix} 7 \\ 5 \end{bmatrix}$$

$$\mathbf{x}(t) = \begin{bmatrix} 3+11t+8t^2 \\ 5+17t+24t^2 \end{bmatrix}$$

35. $\mathbf{A} = \begin{bmatrix} 2 & -5 \\ 1 & -2 \end{bmatrix}, \quad \lambda_1 = -i, \quad \lambda_2 = i$

$$\mathbf{P}_1 = \frac{1}{-2i}(\mathbf{A} - i\mathbf{I}) = \frac{1}{2}\begin{bmatrix} 1+2i & -5i \\ i & 1-2i \end{bmatrix}, \quad \mathbf{P}_2 = \frac{1}{2i}(\mathbf{A} + i\mathbf{I}) = \frac{1}{2}\begin{bmatrix} 1-2i & 5i \\ -i & 1+2i \end{bmatrix}$$

$$e^{\mathbf{A}t} = e^{-it}\mathbf{P}_1 + e^{it}\mathbf{P}_2 = (\cos t - i\sin t)\mathbf{P}_1 + (\cos t + i\sin t)\mathbf{P}_2$$

$$= \begin{bmatrix} \cos t + 2\sin t & -5\sin t \\ \sin t & \cos t - 2\sin t \end{bmatrix}$$

$$\mathbf{x}(0) = \begin{bmatrix} 0 \\ 0 \end{bmatrix}, \qquad \mathbf{f}(t) = \begin{bmatrix} 4t \\ 1 \end{bmatrix}$$

The solution vector $\mathbf{x}(t) = \begin{bmatrix} -1+8t+\cos t - 8\sin t \\ -2+4t+2\cos t - 3\sin t \end{bmatrix}$ is derived by variation of parameters in the solution to Problem 25 of Section 8.2.

37. $\mathbf{A} = \begin{bmatrix} 3 & -1 \\ 9 & -3 \end{bmatrix}, \quad \lambda_1 = 0, \quad \lambda_2 = 0$

$$e^{\mathbf{A}t} = e^{-0t}\left[\mathbf{I} + (\mathbf{A} + 0\mathbf{I})t\right] = \mathbf{I} + \mathbf{A}t = \begin{bmatrix} 1+2t & -4t \\ t & 1-2t \end{bmatrix} \qquad \text{(as in Problems 21–26)}$$

$$\mathbf{x}(0) = \begin{bmatrix} 0 \\ 0 \end{bmatrix}, \qquad \mathbf{f}(t) = \begin{bmatrix} 36t^2 \\ 6t \end{bmatrix}$$

$$\mathbf{x}(t) = \begin{bmatrix} 8t^3 + 6t^4 \\ 3t^2 - 2t^3 + 3t^4 \end{bmatrix}$$

39. $\mathbf{A} = \begin{bmatrix} 0 & -1 \\ 1 & 0 \end{bmatrix}, \quad \lambda_1 = -i, \quad \lambda_2 = i$

$$\mathbf{P}_1 = \frac{1}{-2i}(\mathbf{A} - i\mathbf{I}) = \frac{1}{2}\begin{bmatrix} 1 & -i \\ i & 1 \end{bmatrix}, \quad \mathbf{P}_2 = \frac{1}{2i}(\mathbf{A} + i\mathbf{I}) = \frac{1}{2}\begin{bmatrix} 1 & i \\ -i & 1 \end{bmatrix}$$

$$e^{\mathbf{A}t} = e^{-it}\mathbf{P}_1 + e^{it}\mathbf{P}_2$$

$$= (\cos t - i\sin t)\mathbf{P}_1 + (\cos t + i\sin t)\mathbf{P}_2 = \begin{bmatrix} \cos t & -\sin t \\ \sin t & \cos t \end{bmatrix}$$

$$\mathbf{x}(0) = \begin{bmatrix} 0 \\ 0 \end{bmatrix}, \qquad \mathbf{f}(t) = \begin{bmatrix} \sec t \\ 0 \end{bmatrix}$$

$$e^{-\mathbf{A}s}\mathbf{f}(s) = \begin{bmatrix} \cos s & \sin s \\ -\sin s & \cos s \end{bmatrix}\begin{bmatrix} \sec s \\ 0 \end{bmatrix} = \begin{bmatrix} 1 \\ -\tan s \end{bmatrix}$$

$$\mathbf{x}(0) + \int_0^t e^{-\mathbf{A}s}\mathbf{f}(s)\,ds = \begin{bmatrix} 0 \\ 0 \end{bmatrix} + \int_0^t \begin{bmatrix} 1 \\ -\tan s \end{bmatrix}\,ds = \begin{bmatrix} t \\ \ln(\cos t) \end{bmatrix}$$

$$\mathbf{x}(t) = e^{\mathbf{A}t}\left(\mathbf{x}(0) + \int_0^t e^{-\mathbf{A}s}\mathbf{f}(s)\,ds\right) = \begin{bmatrix} \cos t & -\sin t \\ \sin t & \cos t \end{bmatrix}\begin{bmatrix} t \\ \ln(\cos t) \end{bmatrix}$$

$$\mathbf{x}(t) = \begin{bmatrix} t\cos t - \ln(\cos t)\sin t \\ t\sin t + \ln(\cos t)\cos t \end{bmatrix}$$

In Problems 41-46 we apply the methods of this section to compute the exponential matrices of the coefficient matrices in Problems 23-26, 29, and 31 of Section 7.6.

Each of the 3×3 coefficient matrices in Problems 41 and 42 has a characteristic polynomial of the form $p(\lambda) = (\lambda - \lambda_1)((\lambda - \lambda_2)^2$ yielding an eigenvalue λ_1 of multiplicity 1 and an eigenvalue λ_2 of multiplicity 2. We therefore use the method explained in Problem 19 above, and list here the results of the principal steps in the calculation of the fundamental matrix $e^{\mathbf{A}t}$.

41. $\quad \mathbf{A} = \begin{bmatrix} 39 & 8 & -16 \\ -36 & -5 & 16 \\ 72 & 16 & -29 \end{bmatrix}; \qquad p(\lambda) = (\lambda+1)(\lambda-3)^2, \quad \lambda_1 = -1, \quad \lambda_3 = 3$

$$\frac{1}{p(\lambda)} = \frac{1}{16(\lambda+1)} + \frac{7-\lambda}{16(\lambda-3)^2}; \qquad a_1(\lambda) = \frac{1}{16}, \qquad a_2(\lambda) = \frac{7-\lambda}{16}$$

$$\mathbf{P}_1 = a_1(\mathbf{A})\cdot(\mathbf{A}-\lambda_2\mathbf{I})^2 = \frac{1}{16}(\mathbf{A}-3\mathbf{I})^2 = \begin{bmatrix} -9 & -2 & 4 \\ 9 & 2 & -4 \\ -18 & -4 & 8 \end{bmatrix}$$

$$\mathbf{P}_2 = a_2(\mathbf{A})\cdot(\mathbf{A}-\lambda_1\mathbf{I}) = \frac{1}{16}(7\mathbf{I}-\mathbf{A})(\mathbf{A}+\mathbf{I}) = \begin{bmatrix} 10 & 2 & -4 \\ -9 & -1 & 4 \\ 18 & 4 & -7 \end{bmatrix}$$

$$e^{\mathbf{A}t} = e^{-t}\mathbf{P}_1 + e^{3t}\mathbf{P}_2\left(\mathbf{I}+(\mathbf{A}-3\mathbf{I})\,t\right) = \begin{bmatrix} -9e^{-t}+10e^{3t} & -2e^{-t}+2e^{3t} & 4e^{-t}-4e^{3t} \\ 9e^{-t}-9e^{3t} & 2e^{-t}-e^{3t} & -4e^{-t}+4e^{3t} \\ -18e^{-t}+18e^{3t} & -4e^{-t}+4e^{3t} & 8e^{-t}-7e^{3t} \end{bmatrix}$$

In each of Problems 43 and 44, the given 3×3 coefficient matrix $\mathbf{A}$ has characteristic polynomial of the form $p(\lambda) = (\lambda-\lambda_1)^3$ and thus a single eigenvalue λ_1 of multiplicity 3. Consequently, Equations (25) and (26) in Section 8.3 of the text imply that $a_1(\lambda) = b_1(\lambda) = 1$ and hence that the associated projection matrix $\mathbf{P}_1 = \mathbf{I}$. Therefore Equation (35) in the text reduces (with $q = 1$ and $m_1 = 3$) to

$$e^{\mathbf{A}t} = e^{\lambda_1 t}\left[\mathbf{I}+(\mathbf{A}-\lambda_1\mathbf{I})t+\tfrac{1}{2}(\mathbf{A}-\lambda_1\mathbf{I})^2\,t^2\right]. \tag{8}$$

43. $\mathbf{A} = \begin{bmatrix} -2 & 17 & 4 \\ -1 & 6 & 1 \\ 0 & 1 & 2 \end{bmatrix};$ $p(\lambda) = (\lambda-2)^2,$ $\lambda_1 = 2$

Substitution of $\mathbf{A}$ and $\lambda_1 = 2$ into the formula in (8) above yields

$$e^{\mathbf{A}t} = \frac{1}{2}e^{2t}\begin{bmatrix} -t^2-8t+2 & 4t^2+34t & t^2+8t \\ -2t & 8t+2 & 2t \\ -t^2 & 4t^2+2t & t^2+2 \end{bmatrix}.$$

45. $\mathbf{A} = \begin{bmatrix} -1 & 1 & 1 & -2 \\ 7 & -4 & -6 & 11 \\ 5 & -1 & 1 & 3 \\ 6 & -2 & -2 & 6 \end{bmatrix};$ $p(\lambda) = (\lambda+1)^2(\lambda-2)^2,$ $\lambda_1 = -1,\quad \lambda_2 = 2$

Since $\mathbf{A}$ has two eigenvalues $\lambda_1 = -1$ and $\lambda_2 = 2$ each of multiplicity 2, Equation (35) in the text reduces (with $q = m_1 = m_2 = 2$) to

$$e^{\mathbf{A}t} = e^{\lambda_1 t}\mathbf{P}_1\left[\mathbf{I}+(\mathbf{A}-\lambda_1\mathbf{I})t\right]+e^{\lambda_2 t}\mathbf{P}_2\left[\mathbf{I}+(\mathbf{A}-\lambda_2\mathbf{I})t\right]. \tag{9}$$

To calculate the projection matrices $\mathbf{P}_1$ and $\mathbf{P}_2$, we start with the partial fraction decomposition

$$\frac{1}{p(\lambda)} = \frac{2\lambda+5}{27(\lambda+1)^2}+\frac{7-2\lambda}{27(\lambda-2)^2}.$$

Then Equation (25) in the text implies that $a_1(\lambda) = (2\lambda + 5)/27$ and $a_2(\lambda) = (7 - 2\lambda)/27$. Hence Equation (26) yields

$$\mathbf{P}_1 = \frac{1}{27}(2\mathbf{A} + 5\mathbf{I})(\mathbf{A} + \mathbf{I})^2 = \begin{bmatrix} 1 & 0 & 0 & 0 \\ -3 & 1 & 1 & -2 \\ -1 & 0 & 0 & 0 \\ -2 & 0 & 0 & 0 \end{bmatrix}$$

and

$$\mathbf{P}_2 = \frac{1}{27}(7\mathbf{I} - 2\mathbf{A})(\mathbf{A} - 2\mathbf{I})^2 = \begin{bmatrix} 0 & 0 & 0 & 0 \\ 3 & 0 & -1 & 2 \\ 1 & 0 & 1 & 0 \\ 2 & 0 & 0 & 1 \end{bmatrix}.$$

When we substitute these eigenvalues and projection matrices in Eq. (9) above we get

$$e^{\mathbf{A}t} = \begin{bmatrix} e^{-t} & te^{-t} & te^{-t} & -2te^{-t} \\ -3e^{-t} + (3-2t)e^{2t} & (1-3t)e^{-t} & (1-3t)e^{-t} - e^{2t} & 2(3t-1)e^{-t} + (2-t)e^{2t} \\ -e^{-t} + (1+2t)e^{2t} & -te^{-t} & -te^{-t} + e^{2t} & 2te^{-t} + te^{2t} \\ -2e^{-t} + 2e^{2t} & -2te^{-t} & -2te^{-t} & 4te^{-t} + e^{2t} \end{bmatrix}.$$

In Problems 47-50 we revisit the systems of Problems 2, 4, 5, and 7 in Section 7.5. In each problem we use the given values of $k_1, k_2,$ and k_3 to set up the second-order linear system $\mathbf{x}'' = \mathbf{A}\mathbf{x}$ with coefficient matrix

$$\mathbf{A} = \begin{bmatrix} -k_1 - k_2 & k_2 \\ k_2 & -k_2 - k_3 \end{bmatrix}.$$

We then calculate the particular solution

$$\mathbf{x}(t) = e^{\sqrt{\mathbf{A}}\,t}\mathbf{c}_1 + e^{-\sqrt{\mathbf{A}}\,t}\mathbf{c}_2 \tag{10}$$

given by Theorem 3 with $\mathbf{c}_1 = \begin{bmatrix} 1 & 0 \end{bmatrix}^T$ and $\mathbf{c}_2 = \begin{bmatrix} 0 & 1 \end{bmatrix}^T$. Given the distinct eigenvalues λ_1, λ_2 and the projection matrices $\mathbf{P}_1, \mathbf{P}_2$ of $\mathbf{A}$, the matrix exponential needed in (10) is given by

$$e^{\sqrt{\mathbf{A}}\,t} = e^{\sqrt{\lambda_1}\,t}\mathbf{P}_1 + e^{\sqrt{\lambda_2}\,t}\mathbf{P}_2. \tag{11}$$

47. $A = \begin{bmatrix} -5 & 4 \\ 4 & -5 \end{bmatrix}$; $p(\lambda) = \lambda^2 - 10\lambda + 9$, $\lambda_1 = -1$, $\lambda_2 = -9$

$$P_1 = \frac{1}{-10}(A + 9I) = \frac{1}{2}\begin{bmatrix} 1 & 1 \\ 1 & 1 \end{bmatrix}, \quad P_2 = \frac{1}{10}(A + I) = \frac{1}{2}\begin{bmatrix} 1 & -1 \\ -1 & 1 \end{bmatrix}$$

$$e^{\sqrt{A}\,t} = e^{\sqrt{\lambda_1}\,t}\,P_1 + e^{\sqrt{\lambda_2}\,t}\,P_2 = e^{it}\,P_1 + e^{3it}\,P_2 = \frac{1}{2}\begin{bmatrix} e^{it} + e^{3it} & e^{it} - e^{3it} \\ e^{it} - e^{3it} & e^{it} + e^{3it} \end{bmatrix}$$

$$\mathbf{x}(t) = e^{\sqrt{A}\,t}\mathbf{c}_1 + e^{-\sqrt{A}\,t}\mathbf{c}_2 = \frac{1}{2}\begin{bmatrix} \left(e^{it} + e^{-it}\right) - \left(e^{3it} - e^{3it}\right) \\ \left(e^{it} + e^{-it}\right) + \left(e^{3it} - e^{3it}\right) \end{bmatrix} = \begin{bmatrix} \cos t \\ \cos t \end{bmatrix} + i\begin{bmatrix} -\sin 3t \\ \sin 3t \end{bmatrix}$$

Note that $\mathbf{x}(t)$ is a linear combination of two motions — one in which the two masses move in the same direction with frequency $\omega_1 = 1$ and with equal amplitudes, and one in which they move in opposite directions with frequency $\omega_2 = 3$ and with equal amplitudes.

49. $A = \begin{bmatrix} -3 & 1 \\ 1 & -3 \end{bmatrix}$; $p(\lambda) = \lambda^2 - 6\lambda + 8$, $\lambda_1 = -2$, $\lambda_2 = -4$

$$P_1 = \frac{1}{-6}(A + 4I) = \frac{1}{2}\begin{bmatrix} 1 & 1 \\ 1 & 1 \end{bmatrix}, \quad P_2 = \frac{1}{6}(A + 2I) = \frac{1}{2}\begin{bmatrix} 1 & -1 \\ -1 & 1 \end{bmatrix}$$

$$e^{\sqrt{A}\,t} = e^{\sqrt{\lambda_1}\,t}\,P_1 + e^{\sqrt{\lambda_2}\,t}\,P_2 = e^{\sqrt{2}\,it}\,P_1 + e^{2it}\,P_2 = \frac{1}{2}\begin{bmatrix} e^{\sqrt{2}\,it} + e^{2it} & e^{\sqrt{2}\,it} - e^{2it} \\ e^{\sqrt{2}\,it} - e^{2it} & e^{\sqrt{2}\,it} + e^{2it} \end{bmatrix}$$

$$\mathbf{x}(t) = e^{\sqrt{A}\,t}\mathbf{c}_1 + e^{-\sqrt{A}\,t}\mathbf{c}_2 = \frac{1}{2}\begin{bmatrix} \left(e^{\sqrt{2}\,it} + e^{-\sqrt{2}\,it}\right) - \left(e^{2it} - e^{2it}\right) \\ \left(e^{\sqrt{2}\,it} + e^{-\sqrt{2}\,it}\right) + \left(e^{2it} - e^{2it}\right) \end{bmatrix} = \begin{bmatrix} \cos\sqrt{2}\,t \\ \cos\sqrt{2}\,t \end{bmatrix} + i\begin{bmatrix} -\sin 2t \\ \sin 2t \end{bmatrix}$$

Note that $\mathbf{x}(t)$ is a linear combination of two motions — one in which the two masses move in the same direction with frequency $\omega_1 = \sqrt{2}$ and with equal amplitudes, and one in which they move in opposite directions with frequency $\omega_2 = 2$ and with equal amplitudes.

CHAPTER 9

NONLINEAR SYSTEMS AND PHENOMENA

SECTION 9.1

STABILITY AND THE PHASE PLANE

1. The only solution of the homogeneous system $2x - y = 0$, $x - 3y = 0$ is the origin $(0,0)$. The only figure among Figs. 9.1.11 through 9.1.18 showing a single critical point at the origin is Fig. 9.1.13. Thus the only critical point of the given autonomous system is the saddle point $(0,0)$ shown in Figure 9.1.13 in the text.

3. The only solution of the system $x - 2y + 3 = 0$, $x - y + 2 = 0$ is the point $(-1,1)$. The only figure among Figs. 9.1.11 through 9.1.18 showing a single critical point at $(-1,1)$ is Fig. 9.1.18. Thus the only critical point of the given autonomous system is the stable center $(-1,1)$ shown in Figure 9.1.18 in the text.

5. The first equation $1 - y^2 = 0$ gives $y = 1$ or $y = -1$ at a critical point. Then the second equation $x + 2y = 0$ gives $x = -2$ or $x = 2$, respectively. The only figure among Figs. 9.1.11 through 9.1.18 showing two critical points at $(-2,1)$ and $(2,-1)$ is Fig. 9.1.11. Thus the critical points of the given autonomous system are the spiral point $(-2,1)$ and the saddle point $(2,1)$ shown in Figure 9.1.11 in the text.

7. The first equation $4x - x^3 = 0$ gives $x = -2$, $x = 0$, or $x = 2$ at a critical point. Then the second equation $x - 2y = 0$ gives $y = -1$, $y = 0$, or $y = 1$, respectively. The only figure among Figs. 9.1.11 through 9.1.18 showing three critical points at $(-2,-1)$, $(0,0)$, and $(2,1)$ is Fig. 9.1.14. Thus the critical points of the given autonomous system are the spiral point $(0,0)$ and the saddle points $(-2,-1)$ and $(2,1)$ shown in Figure 9.1.14 in the text.

In each of Problems 9-12 we need only set $x' = x'' = 0$ and solve the resulting equation for x.

9. The equation $4x - x^3 = x(4 - x^2) = 0$ has the three solutions $x = 0, \pm 2$. This gives the three equilibrium solutions $x(t) \equiv 0$, $x(t) \equiv 2$, and $x(t) \equiv -2$ of the given 2nd-order differential equation. A phase plane portrait for the equivalent 1st-order system $x' = y$,

$y' = -4x + x^3$ is shown in the figure. We observe that the critical point $(0,0)$ in the phase plane appears to be a center, whereas the points $(\pm 2, 0)$ appear to be saddle points.

Problem 9

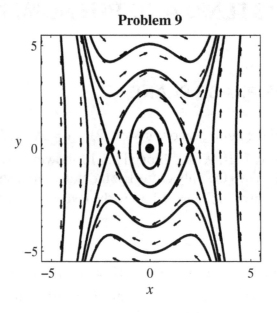

11.   The equation $4\sin x = 0$ is satisfied by $x = n\pi$ for any integer n. Thus the given 2nd-order equation has infinitely many equilibrium solutions: $x(t) = n\pi$ for any integer n. A phase portrait for the equivalent 1st-order system $x' = y$, $y' = -3y - 4\sin x$ is shown. We observe that the critical point $(n\pi, 0)$ in the phase plane looks like a spiral sink if n is even, but a saddle point if n is odd.

Problem 11

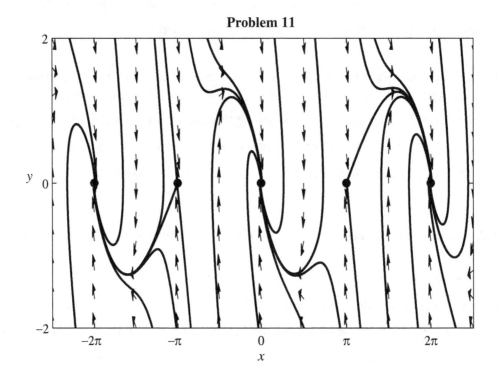

In Problems 13–16, the given x- and y-equations are independent exponential differential equations that we can solve immediately by inspection.

13.   Solution: $x(t) = x_0 e^{-2t}$, $y(t) = y_0 e^{-2t}$.

Then $y = \left(\dfrac{y_0}{x_0} \right) x = kx$, so the trajectories are straight lines through the origin.   Clearly

$x(t), y(t) \to 0$ as $t \to +\infty$, so the origin is a stable proper node like the one shown.

Problem 13

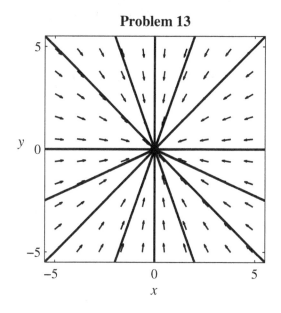

Problem 15

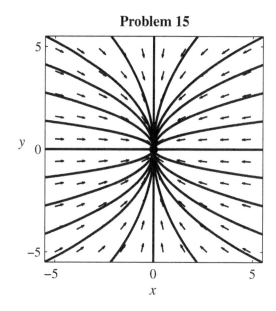

15.   Solution: $x(t) = x_0 e^{-2t}$, $y(t) = y_0 e^{-t}$.

Then $x = \left(\dfrac{x_0}{y_0^2} \right) \left(y_0 e^{-t} \right)^2 = ky^2$, so the trajectories are parabolas of the form $x = ky^2$, and

clearly $x(t), y(t) \to 0$ as $t \to +\infty$. Thus the origin is a stable improper node like the one shown.

17.   Differentiation of the first equation and substitution using the second one gives $x'' = y' = x$, so $x'' + x = 0$. Solving this differential equation for $x(t)$ and using the fact that $y = x'$ lead to the general solution

$$x(t) = A\cos t + B\sin t, \quad y(t) = B\cos t - A\sin t.$$

Then

$$
\begin{aligned}
x^2 + y^2 &= \left(A\cos t + B\sin t \right)^2 + \left(B\cos t - A\sin t \right)^2 \\
&= \left(A^2 + B^2 \right) \cos^2 t + \left(A^2 + B^2 \right) \sin^2 t \\
&= A^2 + B^2.
\end{aligned}
$$

Therefore the trajectories are clockwise-oriented circles centered at the origin, and the origin is a stable center as shown.

Problem 17   **Problem 19**

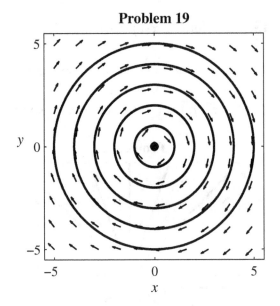

19.   Elimination of y as in Problem 17 gives $x'' + 4x = 0$, and then using $y = \dfrac{1}{2}x'$ we get the general solution

$$x(t) = A\cos 2t + B\sin 2t, \quad y(t) = B\cos 2t - A\sin 2t.$$

Then $x^2 + y^2 = A^2 + B^2$, so the origin is a stable center, and the trajectories are clockwise-oriented circles centered at $(0,0)$, as shown.

21.   We want to solve the system

$$-ky + x\left(1 - x^2 - y^2\right) = 0,$$
$$kx + y\left(1 - x^2 - y^2\right) = 0.$$

If we multiply the first equation by $-y$ and the second one by x, then add the two results, we get $k\left(x^2 + y^2\right) = 0$. It therefore follows that $x = y = 0$.

23.   The equation $\dfrac{dy}{dx} = -\dfrac{x}{y}$ separates to $x\,dx + y\,dy = 0$, so $x^2 + y^2 = C$. Thus the trajectories consist of the origin $(0,0)$ and the circles $x^2 + y^2 = C > 0$, as shown.

Problem 23

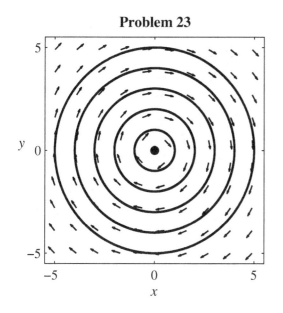

Problem 25

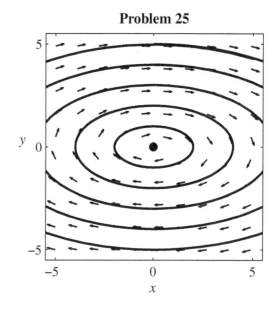

25.   The equation $\dfrac{dy}{dx} = -\dfrac{x}{4y}$ separates to $x\,dx + 4y\,dy = 0$, so $x^2 + 4y^2 = C$. Thus the trajectories consist of the origin $(0,0)$ and the ellipses $x^2 + 4y^2 = C > 0$, as shown.

27.   If $\phi(t) = x(t+\gamma)$ and $\psi(t) = y(t+\gamma)$, then

$$\phi'(t) = x'(t+\gamma) = y(t+\gamma) = \psi(t),$$

but

$$\psi'(t) = y'(t+\gamma) = (t+\gamma)x(t+\gamma) = t\phi(t) + \gamma\phi(t) \neq t\phi(t).$$

SECTION 9.2

LINEAR AND ALMOST LINEAR SYSTEMS

In Problems 1–10 we first find the roots λ_1 and λ_2 of the characteristic equation of the coefficient matrix of the given linear system. We can then read the type and stability of the critical point $(0,0)$ from the table of Figure 9.2.4 in the text.

1.   The roots $\lambda_1 = -1$ and $\lambda_2 = -3$ of the characteristic equation $\lambda^2 + 4\lambda + 3 = 0$ are both negative, so $(0,0)$ is an asymptotically stable node.

Problem 1

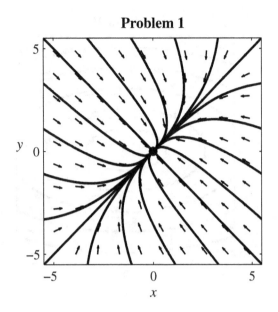

Problem 3

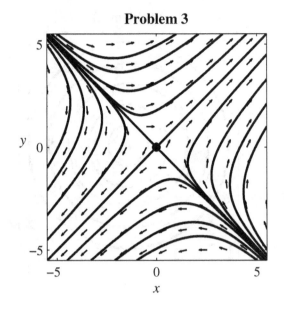

3. The roots $\lambda_1 = -1$ and $\lambda_2 = 3$ of the characteristic equation $\lambda^2 - 2\lambda - 3 = 0$ have different signs, so $(0,0)$ is an unstable saddle point.

5. The roots $\lambda_1 = \lambda_2 = -1$ of the characteristic equation $\lambda^2 + 2\lambda + 1 = 0$ are negative and equal, so $(0,0)$ is an asymptotically stable node.

Problem 5

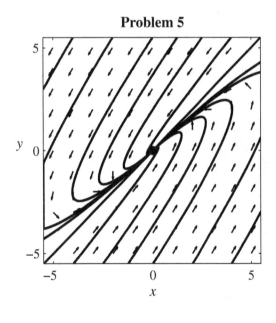

Problem 7

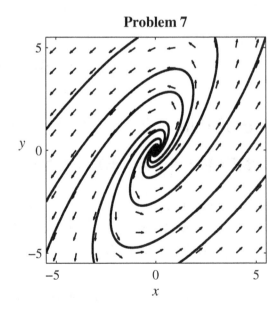

7. The roots $\lambda_1, \lambda_2 = 1 \pm 2i$ of the characteristic equation $\lambda^2 - 2\lambda + 5 = 0$ are complex conjugates with positive real part, so $(0,0)$ is an unstable spiral point as shown in the left-hand figure below.

9.   The roots $\lambda_1, \lambda_2 = \pm 2i$ of the characteristic equation $\lambda^2 + 4 = 0$ are pure imaginary, so $(0,0)$ is a stable (but not asymptotically stable) center.

Problem 9

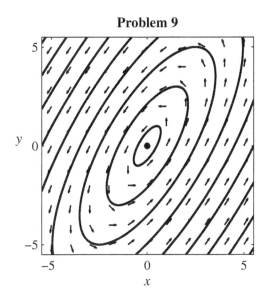

Problem 11

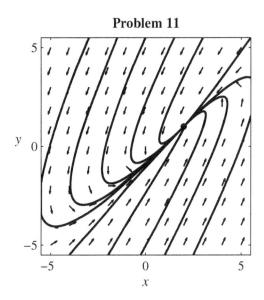

11.   The Jacobian matrix $\mathbf{J} = \begin{bmatrix} 1 & -2 \\ 3 & -4 \end{bmatrix}$ has characteristic equation $\lambda^2 + 3\lambda + 2 = 0$ and eigenvalues $\lambda_1 = -1$, $\lambda_2 = -2$ that are both negative. Hence the critical point $(2,1)$ is an asymptotically stable node.

13.   The Jacobian matrix $\mathbf{J} = \begin{bmatrix} 2 & -1 \\ 3 & -2 \end{bmatrix}$ has characteristic equation $\lambda^2 - 1 = 0$ and eigenvalues $\lambda_1 = -1$, $\lambda_2 = +1$ having different signs. Hence the critical point $(2,2)$ is an unstable saddle point.

Problem 13

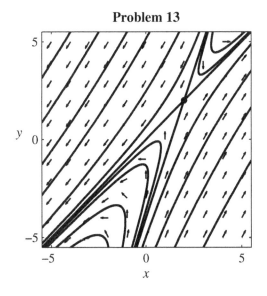

Problem 15

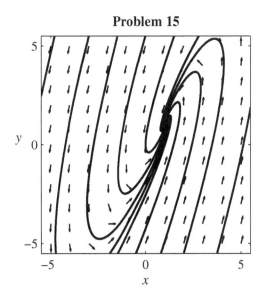

15.   The Jacobian matrix $\mathbf{J} = \begin{bmatrix} 1 & -1 \\ 5 & -3 \end{bmatrix}$ has characteristic equation $\lambda^2 + 2\lambda + 2 = 0$ and eigen-

values $\lambda_1, \lambda_2 = -1 \pm i$ that are complex conjugates with negative real part. Hence the crit-

ical point $(1,1)$ is an asymptotically stable spiral point.

17.   The Jacobian matrix $\mathbf{J} = \begin{bmatrix} 1 & -5 \\ 1 & -1 \end{bmatrix}$ has characteristic equation $\lambda^2 + 4 = 0$ and pure imagi-

nary eigenvalues $\lambda_1, \lambda_2 = \pm 2i$. Hence $\left(\dfrac{5}{2}, -\dfrac{1}{2} \right)$ is a stable (but not asymptotically stable)

center.

Problem 17

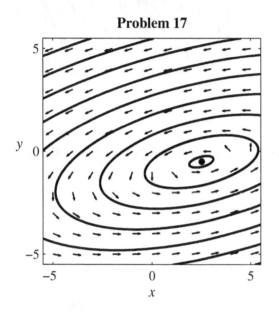

Problem 19

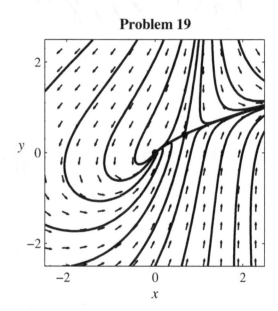

In each of Problems 19-28 we first calculate the Jacobian matrix $\mathbf{J}$ and its eigenvalues at $(0,0)$
and at each of the other critical points we observe in our phase portrait for the given system.
Then we apply Theorem 2 to determine as much as we can about the type and stability of each of
these critical points of the given almost linear system. Finally we draw a phase portrait that
shows typical solution curves for the given system.

19. $\mathbf{J} = \begin{bmatrix} 1+2y & -3+2x \\ 4-y & -6-x \end{bmatrix}$

At $(0,0)$: The Jacobian matrix $\mathbf{J} = \begin{bmatrix} 1 & -3 \\ 4 & -6 \end{bmatrix}$ has characteristic equation $\lambda^2 + 5\lambda + 6 = 0$

and eigenvalues $\lambda_1 = -3$, $\lambda_2 = -2$ that are both negative. Hence $(0,0)$ is an asymptotically stable node of the given almost linear system.

At $(2/3, 2/5)$: The Jacobian matrix $\mathbf{J} = \begin{bmatrix} 9/5 & -5/3 \\ 18/5 & -20/3 \end{bmatrix}$ has characteristic equation

$\lambda^2 + \dfrac{73}{15}\lambda - 6 = 0$ and approximate eigenvalues $\lambda_1 \approx -5.89$, $\lambda_2 \approx 1.02$ with different

signs. Hence $(2/3, 2/5)$ is a saddle point.

21. $\mathbf{J} = \begin{bmatrix} 1+2x & 2+2y \\ 2-3y & -2-3x \end{bmatrix}$

At $(0,0)$: The Jacobian matrix $\mathbf{J} = \begin{bmatrix} 1 & 2 \\ 2 & -2 \end{bmatrix}$ has characteristic equation $\lambda^2 + \lambda - 6 = 0$

and eigenvalues $\lambda_1 = -3$, $\lambda_2 = 2$ with different signs. Hence $(0,0)$ is a saddle point of the given almost linear system.

At $(-0.51, -2.12)$: The Jacobian matrix $\mathbf{J} \approx \begin{bmatrix} -0.014 & -2.236 \\ 8.354 & -0.479 \end{bmatrix}$ has complex conjugate

eigenvalues $\lambda_1, \lambda_2 \approx -0.25 \pm 4.32i$ with negative real parts. Hence $(-0.51, -2.12)$ is a spiral sink.

Problem 21

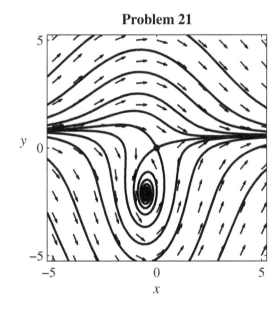

Problem 23

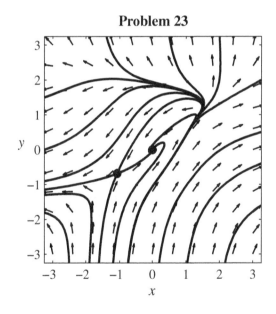

23. $\mathbf{J} = \begin{bmatrix} 2+3x^2 & -5 \\ 4 & -6+4y^3 \end{bmatrix}$

At $(0,0)$: The Jacobian matrix $\mathbf{J} = \begin{bmatrix} 2 & -5 \\ 4 & -6 \end{bmatrix}$ has characteristic equation $\lambda^2 + 4\lambda + 8 = 0$

and complex conjugate eigenvalues $\lambda_1, \lambda_2 = -2 \pm 2i$, with negative real part. Hence $(0,0)$ is a spiral sink of the given almost linear system.

At $(-1.08, -0.68)$: The Jacobian matrix $\mathbf{J} \approx \begin{bmatrix} 5.495 & -5 \\ 4 & -7.276 \end{bmatrix}$ has eigenvalues

$\lambda_1 \approx -5.45$, $\lambda_2 \approx 3.67$ with different signs. Hence $(-1.08, -0.68)$ is a saddle point.

25. $\mathbf{J} = \begin{bmatrix} 1+3y & -2+3x \\ 2-2x & -3-2y \end{bmatrix}$

At $(0,0)$: The Jacobian matrix $\mathbf{J} = \begin{bmatrix} 1 & -2 \\ 2 & -3 \end{bmatrix}$ has characteristic equation $\lambda^2 + 2\lambda + 1 = 0$

and equal negative eigenvalues $\lambda_1 = -1$, $\lambda_1 = -1$. Hence $(0,0)$ is either a nodal sink or a spiral sink of the given almost linear system.

At $(0.74, -3.28)$: The Jacobian matrix $\mathbf{J} \approx \begin{bmatrix} -8.853 & 0.226 \\ 0.516 & 3.568 \end{bmatrix}$ has real eigenvalues

$\lambda_1 \approx -8.86$, $\lambda_2 \approx 3.58$ with different signs. Hence $(0.74, -3.28)$ is a saddle point.

At $(2.47, -0.46)$: The Jacobian matrix $\mathbf{J} \approx \begin{bmatrix} -0.370 & 5.410 \\ -2.940 & -2.087 \end{bmatrix}$ has complex conjugate

eigenvalues $\lambda_1, \lambda_2 \approx -1.23 \pm 3.89i$ with negative real part. Hence $(2.47, -0.46)$ is a spiral sink.

At $(0.121, 0.074)$: The Jacobian matrix $\mathbf{J} \approx \begin{bmatrix} 1.222 & -1.636 \\ 1.758 & -3.148 \end{bmatrix}$ has real eigenvalues

$\lambda_1 \approx -2.34$, $\lambda_2 \approx 0.42$ with different signs. Hence $(0.121, 0.074)$ is a saddle point.

Figure a shows clearly the first three of these critical points. Figure b is a close-up near the origin with the final critical point now visible.

Problem 25a

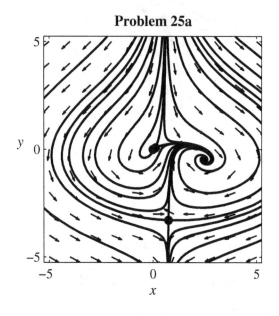

Problem 25b

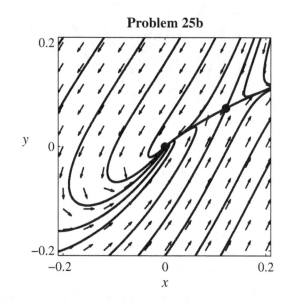

27. $\mathbf{J} = \begin{bmatrix} 1+4x^3 & -1-2y \\ 2-2x & -1+4y^3 \end{bmatrix}$

At $(0,0)$: The Jacobian matrix $\mathbf{J} = \begin{bmatrix} 1 & -1 \\ 2 & -1 \end{bmatrix}$ has characteristic equation $\lambda^2 + 1 = 0$ and equal positive eigenvalues $\lambda_1, \lambda_2 = \pm i$. Hence $(0,0)$ is either a center or a spiral point, but its stability is not determined by Theorem 2.

At $(-0.254, -0.507)$: The Jacobian matrix $\mathbf{J} \approx \begin{bmatrix} 0.934 & 0.014 \\ 2.508 & -1.521 \end{bmatrix}$ has real eigenvalues $\lambda_1 \approx -1.53$, $\lambda_2 \approx 0.95$ with different signs. Hence $(-0.254, -0.507)$ is a saddle point.

At $(-1.557, 1.637)$: The Jacobian matrix $\mathbf{J} \approx \begin{bmatrix} -14.087 & -4.273 \\ 5.113 & 16.532 \end{bmatrix}$ has real eigenvalues $\lambda_1 \approx -13.36$, $\lambda_2 \approx 15.80$ with different signs. Hence $(-1.557, 1.637)$ is a saddle point.

At $(-1.070, -1.202)$: The Jacobian matrix $\mathbf{J} \approx \begin{bmatrix} -3.905 & 1.403 \\ 4.141 & -7.940 \end{bmatrix}$ has unequal negative eigenvalues $\lambda_1 \approx -9.07$, $\lambda_2 \approx -2.78$. Hence $(-1.070, -1.202)$ is a nodal sink.

Figure a shows these four critical points. Figure b, a close-up, suggests that the origin *may* be a stable center; note both the trajectory in grey that seems to spiral inward, and the oval-shaped trajectories that lie closer to the origin.

Problem 27a

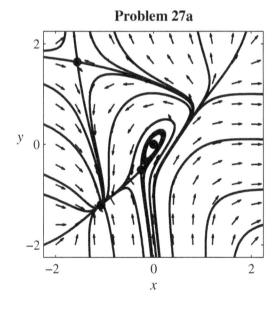

Problem 27b

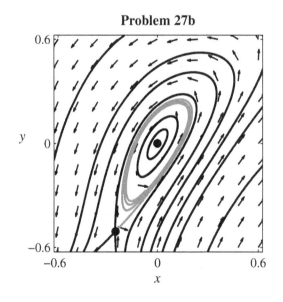

29. $\mathbf{J} = \begin{bmatrix} 1 & -1 \\ 2x & -1 \end{bmatrix}$

At $(0,0)$: The Jacobian matrix $\mathbf{J} = \begin{bmatrix} 1 & -1 \\ 0 & -1 \end{bmatrix}$ has characteristic equation $\lambda^2 - 1 = 0$ and real eigenvalues $\lambda_1 = -1$, $\lambda_2 = +1$ with different signs. Hence $(0,0)$ is a saddle point.

At $(1,1)$: The Jacobian matrix $\mathbf{J} = \begin{bmatrix} 1 & -1 \\ 2 & -1 \end{bmatrix}$ has characteristic equation $\lambda^2 + 1 = 0$ and pure imaginary eigenvalues $\lambda_1, \lambda_2 = \pm i$. Hence $(1,1)$ is either a center or a spiral point, but its stability is not determined by Theorem 2. However the figure suggests that $(1,1)$ is a stable center.

Problem 29

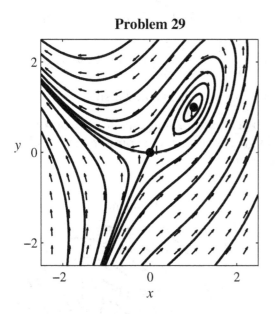

Problem 31

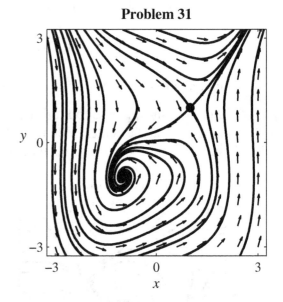

31. $\mathbf{J} = \begin{bmatrix} 0 & 2y \\ 3x^2 & -1 \end{bmatrix}$

At $(1,1)$: The Jacobian matrix $\mathbf{J} = \begin{bmatrix} 0 & 2 \\ 3 & -1 \end{bmatrix}$ has characteristic equation $\lambda^2 + \lambda - 6 = 0$ and real eigenvalues $\lambda_1 = -3$, $\lambda_2 = +2$ with different signs. Hence $(1,1)$ is a saddle point.

At $(-1,-1)$: The Jacobian matrix $\mathbf{J} = \begin{bmatrix} 0 & -2 \\ 3 & -1 \end{bmatrix}$ has characteristic equation $\lambda^2 + \lambda + 6 = 0$ and complex conjugate eigenvalues $\lambda_1, \lambda_2 \approx -0.5 \pm 2.398i$ with negative real part. Hence $(-1,-1)$ is a spiral sink.

33.   The characteristic equation of the given linear system is $(\lambda - \varepsilon)^2 + 1 = 0$, with characteristic roots $\lambda_1, \lambda_2 = \varepsilon \pm i$.

(a) So if $\varepsilon < 0$, then λ_1, λ_2 are complex conjugates with negative real part, and hence $(0,0)$ is an asymptotically stable spiral point.

(b) If $\varepsilon = 0$, then $\lambda_1, \lambda_2 = \pm i$ (pure imaginary), so $(0,0)$ is a stable center.

(c) If $\varepsilon > 0$, the situation is the same as in (a) except that the real part is positive, so $(0,0)$ is an unstable spiral point.

35.   **(a)** If $h = 0$, then we have the familiar system $x' = y$, $y' = -x$ with circular trajectories about the origin, which is therefore a center.

(b) The change to polar coordinates as in Example 6 of Section 9.1 is routine, yielding $r' = hr^3$ and $\theta' = -1$.

(c) If $h = -1$, then $r' = -r^3$ integrates to give $2r^2 = \dfrac{1}{t + C}$, where C is a positive constant, so clearly $r \to 0$ as $t \to +\infty$, and thus the origin is a stable spiral point.

(d) If $h = +1$, then $r' = r^3$ integrates to give $2r^2 = -\dfrac{1}{t + C}$, where $C = -B$ is a positive constant. It follows that $2r^2 = \dfrac{1}{B - t}$, so now r increases as t starts at 0 and increases.

37.   The substitution $y = vx$ in the homogeneous first-order equation

$$\frac{dy}{dx} = \frac{y(2x^3 - y^3)}{x(x^3 - 2y^3)}$$

yields

$$x\frac{dv}{dx} = -\frac{v^4 + v}{2v^3 - 1}.$$

Separating the variables and integrating by partial fractions, we get

$$\int\left(-\frac{1}{v} + \frac{1}{v+1} + \frac{2v-1}{v^2 - v + 1}\right)dv = -\int\frac{dx}{x},$$

or

$$\ln\left[(v+1)(v^2 - v + 1)\right] = \ln v - \ln x + \ln C,$$

or

$$(v+1)(v^2 - v + 1) = \frac{Cv}{x},$$

or

$$v^3 + 1 = \frac{Cv}{x}.$$

Finally, the replacement $v = \dfrac{y}{x}$ yields $x^3 + y^3 = Cxy$.

SECTION 9.3

ECOLOGICAL APPLICATIONS: PREDATORS AND COMPETITORS

1.　$\mathbf{J} = \begin{bmatrix} 200 - 4y & -4x \\ 2y & -150 + 2x \end{bmatrix}$

At $(0,0)$: The Jacobian matrix $\mathbf{J} = \begin{bmatrix} 200 & 0 \\ 0 & -150 \end{bmatrix}$ has characteristic equation

$(200 - \lambda)(-150 - \lambda) = 0$ and real eigenvalues $\lambda_1 = -150$, $\lambda_2 = 200$ with different signs. Hence $(0,0)$ is a saddle point of the linearized system $x' = 200x$, $y' = -150y$ (Figure a).

At $(75,50)$: The Jacobian matrix $\mathbf{J} = \begin{bmatrix} 0 & -300 \\ 100 & 0 \end{bmatrix}$ has characteristic equation

$\lambda^2 + 30000 = 0$ and pure imaginary eigenvalues $\lambda_1, \lambda_2 = \pm 100i\sqrt{3}$. Hence $(75,50)$ is a stable center of the linearization $u' = -300v$, $v' = 100u$ (Figure b).

Problem 1a

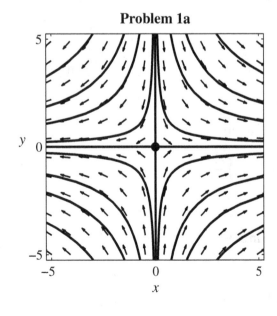

Problem 1b

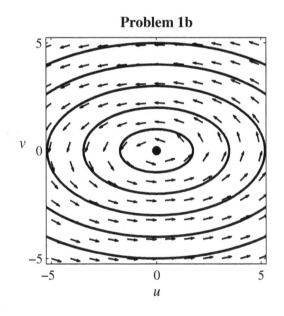

3. The effect of using the insecticide is to replace b by $b+f$ and a by $a-f$ in the predator-prey equations, while leaving p and q unchanged. Hence the new harmful population is

$$\frac{b+f}{q} > \frac{b}{q} = x_E$$

and the new benign population is

$$\frac{a-f}{p} < \frac{a}{p} = y_E \, .$$

Problems 4–7 deal with the competition system

$$x' = 60x - 4x^2 - 3xy, \quad y' = 42y - 2y^2 - 3xy,$$

that has Jacobian matrix $\mathbf{J} = \begin{bmatrix} 60 - 8x - 3y & -3x \\ -3y & 42 - 4y - 3x \end{bmatrix}.$

5. At $(0,21)$ the Jacobian matrix $\mathbf{J} = \begin{bmatrix} -3 & 0 \\ -63 & -42 \end{bmatrix}$ has characteristic equation

$(-3-\lambda)(-42-\lambda) = 0$ and negative real eigenvalues $\lambda_1 = -42$, $\lambda_2 = -3$. Hence $(0,21)$ is a nodal sink of the linearized system $u' = -3u$, $v' = -63u - 42v$.

7. At $(6,12)$ the Jacobian matrix $\mathbf{J} = \begin{bmatrix} -24 & -18 \\ -36 & -24 \end{bmatrix}$ has characteristic equation

$(-24-\lambda)^2 - (-36)(-18) = 0$ and real eigenvalues $\lambda_1 = -24 + 18\sqrt{2} > 0$,

$\lambda_2 = -24 - 18\sqrt{2} < 0$ with different signs. Hence $(6,12)$ is a saddle point of the linearized system $u' = -24u - 18v$, $v' = -36u - 24v$. Figure **a** illustrates this saddle point, whereas Figure **b** shows all four critical points of the system.

Problem 7a

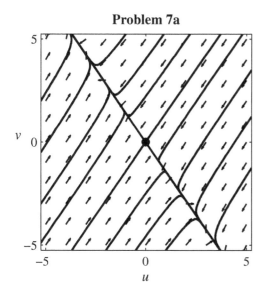

Problem 7b

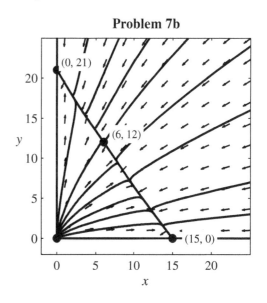

Problems 8–10 deal with the competition system

$$x' = 60x - 3x^2 - 4xy, \quad y' = 42y - 3y^2 - 2xy$$

that has Jacobian matrix $\mathbf{J} = \begin{bmatrix} 60 - 6x - 4y & -4x \\ -2y & 42 - 6y - 2x \end{bmatrix}$.

9.   At $(20,0)$ the Jacobian matrix $\mathbf{J} = \begin{bmatrix} -60 & -80 \\ 0 & 2 \end{bmatrix}$ has characteristic equation

$(-60 - \lambda)(2 - \lambda) = 0$ and real eigenvalues $\lambda_1 = -60$, $\lambda_2 = 2$ with different signs. Hence

$(20,0)$ is a saddle point of the linearized system $u' = -60u - 80v$, $v' = 2v$.

Problems 11–13 deal with the predator-prey system

$$x' = 5x - x^2 - xy, \quad y' = -2y + xy$$

that has Jacobian matrix $\mathbf{J} = \begin{bmatrix} 5 - 2x - y & -x \\ y & -2 + x \end{bmatrix}$.

11.   At $(0,0)$ the Jacobian matrix $\mathbf{J} = \begin{bmatrix} 5 & 0 \\ 0 & -2 \end{bmatrix}$ has characteristic equation

$(5 - \lambda)(-2 - \lambda) = 0$ and real eigenvalues $\lambda_1 = -2$, $\lambda_2 = 5$ with different signs. Hence

$(0,0)$ is a saddle point of the linearized system $x' = 5x$, $y' = -2y$.

13.   At $(2,3)$ the Jacobian matrix $\mathbf{J} = \begin{bmatrix} -2 & -2 \\ 3 & 0 \end{bmatrix}$ has characteristic equation

$(-2 - \lambda)(-\lambda) - 3 \cdot (-2) = \lambda^2 + 2\lambda + 6 = 0$ and complex conjugate eigenvalues

$\lambda_1, \lambda_2 = -1 \pm i\sqrt{5}$ with negative real part. Hence $(2,3)$ is a spiral sink (illustrated in the

figure) of the linearized system $u' = -2u - 2v$, $v' = 3u$.

Problem 13

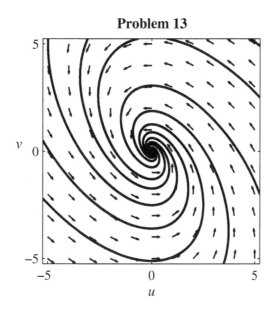

Problems 14–17 deal with the predator-prey system

$$x' = x^2 - 2x - xy, \quad y' = y^2 - 4y + xy$$

that has Jacobian matrix $\mathbf{J} = \begin{bmatrix} 2x - 2 - y & -x \\ y & 2y - 4 + x \end{bmatrix}$.

15.   At $(0,4)$ the Jacobian matrix $\mathbf{J} = \begin{bmatrix} -6 & 0 \\ 4 & 4 \end{bmatrix}$ has characteristic equation

$(-6 - \lambda)(4 - \lambda) = 0$ and real eigenvalues $\lambda_1 = -6$, $\lambda_2 = 4$ with different signs. Hence $(0,4)$ is a saddle point of the linearized system $u' = -6u$, $v' = 4u + 4v$.

17.   At $(3,1)$ the Jacobian matrix $\mathbf{J} = \begin{bmatrix} 3 & -3 \\ 1 & 1 \end{bmatrix}$ has characteristic equation

$(3 - \lambda)(1 - \lambda) - 1 \cdot (-3) = \lambda^2 - 4\lambda + 6 = 0$ and complex conjugate eigenvalues $\lambda_1, \lambda_2 = 2 \pm i\sqrt{2}$ with positive real part. Hence $(3,1)$ is a spiral source of the linearized system $u' = 3u - 3v$, $v' = u + v$ (illustrated in the figure).

Problem 17

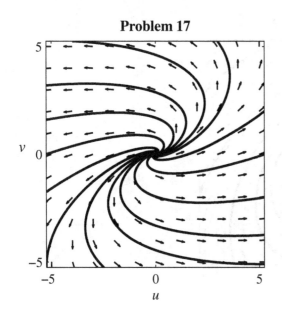

Problem 19

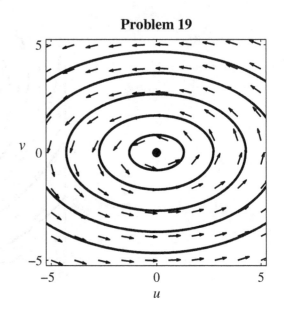

Problems 18 and 19 deal with the predator-prey system

$$x' = 2x - xy, \quad y' = -5y + xy$$

that has Jacobian matrix $\mathbf{J} = \begin{bmatrix} 2-y & -x \\ y & -5+x \end{bmatrix}$.

19.   At $(5,2)$ the Jacobian matrix $\mathbf{J} = \begin{bmatrix} 0 & -5 \\ 2 & 0 \end{bmatrix}$ has characteristic equation

$(-\lambda)(-\lambda) - 2 \cdot (-5) = \lambda^2 + 10 = 0$ and pure imaginary roots $\lambda = \pm i\sqrt{10}$, so the origin is a stable center for the linearized system $u' = -5v$, $v' = 2u$. This is the indeterminate case, but the figure suggests that $(5,2)$ is also a stable center for the original system.

Problems 20–22 deal with the predator-prey system

$$x' = -3x + x^2 - xy, \quad y' = -5y + xy$$

that has Jacobian matrix $\mathbf{J} = \begin{bmatrix} -3+2x-y & -x \\ y & -5+x \end{bmatrix}$.

21.   At $(3,0)$ the Jacobian matrix $\mathbf{J} = \begin{bmatrix} 3 & -3 \\ 0 & -2 \end{bmatrix}$ has characteristic equation

$(3-\lambda)(-2-\lambda) = 0$ and real eigenvalues $\lambda_1 = -2$, $\lambda_2 = 3$ with different signs. Hence $(3,0)$ is a saddle point of the linearized system $u' = 3u - 3v$, $v' = -2v$.

Problems 23–25 deal with the predator-prey system

$$x' = 7x - x^2 - xy, \quad y' = -5y + xy$$

that has Jacobian matrix $\mathbf{J} = \begin{bmatrix} 7 - 2x - y & -x \\ y & -5 + x \end{bmatrix}$.

23.   At $(0,0)$ the Jacobian matrix $\mathbf{J} = \begin{bmatrix} 7 & 0 \\ 0 & -5 \end{bmatrix}$ has characteristic equation

$(7-\lambda)(-5-\lambda) = 0$ and real eigenvalues $\lambda_1 = -5$, $\lambda_2 = 7$ with different signs. Hence $(0,0)$ is a saddle point of the linearized system $x' = 7x$, $y' = -5y$.

25.   At $(5,2)$ the Jacobian matrix $\mathbf{J} = \begin{bmatrix} -5 & -5 \\ 2 & 0 \end{bmatrix}$ has characteristic equation

$(-5-\lambda)(-\lambda) - 2 \cdot (-5) = \lambda^2 + 5\lambda + 10 = 0$ and complex conjugate eigenvalues

$\lambda_1, \lambda_2 = \dfrac{1}{2}\left(-5 \pm i\sqrt{15}\right)$ with negative real part. Hence $(5,2)$ is a spiral sink of the linearized system $u' = -5u - 5v$, $v' = 2u$ (illustrated in the figure).

Problem 25

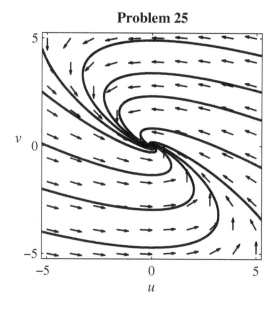

Problem 27

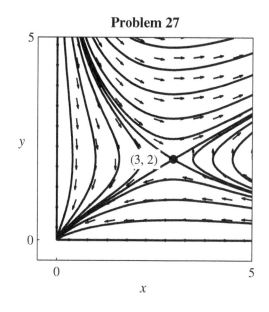

27. $J = \begin{bmatrix} 2y-4 & 2x \\ y & x-3 \end{bmatrix}$

At $(0,0)$: The Jacobian matrix $J = \begin{bmatrix} -4 & 0 \\ 0 & -3 \end{bmatrix}$ has characteristic equation

$\lambda^2 + 7\lambda + 12 = 0$ and negative real eigenvalues $\lambda_1 = -4$, $\lambda_2 = -3$. Hence $(0,0)$ is a nodal sink.

At $(3,2)$: The Jacobian matrix $J = \begin{bmatrix} 0 & 6 \\ 2 & 0 \end{bmatrix}$ has characteristic equation $\lambda^2 - 12 = 0$ and

real eigenvalues $\lambda_1, \lambda_2 = \pm 2\sqrt{3}$ with different signs. Hence $(3,2)$ is a saddle point.

As the figure indicates, if the initial point (x_0, y_0) lies below the northwest-southeast separatrix through $(3,2)$, then $(x(t), y(t)) \to (0,0)$ as $t \to \infty$. But if (x_0, y_0) lies above this separatrix, then $(x(t), y(t)) \to (\infty, \infty)$ as $t \to \infty$.

29. $J = \begin{bmatrix} -2x - \dfrac{1}{2}y + 3 & -\dfrac{1}{2}x \\ -2y & 4-2x \end{bmatrix}$

At $(0,0)$: The Jacobian matrix $J = \begin{bmatrix} 3 & 0 \\ 0 & 4 \end{bmatrix}$ has characteristic equation $\lambda^2 - 7\lambda + 12 = 0$

and positive real eigenvalues $\lambda_1 = 3$, $\lambda_2 = 4$. Hence $(0,0)$ is a nodal source.

At $(3,0)$: The Jacobian matrix $J = \begin{bmatrix} -3 & -3/2 \\ 0 & -2 \end{bmatrix}$ has characteristic equation

$\lambda^2 + 5\lambda + 6 = 0$ and negative real eigenvalues $\lambda_1 = -3$, $\lambda_2 = -2$. Hence $(3,0)$ is a nodal sink.

At $(2,2)$: The Jacobian matrix $J = \begin{bmatrix} -2 & -1 \\ -4 & 0 \end{bmatrix}$ has characteristic equation

$\lambda^2 + 2\lambda - 4 = 0$ and real eigenvalues $\lambda_1 \approx -3.2361$, $\lambda_2 = 1.2361$ with different signs. Hence $(2,2)$ is a saddle point.

As the figure indicates, if the initial point (x_0, y_0) lies above the southwest-northeast separatrix through $(2,2)$, then $(x(t), y(t)) \to (0, \infty)$ as $t \to \infty$. But if (x_0, y_0) lies below this separatrix, then $(x(t), y(t)) \to (3,0)$ as $t \to \infty$.

Problem 29

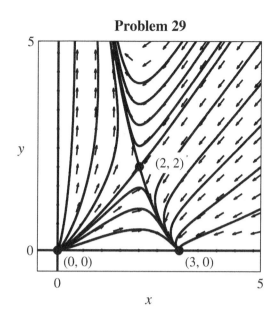

Problem 31

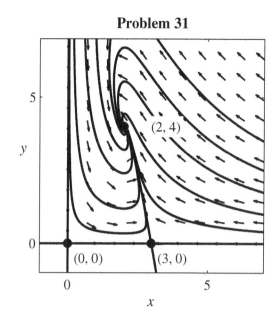

31. $\mathbf{J} = \begin{bmatrix} -2x - \dfrac{1}{4}y + 3 & -\dfrac{1}{4}x \\ y & x - 2 \end{bmatrix}$

At $(0,0)$: The Jacobian matrix $\mathbf{J} = \begin{bmatrix} 3 & 0 \\ 0 & -2 \end{bmatrix}$ has characteristic equation $\lambda^2 - \lambda - 6 = 0$ and real eigenvalues $\lambda_1 = -2$, $\lambda_2 = 3$ of opposite sign. Hence $(0,0)$ is a saddle point.

At $(3,0)$: The Jacobian matrix $\mathbf{J} = \begin{bmatrix} -3 & -3/4 \\ 0 & 1 \end{bmatrix}$ has characteristic equation $\lambda^2 + 2\lambda - 3 = 0$ and real eigenvalues $\lambda_1 = -3$, $\lambda_2 = 1$ of opposite sign. Hence $(3,0)$ is a saddle point.

At $(2,4)$: The Jacobian matrix $\mathbf{J} = \begin{bmatrix} -2 & -1/2 \\ 4 & 0 \end{bmatrix}$ has characteristic equation $\lambda^2 + 2\lambda + 2 = 0$ and complex conjugate eigenvalues $\lambda_1, \lambda_2 = -1 \pm i$ with negative real part. Hence $(2,4)$ is a spiral sink.

As $t \to \infty$, each solution point $(x(t), y(t))$ with nonzero initial conditions approaches the spiral sink $(2,4)$, as indicated by the direction arrows in the figure.

33. $\mathbf{J} = \begin{bmatrix} -6x + y + 30 & x \\ 4y & 4x - 6y + 60 \end{bmatrix}$

At $(0,0)$: The Jacobian matrix $\mathbf{J} = \begin{bmatrix} 30 & 0 \\ 0 & 80 \end{bmatrix}$ has characteristic equation

$\lambda^2 - 110\lambda + 2400 = 0$ and positive real eigenvalues $\lambda_1 = 30$, $\lambda_2 = 80$. Hence $(0,0)$ is a nodal source.

At $(0,20)$: The Jacobian matrix $\mathbf{J} = \begin{bmatrix} 10 & 0 \\ 40 & -80 \end{bmatrix}$ has characteristic equation

$\lambda^2 + 70\lambda - 800 = 0$ and real eigenvalues $\lambda_1 = -80$, $\lambda_2 = 10$ of opposite sign. Hence $(0,20)$ is a saddle point.

At $(15,0)$: The Jacobian matrix $\mathbf{J} = \begin{bmatrix} -30 & 15 \\ 0 & 110 \end{bmatrix}$ has characteristic equation

$\lambda^2 - 80\lambda - 3300 = 0$ and real eigenvalues $\lambda_1 = -30$, $\lambda_2 = 110$ of opposite sign. Hence $(15,0)$ is a saddle point.

At $(4,22)$: The Jacobian matrix $\mathbf{J} = \begin{bmatrix} -8 & -4 \\ 44 & -88 \end{bmatrix}$ has characteristic equation

$\lambda^2 + 96\lambda + 880 = 0$ and negative real eigenvalues $\lambda_1 \approx -85.736$, $\lambda_2 = -10.264$. Hence $(4,22)$ is a nodal sink.

As the figure shows, as $t \to \infty$, each solution point $(x(t), y(t))$ with nonzero initial conditions approaches the nodal sink $(4,22)$.

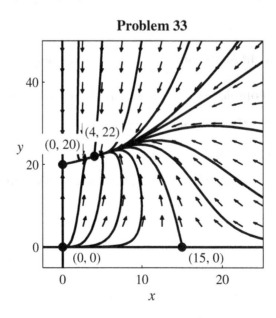

Problem 33

SECTION 9.4

NONLINEAR MECHANICAL SYSTEMS

In each of Problems 1–4 we need only substitute the familiar power series for the exponential, sine, and cosine functions, and then discard all higher-order terms. For each problem we give the corresponding linear system, the eigenvalues λ_1 and λ_2, and the type of this critical point.

1. $x' = 1 - \left(1 + x + \dfrac{1}{2}x^2 + \cdots\right) + 2y \approx -x + 2y$

   $y' = -x - 4\left(y - \dfrac{1}{6}y^3 + \cdots\right) \approx -x - 4y$

   The coefficient matrix $\mathbf{A} = \begin{bmatrix} -1 & 2 \\ -1 & -4 \end{bmatrix}$ has negative eigenvalues $\lambda_1 = -2$ and $\lambda_2 = -3$ indicating a stable nodal sink as illustrated in the figure. Alternatively, we can calculate the Jacobian matrix

   $$\mathbf{J}(x, y) = \begin{bmatrix} -e^x & 2 \\ -1 & -4\cos x \end{bmatrix}, \quad \text{so} \quad \mathbf{J}(0,0) = \begin{bmatrix} -1 & 2 \\ -1 & -4 \end{bmatrix}.$$

Problem 1

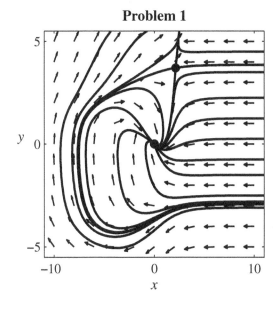

Problem 3

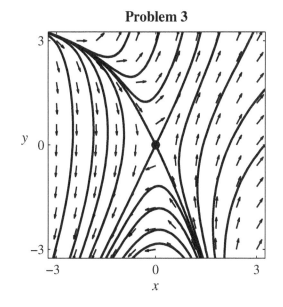

3. $x' = \left(1 + x + \dfrac{1}{2}x^2 + \cdots\right) + 2y - 1 \approx x + 2y$

   $y' = 8x + \left(1 + y + \dfrac{1}{2}y^2 + \cdots\right) - 1 \approx 8x + y$

The coefficient matrix $\mathbf{A} = \begin{bmatrix} 1 & 2 \\ 8 & 1 \end{bmatrix}$ has real eigenvalues $\lambda_1 = -3$ and $\lambda_2 = 5$ of opposite sign, indicating an unstable saddle point as illustrated in the figure. Alternatively, we can calculate the Jacobian matrix

$$\mathbf{J}(x,y) = \begin{bmatrix} e^x & 2 \\ 8 & e^y \end{bmatrix}, \quad \text{so} \quad \mathbf{J}(0,0) = \begin{bmatrix} 1 & 2 \\ 8 & 1 \end{bmatrix}.$$

5. The critical points are of the form $(0, n\pi)$, where n is an integer, so we substitute $x = u$, $y = v + n\pi$. Then

$$u' = x' = -u + \sin(v + n\pi) = -u + (\cos n\pi)v = -u + (-1)^n v.$$

Hence the linearized system at $(0, n\pi)$ is

$$u' = -u \pm v, \quad v' = 2u,$$

where we take the plus sign if n is even, the minus sign if n is odd. If n is even, then the eigenvalues are $\lambda_1 = 1$ and $\lambda_2 = -2$, so $(0, n\pi)$ is an unstable saddle point. If n is odd, then the eigenvalues are $\lambda_1, \lambda_2 = \frac{1}{2}\left(-1 \pm i\sqrt{7}\right)$, so $(0, n\pi)$ is a stable spiral point.

Alternatively, we can start by calculating the Jacobian matrix $\mathbf{J}(x,y) = \begin{bmatrix} -1 & \cos y \\ 2 & 0 \end{bmatrix}$.

At $(0, n\pi)$, n even: The Jacobian matrix $\mathbf{J} = \begin{bmatrix} -1 & 1 \\ 2 & 0 \end{bmatrix}$ has characteristic equation $\lambda^2 + \lambda - 2 = 0$ and real eigenvalues $\lambda_1 = -2$, $\lambda_2 = 1$ of opposite sign. Hence $(0, n\pi)$ is a saddle point if n is even, as we see in the figure above.

At $(0, n\pi)$, n odd: The Jacobian matrix $\mathbf{J} = \begin{bmatrix} -1 & -1 \\ 2 & 0 \end{bmatrix}$ has characteristic equation $\lambda^2 + \lambda + 2 = 0$ and complex conjugate eigenvalues $\lambda_1, \lambda_2 = \frac{1}{2}\left(-1 \pm i\sqrt{7}\right)$ with negative real part. Hence $(0, n\pi)$ is a spiral sink if n is odd, as indicated in the figure.

Problem 5

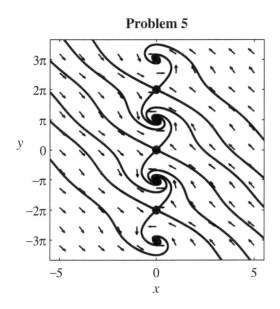

Problem 7

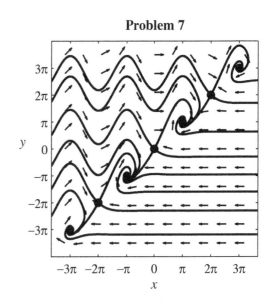

7.   The critical points are of the form $(n\pi, n\pi)$, where n is an integer, so we substitute $x = u + n\pi$, $y = v + n\pi$. Then

$$u' = x' = 1 - e^{u-v} = 1 - \left[1 + (u-v) + \frac{1}{2}(u-v)^2 + \cdots\right] \approx -u + v,$$

$$v' = y' = 2\sin(u + n\pi) = 2\sin u \cos n\pi \approx 2(-1)^n u.$$

Hence the linearized system at $(n\pi, n\pi)$ is

$$u' = -u + v, \quad v' = \pm 2u,$$

and has coefficient matrix $\mathbf{A} = \begin{bmatrix} -1 & 1 \\ \pm 2 & 0 \end{bmatrix}$, where we take the plus sign if n is even, the minus sign if n is odd. With n even, the characteristic equation $\lambda^2 + \lambda - 2 = 0$ has real roots $\lambda_1 = 1$ and $\lambda_2 = -2$ of opposite sign, so $(n\pi, n\pi)$ is an unstable saddle point. With n odd, the characteristic equation $\lambda^2 + \lambda + 2 = 0$ has complex conjugate eigenvalues $\lambda_1, \lambda_2 = \frac{1}{2}\left(-1 \pm i\sqrt{7}\right)$ with negative real part, so $(n\pi, n\pi)$ is a stable spiral point.

Alternatively, we can start by calculating the Jacobian matrix $\mathbf{J}(x,y) = \begin{bmatrix} -e^{x-y} & e^{x-y} \\ 2\cos x & 0 \end{bmatrix}$.

At $(n\pi, n\pi)$, n even: The Jacobian matrix $\mathbf{J} = \begin{bmatrix} -1 & 1 \\ 2 & 0 \end{bmatrix}$ has characteristic equation $\lambda^2 + \lambda - 2 = 0$ and real eigenvalues $\lambda_1 = -2$, $\lambda_2 = 1$ of opposite sign. Hence $(n\pi, n\pi)$ is a saddle point if n is even, as we see in the figure.

At $(n\pi, n\pi)$, n odd: The Jacobian matrix $\mathbf{J} = \begin{bmatrix} -1 & 1 \\ -2 & 0 \end{bmatrix}$ has characteristic equation

$\lambda^2 + \lambda + 2 = 0$ and complex conjugate eigenvalues $\lambda_1, \lambda_2 \approx -0.5 \pm 1.3229i$ with negative real part. Hence $(n\pi, n\pi)$ is a spiral sink if n is odd, as we see in the figure.

As preparation for Problems 9–11, we first calculate the Jacobian matrix

$$\mathbf{J}(x, y) = \begin{bmatrix} 0 & 1 \\ -\omega^2 \cos x & -c \end{bmatrix}$$

of the damped pendulum system in (34) in the text. At the critical point $(n\pi, 0)$ we have

$$\mathbf{J}(n\pi, 0) = \begin{bmatrix} 0 & 1 \\ -\omega^2 \cos n\pi & -c \end{bmatrix} = \begin{bmatrix} 0 & 1 \\ \pm\omega^2 & -c \end{bmatrix},$$

where we take the plus sign if n is odd, the minus sign if n is even.

9. If n is odd then the characteristic equation $\lambda^2 + c\lambda - \omega^2 = 0$ has real roots

$$\lambda_1, \lambda_2 = \frac{-c \pm \sqrt{c^2 + 4\omega^2}}{2}$$

with opposite signs, so $(n\pi, 0)$ is an unstable saddle point.

11. If n is even and $c^2 < 4\omega^2$, then the two eigenvalues

$$\lambda_1, \lambda_2 = \frac{-c \pm \sqrt{c^2 - 4\omega^2}}{2} = -\frac{c}{2} \pm \frac{i}{2}\sqrt{4\omega^2 - c^2}$$

are complex conjugates with negative real part, so $(n\pi, 0)$ is a stable spiral point.

Problems 12-16 call for us to find and classify the critical points of the first order-system $x' = y$, $y' = -f(x, y)$ that corresponds to the given equation $x'' + f(x, x') = 0$. After finding the critical points $(x, 0)$ where $f(x, 0) = 0$, we first calculate the Jacobian matrix $\mathbf{J}(x, y)$.

13. $\mathbf{J}(x, y) = \begin{bmatrix} 0 & 1 \\ 15x^2 - 20 & -2 \end{bmatrix}$.

At $(0,0)$: The Jacobian matrix $\mathbf{J} = \begin{bmatrix} 0 & 1 \\ -20 & -2 \end{bmatrix}$ has characteristic equation

$\lambda^2 + 2\lambda + 20 = 0$ and complex conjugate eigenvalues $\lambda_1, \lambda_2 = -1 \pm i\sqrt{19}$ consistent with the spiral node we see at $(0, 0)$ in Fig. 9.4.6 in the textbook.

At $(\pm2, 0)$: The Jacobian matrix $\mathbf{J} = \begin{bmatrix} 0 & 1 \\ 40 & -2 \end{bmatrix}$ has characteristic equation

$\lambda^2 + 2\lambda - 40 = 0$ and real eigenvalues $\lambda_1, \lambda_2 = -1 \pm \sqrt{41}$ of opposite sign, consistent with the saddle points we see at $(\pm2, 0)$ in Fig. 9.4.6.

15. $\mathbf{J}(x, y) = \begin{bmatrix} 0 & 1 \\ 2x - 4 & 0 \end{bmatrix}$.

At $(0, 0)$: The Jacobian matrix $\mathbf{J} = \begin{bmatrix} 0 & 1 \\ -4 & 0 \end{bmatrix}$ has characteristic equation $\lambda^2 + 4 = 0$ and

pure imaginary eigenvalues $\lambda_1, \lambda_2 = \pm 2i$ consistent with the stable center we see at $(0, 0)$ in Fig. 9.4.13 in the textbook.

At $(4, 0)$: The Jacobian matrix $\mathbf{J} = \begin{bmatrix} 0 & 1 \\ 4 & 0 \end{bmatrix}$ has characteristic equation $\lambda^2 - 4 = 0$ and

real eigenvalues $\lambda_1, \lambda_2 = \pm 2$ of opposite sign, consistent with the saddle point we see at $(4, 0)$ in Fig. 9.4.13.

17. $\mathbf{J}(x, y) = \begin{bmatrix} 0 & 1 \\ -5 - \dfrac{15}{4}x^2 & -2 \end{bmatrix}$.

At $(0, 0)$: The Jacobian matrix $\mathbf{J} = \begin{bmatrix} 0 & 1 \\ -5 & -2 \end{bmatrix}$ has characteristic equation

$\lambda^2 + 2\lambda + 5 = 0$ and complex conjugate eigenvalues $\lambda_1, \lambda_2 = -1 \pm 2i$ with negative real part, consistent with the spiral sink we see in the figure.

Problem 17

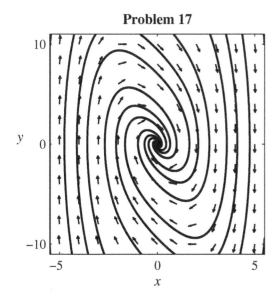

Problem 19

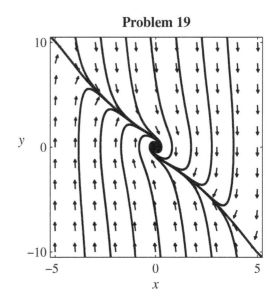

19. $\mathbf{J}(x,y) = \begin{bmatrix} 0 & 1 \\ -5 - \dfrac{15}{4}x^2 & -4|y| \end{bmatrix}$.

At $(0,0)$: The Jacobian matrix $\mathbf{J} = \begin{bmatrix} 0 & 1 \\ -5 & 0 \end{bmatrix}$ has characteristic equation $\lambda^2 + 5 = 0$ and

pure imaginary eigenvalues $\lambda_1, \lambda_2 = \pm i\sqrt{5}$. This corresponds to the indeterminate case of Theorem 2 in Section 9.3, but is not inconsistent with the spiral sink we see in the figure.

The statements of Problems 21–26 in the text include their answers and rather fully outline their solutions, which therefore are omitted here.

CHAPTER 10

LAPLACE TRANSFORM METHODS

SECTION 10.1

LAPLACE TRANSFORMS AND INVERSE TRANSFORMS

The objectives of this section are especially clear cut. They include familiarity with the definition of the Laplace transform $\mathcal{L}\{f(t)\} = F(s)$ that is given in Equation (1) in the textbook, the direct application of this definition to calculate Laplace transforms of simple functions (as in Examples 1-3), and the use of known transforms (those listed in Figure 10.1.2) to find Laplace transforms and inverse transforms (as in Examples 4-6). Perhaps students need to be told explicitly to memorize the transforms that are listed in the short table that appears in Figure 10.1.2.

1. $\quad \mathcal{L}\{t\} = \displaystyle\int_0^\infty e^{-st} t \, dt \qquad\qquad (u = -st, \quad du = -s \, dt)$

$$= \int_0^{-\infty} \left[\frac{1}{s^2}\right] u e^u \, du = \frac{1}{s^2}\left[(u-1)e^u\right]_0^{-\infty} = \frac{1}{s^2}$$

3. $\quad \mathcal{L}\left\{e^{3t+1}\right\} = \displaystyle\int_0^\infty e^{-st} e^{3t+1} \, dt = e\int_0^\infty e^{-(s-3)t} \, dt = \frac{e}{s-3}$

5. $\quad \mathcal{L}\{\sinh t\} = \tfrac{1}{2}\mathcal{L}\left\{e^t - e^{-t}\right\} = \tfrac{1}{2}\displaystyle\int_0^\infty e^{-st}\left(e^t - e^{-t}\right)dt = \tfrac{1}{2}\int_0^\infty \left(e^{-(s-1)t} - e^{-(s+1)t}\right)dt$

$$= \frac{1}{2}\left[\frac{1}{s-1} - \frac{1}{s+1}\right] = \frac{1}{s^2-1}$$

7. $\quad \mathcal{L}\{f(t)\} = \displaystyle\int_0^1 e^{-st} \, dt = \left[-\frac{1}{s}e^{-st}\right]_0^1 = \frac{1-e^{-s}}{s}$

9. $\quad \mathcal{L}\{f(t)\} = \displaystyle\int_0^1 e^{-st} t \, dt = \frac{1-e^{-s}-se^{-s}}{s^2}$

11. $\quad \mathcal{L}\left\{\sqrt{t} + 3t\right\} = \dfrac{\Gamma(3/2)}{s^{3/2}} + 3\cdot\dfrac{1}{s^2} = \dfrac{\sqrt{\pi}}{2s^{3/2}} + \dfrac{3}{s^2}$

13. $\mathcal{L}\{t - 2e^{3t}\} = \dfrac{1}{s^2} - \dfrac{2}{s-3}$

15. $\mathcal{L}\{1 + \cosh 5t\} = \dfrac{1}{s} + \dfrac{s}{s^2 - 25}$

17. $\mathcal{L}\{\cos^2 2t\} = \dfrac{1}{2}\mathcal{L}\{1 + \cos 4t\} = \dfrac{1}{2}\left(\dfrac{1}{s} + \dfrac{s}{s^2 + 16}\right)$

19. $\mathcal{L}\{(1+t)^3\} = \mathcal{L}\{1 + 3t + 3t^2 + t^3\} = \dfrac{1}{s} + 3 \cdot \dfrac{1!}{s^2} + 3 \cdot \dfrac{2!}{s^3} + \dfrac{3!}{s^4} = \dfrac{1}{s} + \dfrac{3}{s^2} + \dfrac{6}{s^3} + \dfrac{6}{s^4}$

21. Integration by parts with $u = t$ and $dv = e^{-st}\cos 2t \, dt$ yields

$$\mathcal{L}\{t \cos 2t\} = \int_0^\infty t e^{-st}\cos 2t \, dt = -\dfrac{1}{s^2 + 4}\int_0^\infty e^{-st}\left(-s\cos 2t + 2\sin 2t\right) dt$$

$$= -\dfrac{1}{s^2 + 4}\Big[-s\,\mathcal{L}\{\cos 2t\} + 2\,\mathcal{L}\{\sin 2t\}\Big]$$

$$= -\dfrac{1}{s^2 + 4}\left[\dfrac{-s^2}{s^2 + 4} + \dfrac{4}{s^2 + 4}\right] = \dfrac{s^2 - 4}{\left(s^2 + 4\right)^2}.$$

23. $\mathcal{L}^{-1}\left\{\dfrac{3}{s^4}\right\} = \mathcal{L}^{-1}\left\{\dfrac{1}{2} \cdot \dfrac{6}{s^4}\right\} = \dfrac{1}{2}t^3$

25. $\mathcal{L}^{-1}\left\{\dfrac{1}{s} - \dfrac{2}{s^{5/2}}\right\} = \mathcal{L}^{-1}\left\{\dfrac{1}{s} - \dfrac{2}{\Gamma(5/2)} \cdot \dfrac{\Gamma(5/2)}{s^{5/2}}\right\} = 1 - \dfrac{2}{\frac{3}{2} \cdot \frac{1}{2}\sqrt{\pi}} \cdot t^{3/2} = 1 - \dfrac{8t^{3/2}}{3\sqrt{\pi}}$

27. $\mathcal{L}^{-1}\left\{\dfrac{3}{s-4}\right\} = 3 \cdot \mathcal{L}^{-1}\left\{\dfrac{1}{s-4}\right\} = 3e^{4t}$

29. $\mathcal{L}^{-1}\left\{\dfrac{5 - 3s}{s^2 + 9}\right\} = \dfrac{5}{3} \cdot \mathcal{L}^{-1}\left\{\dfrac{3}{s^2 + 9}\right\} - 3 \cdot \mathcal{L}^{-1}\left\{\dfrac{s}{s^2 + 9}\right\} = \dfrac{5}{3}\sin 3t - 3\cos 3t$

31. $\mathcal{L}^{-1}\left\{\dfrac{10s - 3}{25 - s^2}\right\} = -10 \cdot \mathcal{L}^{-1}\left\{\dfrac{s}{s^2 - 25}\right\} + \dfrac{3}{5} \cdot \mathcal{L}^{-1}\left\{\dfrac{5}{s^2 - 25}\right\} = -10\cosh 5t + \dfrac{3}{5}\sinh 5t$

33. $\mathcal{L}\{\sin kt\} = \mathcal{L}\left\{\dfrac{e^{ikt} - e^{-ikt}}{2i}\right\} = \dfrac{1}{2i}\left(\dfrac{1}{s - ik} - \dfrac{1}{s + ik}\right)$

$$= \dfrac{1}{2i} \cdot \dfrac{2ik}{(s - ik)(s - ik)} = \dfrac{k}{s^2 + k^2} \qquad \text{(because } i^2 = -1\text{)}$$

35. Using the given tabulated integral with $a = -s$ and $b = k$, we find that

$$\mathcal{L}\{\cos kt\} = \int_0^\infty e^{-st} \cos kt \, dt = \left[\frac{e^{-st}}{s^2 + k^2} (-s \cos kt + k \sin kt) \right]_{t=0}^\infty$$

$$= \lim_{t \to \infty} \left(\frac{e^{-st}}{s^2 + k^2} (-s \cos kt + k \sin kt) \right) - \frac{e^0}{s^2 + k^2} (-s \cdot 1 + k \cdot 0) = \frac{s}{s^2 + k^2}.$$

37. $f(t) = 1 - u_a(t) = 1 - u(t - a)$ so

$$\mathcal{L}\{f(t)\} = \mathcal{L}\{1\} - \mathcal{L}\{u_a(t)\} = \frac{1}{s} - \frac{e^{-as}}{s} = s^{-1}(1 - e^{-as}).$$

For the graph of f, note that $f(a) = 1 - u(a) = 1 - 1 = 0$.

39. Use of the geometric series gives

$$\mathcal{L}\{f(t)\} = \sum_{n=0}^\infty \mathcal{L}\{u(t-n)\} = \sum_{n=0}^\infty \frac{e^{-ns}}{s} = \frac{1}{s}\left(1 + e^{-s} + e^{-2s} + e^{-3s} + \cdots\right)$$

$$= \frac{1}{s}\left(1 + (e^{-s}) + (e^{-s})^2 + (e^{-s})^3 + \cdots\right) = \frac{1}{s} \cdot \frac{1}{1 - e^{-s}} = \frac{1}{s(1 - e^{-s})}.$$

41. By checking values at sample points, you can verify that $g(t) = 2f(t) - 1$ in terms of the square wave function $f(t)$ of Problem 40. Hence

$$\mathcal{L}\{g(t)\} = \mathcal{L}\{2f(t) - 1\} = \frac{2}{s(1 + e^{-s})} - \frac{1}{s} = \frac{1}{s}\left(\frac{2}{1 + e^{-s}} - 1\right) = \frac{1}{s} \cdot \frac{1 - e^{-s}}{1 + e^{-s}}$$

$$= \frac{1}{s} \cdot \frac{1 - e^{-s}}{1 + e^{-s}} \cdot \frac{e^{s/2}}{e^{s/2}} = \frac{1}{s} \cdot \frac{e^{s/2} - e^{-s/2}}{e^{s/2} + e^{-s/2}} = \frac{1}{s} \cdot \frac{\frac{1}{2}\left(e^{s/2} - e^{-s/2}\right)}{\frac{1}{2}\left(e^{s/2} + e^{-s/2}\right)}$$

$$= \frac{1}{s} \cdot \frac{\sinh(s/2)}{\cosh(s/2)} = \frac{1}{s} \tanh \frac{s}{2}.$$

SECTION 10.2

TRANSFORMATION OF INITIAL VALUE PROBLEMS

The focus of this section is on the use of transforms of derivatives (Theorem 1) to solve initial value problems (as in Examples 1 and 2). Transforms of integrals (Theorem 2) appear less frequently in practice, and the extension of Theorem 1 at the end of Section 10.2 may be considered entirely optional (except perhaps for electrical engineering students).

In Problems 1–10 we give first the transformed differential equation, then the transform $X(s)$ of the solution, and finally the inverse transform $x(t)$ of $X(s)$.

1. $[s^2X(s) - 5s] + 4\{X(s)\} = 0$

$$X(s) = \frac{5s}{s^2+4} = 5 \cdot \frac{s}{s^2+4}$$

$x(t) = \mathcal{L}^{-1}\{X(s)\} = 5\cos 2t$

3. $[s^2X(s) - 2] - [sX(s)] - 2[X(s)] = 0$

$$X(s) = \frac{2}{s^2-s-2} = \frac{2}{(s-2)(s+1)} = \frac{2}{3}\left(\frac{1}{s-2} - \frac{1}{s+1}\right)$$

$x(t) = (2/3)(e^{2t} - e^{-t})$

5. $[s^2X(s)] + [X(s)] = 2/(s^2+4)$

$$X(s) = \frac{2}{(s^2+1)(s^2+4)} = \frac{2}{3} \cdot \frac{1}{s^2+1} - \frac{1}{3} \cdot \frac{2}{s^2+4}$$

$x(t) = (2\sin t - \sin 2t)/3$

7. $[s^2X(s) - s] + [X(s)] = s/s^2 + 9)$

$(s^2 + 1)X(s) = s + s/(s^2 + 9) = (s^3 + 10s)/(s^2 + 9)$

$$X(s) = \frac{s^2+10s}{(s^2+1)(s^2+9)} = \frac{9}{9} \cdot \frac{s}{s^2+1} - \frac{1}{8} \cdot \frac{s}{s^2+9}$$

$x(t) = (9\cos t - \cos 3t)/8$

9. $s^2 X(s) + 4sX(s) + 3X(s) = 1/s$

$$X(s) = \frac{1}{s(s^2 + 4s + 3)} = \frac{1}{s(s+1)(s+3)} = \frac{1}{3} \cdot \frac{1}{s} - \frac{1}{2} \cdot \frac{1}{s+1} + \frac{1}{6} \cdot \frac{1}{s+3}$$

$x(t) = (2 - 3e^{-t} + e^{-3t})/6$

11.   The transformed equations are

$$sX(s) - 1 = 2X(s) + Y(s)$$
$$sY(s) + 2 = 6X(s) + 3Y(s).$$

We solve for the Laplace transforms

$$X(s) = \frac{s-5}{s(s-5)} = \frac{1}{s}$$

$$Y(s) = X(s) = \frac{-2s+10}{s(s-5)} = -\frac{2}{s}.$$

Hence the solution is given by

$$x(t) = 1, \qquad\qquad y(t) = -2.$$

13.   The transformed equations are

$$sX(s) + 2[sY(s) - 1] + X(s) = 0$$
$$sX(s) - [sY(s) - 1] + Y(s) = 0,$$

which we solve for the transforms

$$X(s) = -\frac{2}{3s^2 - 1} = -\frac{2}{3} \cdot \frac{1}{s^2 - 1/3} = -\frac{2}{\sqrt{3}} \cdot \frac{1/\sqrt{3}}{s^2 - \left(1/\sqrt{3}\right)^2}$$

$$Y(s) = \frac{3s+1}{3s^2 - 1} = \frac{s+1/3}{s^2 - 1/3} = \frac{s}{s^2 - \left(1/\sqrt{3}\right)^2} + \frac{1}{\sqrt{3}} \cdot \frac{1/\sqrt{3}}{s^2 - \left(1/\sqrt{3}\right)^2}.$$

Hence the solution is

$$x(t) = -\left(2/\sqrt{3}\right) \sinh\left(t/\sqrt{3}\right)$$
$$y(t) = \cosh\left(t/\sqrt{3}\right) + \left(1/\sqrt{3}\right) \sinh\left(t/\sqrt{3}\right).$$

15. The transformed equations are

$$[s^2X - s] + [sX - 1] + [sY - 1] + 2X - Y = 0$$
$$[s^2Y - s] + [sX - 1] + [sY - 1] + 4X - 2Y = 0,$$

which we solve for

$$X(s) = \frac{s^2 + 3s + 2}{s^3 + 3s^2 + 3s} = \frac{1}{3}\left(\frac{2}{s} + \frac{s+3}{s^2 + 3s + 3}\right) = \frac{1}{3}\left(\frac{2}{s} + \frac{s+3}{(s+3/2)^2 + (3/4)}\right)$$

$$= \frac{1}{3}\left(\frac{2}{s} + \frac{s+3/2}{(s+3/2)^2 + (\sqrt{3}/2)^2} + \sqrt{3}\cdot\frac{\sqrt{3}/2}{(s+3/2)^2 + (\sqrt{3}/2)^2}\right)$$

$$Y(s) = \frac{-s^3 - 2s^2 + 2s + 4}{s^3 + 3s^2 + 3s} = \frac{1}{21}\left(\frac{28}{s} - \frac{9}{s-1} + \frac{2s+15}{s^2 + 3s + 3}\right)$$

$$= \frac{1}{21}\left(\frac{28}{s} - \frac{9}{s-1} + \frac{2s+15}{(s+3/2)^2 + 3/4}\right)$$

$$= \frac{1}{21}\left(\frac{28}{s} - \frac{9}{s-1} + 2\cdot\frac{s+3/2}{(s+3/2)^2 + (\sqrt{3}/2)^2} + 8\sqrt{3}\cdot\frac{\sqrt{3}/2}{(s+3/2)^2 + (\sqrt{3}/2)^2}\right).$$

Here we've used some fairly heavy-duty partial fractions (Section 10.3). The transforms

$$\mathcal{L}\{e^{at}\cos kt\} = \frac{s-a}{(s-a)^2 + k^2}, \quad \mathcal{L}\{e^{at}\sin kt\} = \frac{k}{(s-a)^2 + k^2}$$

from the inside-front-cover table (with $a = -3/2$, $k = \sqrt{3}/2$) finally yield

$$x(t) = \frac{1}{3}\left\{2 + e^{-3t/2}\left[\cos\left(\sqrt{3}t/2\right) + \sqrt{3}\sin\left(\sqrt{3}t/2\right)\right]\right\}$$

$$y(t) = \frac{1}{21}\left\{28 - 9e^t + e^{-3t/2}\left[2\cos\left(\sqrt{3}t/2\right) + 8\sqrt{3}\sin\left(\sqrt{3}t/2\right)\right]\right\}.$$

17. $\quad f(t) = \int_0^t e^{3\tau}\,d\tau = \left[\frac{1}{3}e^{3\tau}\right]_{\tau=0}^t = \frac{1}{3}\left(e^{3t} - 1\right)$

19. $\quad f(t) = \int_0^t \frac{1}{2}\sin 2\tau\,d\tau = \left[-\frac{1}{4}\cos 2\tau\right]_{\tau=0}^t = \frac{1}{4}(1 - \cos 2t)$

21. $\quad f(t) = \int_0^t\left[\int_0^\tau \sin t\,dt\right]d\tau = \int_0^t (1 - \cos\tau)\,d\tau = \left[\tau - \sin\tau\right]_{\tau=0}^t = t - \sin t$

23. $f(t) = \int_0^t \left[\int_0^\tau \sinh t \, dt \right] d\tau = \int_0^t (\cosh \tau - 1) \, d\tau = \left[\sinh \tau - \tau \right]_{\tau=0}^t = \sinh t - t$

25.   With $f(t) = \cos kt$ and $F(s) = s/(s^2 + k^2)$, Theorem 1 in this section yields

$$\mathcal{L}\{-k \sin kt\} = \mathcal{L}\{f'(t)\} = sF(s) - 1 = s \cdot \frac{s}{s^2 + k^2} - 1 = -\frac{k^2}{s^2 + k^2},$$

so division by $-k$ yields $\mathcal{L}\{\sin kt\} = k/(s^2 + k^2)$.

27.   **(a)**   With $f(t) = t^n e^{at}$ and $f'(t) = nt^{n-1}e^{at} + at^n e^{at}$, Theorem 1 yields

$$\mathcal{L}\{nt^{n-1}e^{at} + at^n e^{at}\} = s \, \mathcal{L}\{t^n e^{at}\}$$

so

$$n \, \mathcal{L}\{t^{n-1}e^{at}\} = (s - a)\mathcal{L}\{t^n e^{at}\}$$

and hence

$$\mathcal{L}\{t^n e^{at}\} = \frac{n}{s-a} \mathcal{L}\{t^{n-1}e^{at}\}.$$

(b) $n = 1$: $\mathcal{L}\{t e^{at}\} = \frac{1}{s-a} \mathcal{L}\{e^{at}\} = \frac{1}{s-a} \cdot \frac{1}{s-a} = \frac{1}{(s-a)^2}$

$n = 2$: $\mathcal{L}\{t^2 e^{at}\} = \frac{2}{s-a} \mathcal{L}\{t e^{at}\} = \frac{2}{s-a} \cdot \frac{1}{(s-a)^2} = \frac{2!}{(s-a)^3}$

$n = 3$: $\mathcal{L}\{t^3 e^{at}\} = \frac{3}{s-a} \mathcal{L}\{t^2 e^{at}\} = \frac{3}{s-a} \cdot \frac{2!}{(s-a)^3} = \frac{3!}{(s-a)^4}$

And so forth.

29.   Let $f(t) = t \sinh kt$, so $f(0) = 0$. Then

$$f'(t) = \sinh kt + kt \cosh kt$$
$$f''(t) = 2k \cosh kt + k^2 t \sinh kt,$$

and thus $f'(0) = 0$, so Formula (5) in this section yields

$$\mathcal{L}\{2k \cosh kt + k^2 t \sinh kt\} = s^2 \mathcal{L}\{t \sinh kt\},$$

$$2k \cdot \frac{s}{s^2 - k^2} + k^2 F(s) = s^2 F(s).$$

We readily solve this last equation for

$$\mathcal{L}\{t \sinh kt\} = F(s) = \frac{2ks}{\left(s^2 - k^2\right)^2}.$$

31.   Using the known transform of $\sin kt$ and the Problem 28 transform of $t \cos kt$, we obtain

$$\mathcal{L}\left\{\frac{1}{2k^3}\left(\sin kt - kt \cos kt\right)\right\} = \frac{1}{2k^3} \cdot \frac{k}{s^2 + k^2} - \frac{k}{2k^3} \cdot \frac{s^2 - k^2}{\left(s^2 + k^2\right)^2}$$

$$= \frac{1}{2k^2}\left(\frac{1}{s^2 + k^2} - \frac{s^2 - k^2}{\left(s^2 + k^2\right)^2}\right) = \frac{1}{2k^2} \cdot \frac{2k^2}{\left(s^2 + k^2\right)^2} = \frac{1}{\left(s^2 + k^2\right)^2}$$

33. $f(t) = u_a(t) - u_b(t) = u(t - a) - u(t - b),$ so the result of Problem 32 gives

$$\mathcal{L}\{f(t)\} = \mathcal{L}\{u(t - a)\} - \mathcal{L}\{u(t - b)\} = \frac{e^{-as}}{s} - \frac{e^{-bs}}{s} = \frac{e^{-as} - e^{-bs}}{s}.$$

35.   Let's write $g(t)$ for the on-off function of this problem to distinguish it from the square wave function of Problem 34. Then comparison of Figures 10.2.9 and 10.2.10 makes it clear that $g(t) = \frac{1}{2}\left(1 + f(t)\right),$ so (using the result of Problem 34) we obtain

$$G(s) = \frac{1}{2s} + \frac{1}{2}F(s) = \frac{1}{2s} + \frac{1}{2s}\tanh\frac{s}{2} = \frac{1}{2s}\left(1 + \frac{e^{s/2} - e^{-s/2}}{e^{s/2} + e^{-s/2}} \cdot \frac{e^{-s/2}}{e^{-s/2}}\right)$$

$$= \frac{1}{2s}\left(1 + \frac{1 - e^{-s}}{1 + e^{-s}}\right) = \frac{1}{2s} \cdot \frac{2}{1 + e^{-s}} = \frac{1}{s\left(1 + e^{-s}\right)}.$$

37.   We observe that $f(0) = 0$ and that the sawtooth function has jump -1 at each of the points $t_n = n = 1, 2, 3, \cdots$. Also, $f'(t) \equiv 1$ wherever the derivative is defined. Hence Eq. (22) in this section gives

$$\frac{1}{s} = sF(s) + \sum_{n=1}^{\infty} e^{-ns} = sF(s) - 1 + \sum_{n=0}^{\infty} e^{-ns} = sF(s) - 1 + \frac{1}{1 - e^{-ns}},$$

using the geometric series $\sum_{n=0}^{\infty} x^n = 1/(1 - x)$ with $x = e^{-s}$. Solution for $F(s)$ gives

$$F(s) = \frac{1}{s^2} + \frac{1}{s} - \frac{1}{s\left(1 - e^{-s}\right)} = \frac{1}{s^2} - \frac{e^{-s}}{s\left(1 - e^{-s}\right)}.$$

SECTION 10.3

TRANSLATION AND PARTIAL FRACTIONS

This section is devoted to the computational nuts and bolts of the staple technique for the inversion of Laplace transforms — partial fraction decompositions. If time does not permit going further in this chapter, Sections 10.1–10.3 provide a self-contained introduction to Laplace transforms that suffices for the most common elementary applications.

1. $\mathcal{L}\{t^4\} = \dfrac{24}{s^5}$,   so $\mathcal{L}\{t^4 e^{\pi t}\} = \dfrac{24}{(s-\pi)^5}$

3. $\mathcal{L}\{\sin 3\pi t\} = \dfrac{3\pi}{s^2 + 9\pi^2}$,   so $\mathcal{L}\{e^{-2t}\sin 3\pi t\} = \dfrac{3\pi}{(s+2)^2 + 9\pi^2}$.

5. $F(s) = \dfrac{3}{2s-4} = \dfrac{3}{2}\cdot\dfrac{1}{s-2}$,   so $f(t) = \dfrac{3}{2}e^{2t}$

7. $F(s) = \dfrac{1}{(s+2)^2}$,   so $f(t) = t\,e^{-2t}$

9. $F(s) = 3\cdot\dfrac{s-3}{(s-3)^2 + 16} + \dfrac{7}{2}\cdot\dfrac{4}{(s-3)^2 + 16}$,   so $f(t) = e^{3t}[3\cos 4t + (7/2)\sin 4t]$

11. $F(s) = \dfrac{1}{4}\cdot\dfrac{1}{s-2} - \dfrac{1}{4}\cdot\dfrac{1}{s+2}$,   so $f(t) = \dfrac{1}{4}\left(e^{2t} - e^{-2t}\right) = \dfrac{1}{2}\sinh 2t$

13. $F(s) = 3\cdot\dfrac{1}{s+2} - 5\cdot\dfrac{1}{s+5}$,   so $f(t) = 3e^{-2t} - 5e^{-5t}$

15. $F(s) = \dfrac{1}{25}\left(-1\cdot\dfrac{1}{s} - 5\cdot\dfrac{1}{s^2} + \dfrac{1}{s-5}\right)$,   so $f(t) = \dfrac{1}{25}\left(-1 - 5t + e^{5t}\right)$

17. $F(s) = \dfrac{1}{8}\left(\dfrac{1}{s^2-4} - \dfrac{1}{s^2+4}\right) = \dfrac{1}{16}\left(\dfrac{2}{s^2-4} - \dfrac{2}{s^2+4}\right)$

$f(t) = \dfrac{1}{16}\left(\sinh 2t - \sin 2t\right)$

19. $F(s) = \dfrac{s^2 - 2s}{(s^2+1)(s^2+4)} = \dfrac{1}{3}\left(\dfrac{-2s-1}{s^2+1} + \dfrac{2s+4}{s^2+4}\right)$

$$f(t) = \frac{1}{3}\left(-2\cos t - \sin t + 2\cos 2t + 2\sin 2t\right)$$

21. First we need to find A, B, C, D so that

$$\frac{s^2+3}{\left(s^2+2s+2\right)^2} = \frac{As+B}{s^2+2s+2} + \frac{Cs+D}{\left(s^2+2s+2\right)^2}.$$

When we multiply both sides by the quadratic factor s^2+2s+2 and collect coefficients, we get the linear equations

$$-2B - D + 3 = 0$$
$$-2A - 2B - C = 0$$
$$-2A - B + 1 = 0$$
$$-A = 0$$

which we solve for $A=0$, $B=1$, $C=-2$, $D=1$. Thus

$$F(s) = \frac{1}{\left(s+1\right)^2+1} + \frac{-2s+1}{\left[\left(s+1\right)^2+1\right]^2} = \frac{1}{\left(s+1\right)^2+1} - 2\cdot\frac{s+1}{\left[\left(s+1\right)^2+1\right]^2} + 3\cdot\frac{1}{\left[\left(s+1\right)^2+1\right]^2}.$$

We now use the inverse Laplace transforms given in Eq. (16) and (17) of Section 10.3 — supplying the factor e^{-t} corresponding to the translation $s \to s+1$ — and get

$$f(t) = e^{-t}\left[\sin t - 2\cdot\frac{1}{2}t\sin t + 3\cdot\frac{1}{2}\left(\sin t - t\cos t\right)\right] = \frac{1}{2}e^{-t}\left(5\sin t - 2t\sin t - 3t\cos t\right).$$

23. $$\frac{s^3}{s^4+4a^4} = \frac{1}{2}\left(\frac{s-a}{s^2-2as+2a^2} + \frac{s+a}{s^2+2as+2a^2}\right),$$

and $s^2 \pm 2as + 2a^2 = (s \pm a)^2 + a^2$, so it follows that

$$\mathcal{L}^{-1}\left\{\frac{s^3}{s^4+4a^4}\right\} = \frac{1}{2}\left(e^{at} + e^{-at}\right)\cos at = \cosh at \cos at.$$

25. $$\frac{s}{s^4+4a^4} = \frac{1}{4a}\left(\frac{s}{s^2-2as+2a^2} - \frac{s}{s^2+2as+2a^2}\right)$$

$$= \frac{1}{4a}\left(\frac{s-a}{s^2-2as+2a^2} + \frac{a}{s^2-2as+2a^2} - \frac{s+a}{s^2+2as+2a^2} + \frac{a}{s^2+2as+2a^2}\right),$$

and $s^2 \pm 2as + 2a^2 = (s \pm a)^2 + a^2$, so it follows that

$$\mathcal{L}^{-1}\left\{\frac{s}{s^4+4a^4}\right\} = \frac{1}{4a}\left[e^{at}(\cos at + \sin at) - e^{-at}(\cos at - \sin at)\right]$$

$$= \frac{1}{2a}\left[\frac{1}{2}\left(e^{at}+e^{-at}\right)\sin at + \frac{1}{2}\left(e^{at}-e^{-at}\right)\cos at\right]$$

$$= \frac{1}{2a}\left(\cosh at \sin at + \sinh at \cos at\right).$$

In Problems 27–40 we give first the transformed equation, then the Laplace transform $X(s)$ of the solution, and finally the desired solution $x(t)$.

27. $[s^2 X(s) - 2s - 3] + 6[sX(s) - 2] + 25X(s) = 0$

$$X(s) = \frac{2s+15}{s^2+6s+25} = 2\cdot\frac{s+3}{(s+3)^2+16} + \frac{9}{4}\cdot\frac{4}{(s+3)^2+16}$$

$x(t) = e^{-3t}[2\cos 4t + (9/4)\sin 4t]$

29. $s^2 X(s) - 4X(s) = \dfrac{3}{s^2}$

$$X(s) = \frac{3}{s^2(s^2-4)} = \frac{3}{4}\left(\frac{1}{s^2-4} - \frac{1}{s^2}\right)$$

$$x(t) = \frac{3}{8}\sinh 2t - \frac{3}{4}t = \frac{3}{8}\left(\sinh 2t - 2t\right)$$

31. $[s^3 X(s) - s - 1] + [s^2 X(s) - 1] - 6[sX(s)] = 0$

$$X(s) = \frac{s+2}{s^3+s^2-6s} = \frac{1}{15}\left(-\frac{5}{s} - \frac{1}{s+3} + \frac{6}{s-2}\right)$$

$$x(t) = \frac{1}{15}\left(-5 - e^{-3t} + 6e^{2t}\right)$$

33. $[s^4 X(s) - 1] + X(s) = 0$

$$X(s) = \frac{1}{s^4+1}$$

It therefore follows from Problem 26 with $a = \sqrt[4]{1/4} = 1/\sqrt{2}$ that

$$x(t) = \frac{1}{\sqrt{2}}\left(\cosh\frac{t}{\sqrt{2}}\sin\frac{t}{\sqrt{2}} - \sinh\frac{t}{\sqrt{2}}\cos\frac{t}{\sqrt{2}}\right).$$

35. $\left[s^4 X(s) - 1 \right] + 8s^2 X(s) + 16 X(s) = 0$

$$X(s) = \frac{1}{s^4 + 8s^2 + 16} = \frac{1}{\left(s^2 + 4\right)^2}$$

$$x(t) = \frac{1}{16}\left(\sin 2t - 2t\cos 2t\right) \qquad \text{(by Eq. (17) in Section 10.3)}$$

37. $\left[s^2 X(s) - 2 \right] + 4s X(s) + 13 X(s) = \dfrac{1}{\left(s+1\right)^2}$

$$X(s) = \frac{2 + 1/(s+1)^2}{s^2 + 4s + 13} = \frac{2s^2 + 4s + 13}{(s+1)^2 \left(s^2 + 4s + 3\right)}$$

$$= \frac{1}{50}\left[-\frac{1}{s+1} + \frac{5}{(s+1)^2} + \frac{s+98}{(s+2)^2 + 9} \right]$$

$$= \frac{1}{50}\left[-\frac{1}{s+1} + \frac{5}{(s+1)^2} + \frac{s+2}{(s+2)^2 + 9} + 32 \cdot \frac{3}{(s+2)^2 + 9} \right]$$

$$x(t) = \frac{1}{50}\left[(-1 + 5t)e^{-t} + e^{-2t}(\cos 3t + 32\sin 3t) \right]$$

39. $x'' + 9x = 6\cos 3t, \qquad x(0) = x'(0) = 0$

$$s^2 X(s) + 9 X(s) = \frac{6s}{s^2 + 9}$$

$$X(s) = \frac{6s}{\left(s^2 + 9\right)^2}$$

$$x(t) = 6 \cdot \frac{1}{2 \cdot 3} t \sin 3t = t \sin 3t \qquad \text{(by Eq. (16) in Section 10.3)}$$

The graph of this resonance is shown in the figure below.

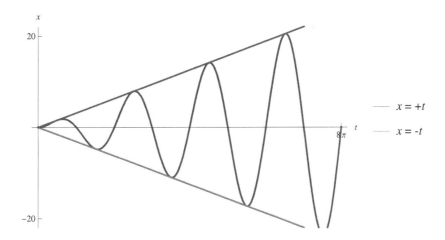

SECTION 10.4

DERIVATIVES, INTEGRALS, AND PRODUCTS OF TRANSFORMS

This section completes the presentation of the standard "operational properties" of Laplace transforms, the most important one here being the convolution property $\mathcal{L}\{f*g\} = \mathcal{L}\{f\}\cdot\mathcal{L}\{g\}$, where the **convolution** $f*g$ is defined by

$$f * g(t) = \int_0^t f(x)g(t-x)\,dx.$$

Here we use x rather than τ as the variable of integration; compare with Eq. (3) in Section 10.4 of the textbook.

1. With $f(t) = t$ and $g(t) = 1$ we calculate

$$t * 1 = \int_0^t x \cdot 1\,dx = \left[\frac{1}{2}x^2\right]_{x=0}^{x=t} = \frac{1}{2}t^2.$$

3. To compute $(\sin t)*(\sin t) = \int_0^t \sin x \sin(t-x)\,dx,$ we first apply the identity $\sin A \sin B = [\cos(A - B) - \cos(A + B)]/2.$ This gives

$$(\sin t) * (\sin t) = \int_0^t \sin x \sin(t-x)\, dx$$

$$= \frac{1}{2}\int_0^t [\cos(2x-t) - \cos t]\, dx$$

$$= \frac{1}{2}\left[\frac{1}{2}\sin(2x-t) - x\cos t\right]_{x=0}^{x=t}$$

$$(\sin t) * (\sin t) = \frac{1}{2}(\sin t - t\cos t).$$

5. $\quad e^{at} * e^{at} = \int_0^t e^{ax} e^{a(t-x)}\, dx = \int_0^t e^{at}\, dx = e^{at}[x]_{x=0}^{x=t} = t e^{at}$

7. $\quad f(t) = 1 * e^{3t} = e^{3t} * 1 = \int_0^t e^{3x} \cdot 1\, dx = \frac{1}{3}(e^{3t} - 1)$

9. $\quad f(t) = \frac{1}{9}\sin 3t * \sin 3t = \frac{1}{9}\int_0^t \sin 3x \sin 3(t-x)\, dx$

$$= \frac{1}{9}\int_0^t \sin 3x [\sin 3t \cos 3x - \cos 3t \sin 3x]\, dx$$

$$= \frac{1}{9}\sin 3t \int_0^t \sin 3x \cos 3x\, dx - \frac{1}{9}\cos 3t \int_0^t \sin^2 3x\, dx$$

$$= \frac{1}{9}\sin 3t \left[\frac{1}{6}\sin^2 3x\right]_{x=0}^{x=t} - \frac{1}{9}\cos 3t \left[\frac{1}{2}\left(x - \frac{1}{6}\sin 6x\right)\right]_{x=0}^{x=t}$$

$$f(t) = \frac{1}{54}(\sin 3t - 3t\cos 3t)$$

11. $\quad f(t) = \cos 2t * \cos 2t = \int_0^t \cos 2x \cos 2(t-x)\, dx$

$$= \int_0^t \cos 2x (\cos 2t \cos 2x + \sin 2t \sin 2x)\, dx$$

$$= (\cos 2t)\int_0^t \cos^2 2x\, dx + (\sin 2t)\int_0^t \cos 2x \sin 2x\, dx$$

$$= (\cos 2t)\left[\frac{1}{2}\left(x + \frac{1}{4}\sin 4x\right)\right]_{x=0}^{x=t} + (\sin 2t)\left[\frac{1}{4}\sin^2 2x\right]_{x=0}^{x=t}$$

$$f(t) = \frac{1}{4}(\sin 2t + 2t\cos 2t)$$

13. $f(t) = e^{3t} * \cos t = \int_0^t (\cos x) e^{3(t-x)} dx$

$= e^{3t} \int_0^t e^{-3x} \cos x \, dx$

$= e^{3t} \left[\dfrac{e^{-3x}}{10} (-3\cos x + \sin x) \right]_{x=0}^{x=t}$ (by integral formula #50)

$f(t) = \dfrac{1}{10} \left(3e^{3t} - 3\cos t + \sin t \right)$

15. $\mathcal{L}\{t \sin 3t\} = -\dfrac{d}{ds} \left(\mathcal{L}\{\sin 3t\} \right) = -\dfrac{d}{ds} \left(\dfrac{3}{s^2+9} \right) = \dfrac{6s}{\left(s^2+9 \right)^2}$

17. $\mathcal{L}\{e^{2t} \cos 3t\} = (s-2)/(s^2 - 4s + 13)$

$\mathcal{L}\{te^{2t} \cos 3t\} = -(d/ds)[(s-2)/(s^2 - 4s + 13)] = (s^2 - 4s - 5)/(s^2 - 4s + 13)^2$

19. $\mathcal{L}\left\{ \dfrac{\sin t}{t} \right\} = \int_s^\infty \dfrac{ds}{s^2+1} = \left[\tan^{-1} s \right]_s^\infty = \dfrac{\pi}{2} - \tan^{-1} s = \tan^{-1}\left(\dfrac{1}{s} \right)$

21. $\mathcal{L}\left\{ e^{3t} - 1 \right\} = \dfrac{1}{s-3} - \dfrac{1}{s}$, so

$\mathcal{L}\left\{ \dfrac{e^{3t}-1}{t} \right\} = \int_s^\infty \left(\dfrac{1}{s-3} - \dfrac{1}{s} \right) ds = \left[\ln\left(\dfrac{s-3}{s} \right) \right]_s^\infty = \ln\left(\dfrac{s}{s-3} \right)$

23. $f(t) = -\dfrac{1}{t} \mathcal{L}^{-1}\{F'(s)\} = -\dfrac{1}{t} \mathcal{L}^{-1}\left\{ \dfrac{1}{s-2} - \dfrac{1}{s+2} \right\} = -\dfrac{1}{t}\left(e^{2t} - e^{-2t} \right) = -\dfrac{2\sinh 2t}{t}$

25. $f(t) = -\dfrac{1}{t} \mathcal{L}^{-1}\{F'(s)\} = -\dfrac{1}{t} \mathcal{L}^{-1}\left\{ \dfrac{2s}{s^2+1} - \dfrac{1}{s+2} - \dfrac{1}{s-3} \right\} = \dfrac{1}{t}\left(e^{-2t} + e^{3t} - 2\cos t \right)$

27. $f(t) = -\dfrac{1}{t} \mathcal{L}^{-1}\{F'(s)\} = -\dfrac{1}{t} \mathcal{L}^{-1}\left\{ \dfrac{-2/s^3}{1+1/s^2} \right\}$

$= \dfrac{2}{t} \mathcal{L}^{-1}\left\{ \dfrac{1}{s^3+s} \right\} = \dfrac{2}{t} \mathcal{L}^{-1}\left\{ \dfrac{1}{s} - \dfrac{s}{s^2+1} \right\} = \dfrac{2}{t}(1 - \cos t)$

29. $-[s^2 X(s) - x'(0)]' - [s X(s)]' - 2[s X(s)] + X(s) = 0$

$s(s+1)X'(s) + 4s X(s) = 0$ (separable)

$$X(s) = \frac{A}{(s+1)^4} \quad \text{with} \ A \neq 0$$

$$x(t) = Ct^3e^{-t} \quad \text{with} \ C \neq 0$$

31. $-[s^2X(s) - x'(0)]' + 4[s\,X(s)]' - [s\,X(s)] - 4[X(s)]' + 2X(s) = 0$

$(s^2 - 4s + 4)X'(s) + (3s - 6)X(s) = 0 \quad$ (separable)

$(s - 2)X'(s) + 3X(s) = 0$

$$X(s) = \frac{A}{(s-2)^3} \quad \text{with} \ A \neq 0$$

$$x(t) = Ct^2e^{2t} \quad \text{with} \ C \neq 0$$

33. $-[s^2X(s) - x(0)]' - 2[s\,X(s)] - [X(s)]' = 0$

$(s^2 + 1)X'(s) + 4s\,X(s) = 0 \quad$ (separable)

$$X(s) = \frac{A}{(s^2+1)^2} \quad \text{with} \ A \neq 0$$

$$x(t) = C(\sin t - t\cos t) \quad \text{with} \ C \neq 0$$

35. $\mathcal{L}^{-1}\left\{\dfrac{1}{(s-1)\sqrt{s}}\right\} = e^t * \dfrac{1}{\sqrt{\pi t}} = \displaystyle\int_0^t \dfrac{1}{\sqrt{\pi x}} \cdot e^{t-x}\,dx$

$$= \frac{e^t}{\sqrt{\pi}}\int_0^{\sqrt{t}} \frac{1}{u} \cdot e^{-u^2} \cdot 2u\,du = \frac{2e^t}{\sqrt{\pi}}\int_0^{\sqrt{t}} e^{-u^2}\,du = e^t\,\mathrm{erf}\left(\sqrt{t}\right)$$

37. $s^2X(s) + 2sX(s) + X(s) = F(s)$

$$X(s) = F(s) \cdot \frac{1}{(s+1)^2}$$

$$x(t) = te^{-t} * f(t) = \int_0^t \tau e^{-\tau} f(t-\tau)\,d\tau$$

SECTION 10.5

PERIODIC AND PIECEWISE CONTINUOUS INPUT FUNCTIONS

In Problems 1 through 10, we first derive the inverse Laplace transform $f(t)$ of $F(s)$ and then show the graph of $f(t)$.

1. $F(s) = e^{-3s}\mathcal{L}\{t\}$ so Eq. (3b) in Theorem 1 gives

$$f(t) = u(t-3)\cdot(t-3) = \begin{cases} 0 \text{ if } t<3, \\ t-3 \text{ if } t\geq 3. \end{cases}$$

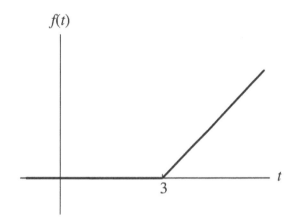

3. $F(s) = e^{-s}\mathcal{L}\{e^{-2t}\}$ so $f(t) = u(t-1)\cdot e^{-2(t-1)} = \begin{cases} 0 \text{ if } t<1, \\ e^{-2(t-1)} \text{ if } t\geq 1. \end{cases}$

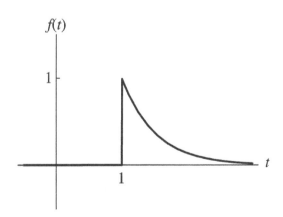

5. $F(s) = e^{-\pi s} \mathcal{L}\{\sin t\}$ so

$$f(t) = u(t-\pi) \cdot \sin(t-\pi) = -u(t-\pi)\sin t = \begin{cases} 0 \text{ if } t<\pi, \\ -\sin t \text{ if } t\ge \pi. \end{cases}$$

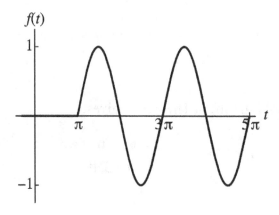

7. $F(s) = \mathcal{L}\{\sin t\} - e^{-2\pi s} \mathcal{L}\{\sin t\}$ so

$$f(t) = \sin t - u(t-2\pi)\sin(t-2\pi) = [1 - u(t-2\pi)]\sin t = \begin{cases} \sin t \text{ if } t<2\pi, \\ 0 \text{ if } t\ge 2\pi. \end{cases}$$

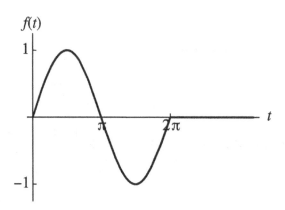

9. $F(s) = \mathcal{L}\{\cos \pi t\} + e^{-3s} \mathcal{L}\{\cos \pi t\}$ so

$$f(t) = \cos \pi t + u(t-3)\cos \pi(t-3) = [1 - u(t-3)]\cos \pi t = \begin{cases} \cos \pi t \text{ if } t<3, \\ 0 \text{ if } t\ge 3. \end{cases}$$

The graph of $f(t)$ is shown at the top of the next page.

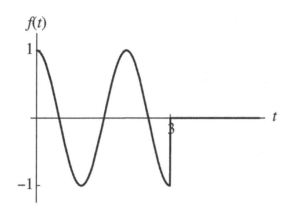

11. $f(t) = 2 - u(t-3) \cdot 2$ so $F(s) = \dfrac{2}{s} - e^{-3s} \dfrac{2}{s} = \dfrac{2}{s}\left(1 - e^{-3s}\right).$

13. $f(t) = [1 - u(t - 2\pi)]\sin t = \sin t - u(t - 2\pi)\sin(t - 2\pi)$ so

$$F(s) = \frac{1}{s^2+1} - e^{-2\pi s} \cdot \frac{1}{s^2+1} = \frac{1 - e^{-2\pi s}}{s^2+1}.$$

15. $f(t) = [1 - u(t - 3\pi)]\sin t = \sin t + u(t - 3\pi)]\sin(t - 3\pi)$ so

$$F(s) = \frac{1}{s^2+1} + \frac{e^{-3\pi s}}{s^2+1} = \frac{1 + e^{-3\pi s}}{s^2+1}.$$

17. $f(t) = [u(t-2) - u(t-3)]\sin \pi t = u(t-2)\sin \pi(t-2) + u(t-3)\sin \pi(t-3)$ so

$$F(s) = \left(e^{-2s} + e^{-3s}\right) \cdot \frac{\pi}{s^2+\pi^2} = \frac{\pi\left(e^{-2s} + e^{-3s}\right)}{s^2+\pi^2}.$$

19. If $g(t) = t + 1$ then $f(t) = u(t-1) \cdot t = u(t-1) \cdot g(t-1)$ so

$$F(s) = e^{-s}G(s) = e^{-s}L\{t+1\} = e^{-s} \cdot \left(\frac{1}{s^2} + \frac{1}{s}\right) = \frac{e^{-s}(s+1)}{s^2}.$$

21. If $g(t) = t + 1$ and $h(t) = t + 2$ then

$$\begin{aligned}
f(t) &= t[1 - u(t-1)] + (2 - t)[u(t-1) - u(t-2)] \\
&= t - 2t\,u(t-1) + 2u(t-1) - 2u(t-2) + t\,u(t-2) \\
&= t - 2u(t-1)g(t-1) + 2u(t-1) - 2u(t-2) + u(t-2)h(t-2)
\end{aligned}$$

so

$$F(s) = \frac{1}{s^2} - 2e^{-s}\left(\frac{1}{s^2} + \frac{1}{s}\right) + \frac{2e^{-s}}{s} - \frac{2e^{-2s}}{s} + e^{-2s}\left(\frac{1}{s^2} + \frac{2}{s}\right) = \frac{\left(1 - e^{-s}\right)^2}{s^2}.$$

23. With $f(t) = 1$ and $p = 1$, Formula (12) in the text gives

$$\mathcal{L}\{1\} = \frac{1}{1-e^{-s}} \int_0^1 e^{-st} \cdot 1 \, dt = \frac{1}{1-e^{-s}} \left[-\frac{e^{-st}}{s} \right]_{t=0}^{t=1} = \frac{1}{s}.$$

25. With $p = 2a$ and $f(t) = 1$ if $0 \le t \le a$, $f(t) = 0$ if $a < t \le 2a$, Formula (12) gives

$$\mathcal{L}\{f(t)\} = \frac{1}{1-e^{-2as}} \int_0^a e^{-st} \cdot 1 \, dt = \frac{1}{1-e^{-2as}} \left[-\frac{e^{-st}}{s} \right]_{t=0}^{t=a}$$

$$= \frac{1-e^{-as}}{s\left(1-e^{-as}\right)\left(1+e^{-as}\right)} = \frac{1}{s\left(1+e^{-as}\right)}.$$

27. $G(s) = \mathcal{L}\{t/a - f(t)\} = (1/as^2) - F(s)$. Now substitution of the result of Problem 26 in place of $F(s)$ immediately gives the desired transform.

29. With $p = 2\pi/k$ and $f(t) = \sin kt$ for $0 \le t \le \pi/k$ while $f(t) = 0$ for $\pi/k \le t \le 2\pi/k$, Formula (12) the integral formula

$$\int e^{at} \sin bt \, dt = e^{at} \left[\frac{a \sin bt - b \cos bt}{a^2 + b^2} \right] + C$$

give

$$\mathcal{L}\{f(t)\} = \frac{1}{1-e^{-2\pi s/k}} \int_0^{\pi/k} e^{-st} \cdot \sin kt \, dt$$

$$= \frac{1}{1-e^{-2\pi s/k}} \left[e^{-st} \left(\frac{-s \sin kt - k \cos kt}{s^2 + k^2} \right) \right]_{t=0}^{t=\pi/k}$$

$$= \frac{1}{1-e^{-2\pi s/k}} \left[\frac{e^{-\pi s/k}\left(k\right)-\left(-k\right)}{s^2 + k^2} \right]$$

$$= \frac{k\left(1+e^{-\pi s/k}\right)}{\left(1-e^{-\pi s/k}\right)\left(1+e^{-\pi s/k}\right)\left(s^2 + k^2\right)} = \frac{k}{\left(s^2 + k^2\right)\left(1-e^{-\pi s/k}\right)}.$$

In Problems 31-42, we first write and transform the appropriate differential equation. Then we solve for the transform of the solution, and finally inverse transform to find the desired solution.

31. $x'' + 4x = 1 - u(t - \pi)$

$$s^2 X(s) + 4X(s) = \frac{1 - e^{-\pi s}}{s}$$

$$X(s) = \frac{1 - e^{-\pi s}}{s(s^2 + 4)} = \frac{1}{4}(1 - e^{-\pi s})\left(\frac{1}{s} - \frac{s}{s^2 + 4}\right)$$

$$x(t) = (1/4)[1 - u(t - \pi)][1 - \cos 2(t - \pi)] = (1/2)[1 - u(t - \pi)]\sin^2 t$$

The graph of the position function $x(t)$ is shown below.

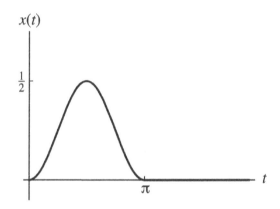

33. $x'' + 9x = [1 - u(t - 2\pi)]\sin t$

$$X(s) = \frac{1 - e^{-2\pi s}}{(s^2 + 1)(s^2 + 4)} = \frac{1}{8}(1 - e^{-2\pi s})\left(\frac{1}{s^2 + 1} - \frac{1}{s^2 + 9}\right)$$

$$x(t) = \frac{1}{8}[1 - u(t - 2\pi)]\left(\sin t - \frac{1}{3}\sin 3t\right)$$

The figure below shows the graph of this position function.

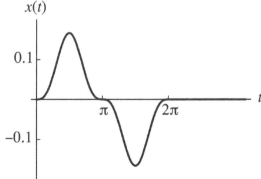

35. $x'' + 4x' + 4x = [1 - u(t-2)]t = t - u(t-2)g(t-2)$ where $g(t) = t + 2$

$$(s+2)^2 X(s) = \frac{1}{s^2} - e^{-2s}\left(\frac{2}{s} + \frac{1}{s^2}\right)$$

$$X(s) = \frac{1}{s^2(s+2)^2} - e^{-2s}\frac{2s+1}{s^2(s+2)^2}$$

$$= \frac{1}{4}\left(-\frac{1}{s} + \frac{1}{s^2} + \frac{1}{s+2} + \frac{1}{(s+2)^2}\right) - \frac{1}{4}e^{-2s}\left(\frac{1}{s} + \frac{1}{s^2} - \frac{1}{s+2} - \frac{3}{(s+2)^2}\right)$$

$$x(t) = (1/4)\{-1 + t + (1 + t)e^{-2t} + u(t-2)[1 - t + (3t - 5)e^{-2(t-2)}]\}$$

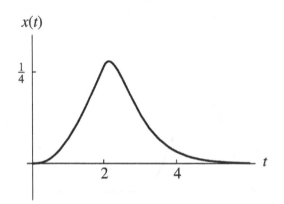

37. $x'' + 2x' + 10x = f(t),$ $x(0) = x'(0) = 0$

As in the solution of Example 7 we find first that

$$\left(s^2 + 2s + 10\right)X(s) = \frac{10}{s} + \frac{20}{s}\sum_{n=1}^{\infty}(-1)^n e^{-n\pi s},$$

so

$$X(s) = \frac{10}{s(s^2 + 2s + 10)} + 2\sum_{n=1}^{\infty}\frac{10(-1)^n e^{-n\pi s}}{s(s^2 + 2s + 10)}.$$

If

$$g(t) = \mathcal{L}^{-1}\left\{\frac{10}{s\left[(s+1)^2 + 9\right]}\right\} = 1 - \frac{1}{3}e^{-t}\left(3\cos 3t + \sin 3t\right),$$

then it follows that

$$x(t) = g(t) + 2\sum_{n=1}^{\infty}(-1)^n u_{n\pi}(t)g(t - n\pi).$$

And here is the graph of $x(t)$:

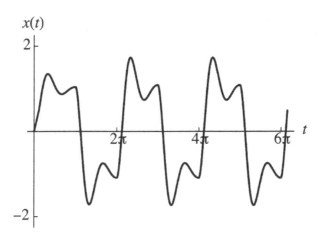

39.   When we substitute the inverse Laplace transforms $a(t)$, $b(t)$, $c(t)$ given at the beginning of part (c), we get

$$v(t) = b_0 a(t) + b_1 b(t) + c(t)$$
$$= \tfrac{1}{4} e^{-2t} \left[b_0 \left(4\cos 4t + 2\sin 4t \right) + b_1 \sin 4t + \left(4 - 4\cos 4t - 2\sin 4t \right) \right].$$

Similarly, Theorem 1 in this section gives

$$w(t) = L^{-1} \left\{ e^{-\pi s} \left[c_0 A(s) + c_1 B(s) - C(s) \right] \right\}$$
$$= u(t - \pi) \left[c_0 a(t - \pi) + c_1 b(t - \pi) - c(t - \pi) \right]$$
$$= \tfrac{1}{4} e^{-2(t - \pi)} \left[c_0 \left(4\cos 4(t - \pi) + 2\sin 4(t - \pi) \right) + c_1 \sin 4(t - \pi) - \right.$$
$$\left. \left(4 e^{2(t - \pi)} - 4\cos 4(t - \pi) - 2\sin 4(t - \pi) \right) \right] \cdot u(t - \pi).$$

where as usual $u(t - \pi) = u_\pi(t)$ denotes the unit stop function at π. Then the four continuity equations listed in part (c) yield the equations

$$e^{-2\pi} b_0 + 1 - e^{-2\pi} = c_0, \qquad e^{-2\pi} b_1 = c_1,$$
$$e^{-2\pi} c_0 - 1 + e^{-2\pi} = b_0, \qquad e^{-2\pi} c_1 = b_1$$

that we solve readily for

$$b_0 = -c_0 = \frac{1 - e^{2\pi}}{1 + e^{2\pi}} \approx -0.996372, \qquad b_1 = c_1 = 0.$$

Finally, these values for the coefficients yield

$$v(t) = 1 - \frac{e^{-2(t-\pi)}}{1+e^{2\pi}}\left(2\cos 4t + \sin 4t\right) \approx 1 - 0.9981 e^{-2t}\left(2\cos 4t + \sin 4t\right),$$

$$w(t) = -\left[1 - \frac{e^{-2(t-2\pi)}}{1+e^{2\pi}}\left(2\cos 4t + \sin 4t\right)\right]u(t-\pi)$$

$$\approx -\left[1 - 0.9981 e^{-2(t-\pi)}\left(2\cos 4t + \sin 4t\right)\right]u(t-\pi).$$

CHAPTER 11

POWER SERIES METHODS

SECTION 11.1

INTRODUCTION AND REVIEW OF POWER SERIES

The power series method consists of substituting a series $y = \Sigma c_n x^n$ into a given differential equation in order to determine what the coefficients $\{c_n\}$ must be in order that the power series will satisfy the equation. It might be pointed out that, if we find a recurrence relation in the form $c_{n+1} = \phi(n)c_n$, then we can determine the radius of convergence ρ of the series solution directly from the recurrence relation

$$\rho = \lim_{n\to\infty}\left|\frac{c_n}{c_{n+1}}\right| = \lim_{n\to\infty}\left|\frac{1}{\phi(n)}\right|.$$

In Problems 1–10 we give first a recurrence relation that can be used to find the radius of convergence and to calculate the succeeding coefficients $c_1, c_2, c_3, \cdots$ in terms of the arbitrary constant c_0. Then we give the series itself.

1. $\quad c_{n+1} = \dfrac{c_n}{n+1};\quad$ it follows that $\quad c_n = \dfrac{c_0}{n!}\quad$ and $\quad \rho = \lim_{n\to\infty}(n+1) = \infty.$

$$y(x) = c_0\left(1 + x + \frac{x^2}{2} + \frac{x^3}{6} + \frac{x^4}{24} + \cdots\right) = c_0\left(1 + \frac{x}{1!} + \frac{x^2}{2!} + \frac{x^3}{3!} + \frac{x^4}{4!} + \cdots\right) = c_0 e^x$$

3. $\quad c_{n+1} = -\dfrac{3c_n}{2(n+1)};\quad$ it follows that $\quad c_n = \dfrac{(-1)^n\, 3^n c_0}{2^n n!}\quad$ and $\quad \rho = \lim_{n\to\infty}\dfrac{2(n+1)}{3} = \infty.$

$$y(x) = c_0\left(1 - \frac{3x}{2} + \frac{9x^2}{8} - \frac{9x^3}{16} + \frac{27x^4}{128} - \cdots\right)$$

$$= c_0\left(1 - \frac{3x}{1!2} + \frac{3^2 x^2}{2!2^2} - \frac{3^3 x^3}{3!2^3} + \frac{3^4 x^4}{4!2^4} - \cdots\right) = c_0 e^{-3x/2}$$

5.   When we substitute $y = \Sigma c_n x^n$ into the equation $y' = x^2 y$, we find that

$$c_1 + 2c_2 x + \sum_{n=0}^{\infty} \left[(n+3)c_{n+3} - c_n \right] x^{n+2} = 0.$$

Hence $c_1 = c_2 = 0$ — which we see by equating constant terms and x-terms on the two

sides of this equation — and $c_{n+3} = \dfrac{c_n}{n+3}$. It follows that

$$c_{3k+1} = c_{3k+2} = 0 \quad \text{and} \quad c_{3k} = \dfrac{c_0}{3 \cdot 6 \cdots (3k)} = \dfrac{c_0}{k! 3^k}.$$

Hence

$$y(x) = c_0 \left(1 + \dfrac{x^3}{3} + \dfrac{x^6}{18} + \dfrac{x^9}{162} + \cdots \right) = c_0 \left(1 + \dfrac{x^3}{1! 3} + \dfrac{x^6}{2! 3^2} + \dfrac{x^9}{3! 3^3} + \cdots \right) = c_0 e^{(x^3/3)}$$

and $\rho = \infty$.

7. $c_{n+1} = 2c_n$;   it follows that $c_n = 2^n c_0$ and $\rho = \lim\limits_{n \to \infty} \dfrac{1}{2} = \dfrac{1}{2}$.

$$y(x) = c_0 \left(1 + 2x + 4x^2 + 8x^3 + 16x^4 + \cdots \right)$$

$$= c_0 \left[1 + (2x) + (2x)^2 + (2x)^3 + (2x)^4 + \cdots \right] = \dfrac{c_0}{1 - 2x}$$

9. $c_{n+1} = \dfrac{(n+2)c_n}{n+1}$;   it follows that $c_n = (n+1)c_0$ and $\rho = \lim\limits_{n \to \infty} \dfrac{n+1}{n+2} = 1$.

$$y(x) = c_0 \left(1 + 2x + 3x^2 + 4x^3 + 5x^4 + \cdots \right)$$

Separation of variables gives $y(x) = \dfrac{c_0}{(1-x)^2}$.

In Problems 11–14 the differential equations are second-order, and we find that the two initial coefficients c_0 and c_1 are both arbitrary. In each case we find the even-degree coefficients in terms of c_0 and the odd-degree coefficients in terms of c_1. The solution series in these problems are all recognizable power series that have infinite radii of convergence.

11. $c_{n+1} = \dfrac{c_n}{(n+1)(n+2)}$;   it follows that $c_{2k} = \dfrac{c_0}{(2k)!}$ and $c_{2k+1} = \dfrac{c_1}{(2k+1)!}$.

$$y(x) = c_0 \left(1 + \dfrac{x^2}{2!} + \dfrac{x^4}{4!} + \dfrac{x^6}{6!} + \cdots \right) + c_1 \left(x + \dfrac{x^3}{3!} + \dfrac{x^5}{5!} + \dfrac{x^7}{7!} + \cdots \right) = c_0 \cosh x + c_1 \sinh x$$

13. $c_{n+2} = -\dfrac{9c_n}{(n+1)(n+2)}$; it follows that $c_{2k} = \dfrac{(-1)^k 3^{2k} c_0}{(2k)!}$ and $c_{2k+1} = \dfrac{(-1)^k 3^{2k} c_1}{(2k+1)!}$.

$$y(x) = c_0\left(1 - \frac{9x^2}{2} + \frac{27x^4}{8} - \frac{81x^6}{80} + \cdots\right) + c_1\left(x - \frac{3x^3}{2} + \frac{27x^5}{40} - \frac{81x^7}{560} + \cdots\right)$$

$$= c_0\left(1 - \frac{(3x)^2}{2!} + \frac{(3x)^4}{4!} - \frac{(3x)^6}{6!} + \cdots\right) + \frac{c_1}{3}\left((3x) - \frac{(3x)^3}{3!} + \frac{(3x)^5}{5!} - \frac{(3x)^7}{7!} + \cdots\right)$$

$$= c_0 \cos 3x + \frac{c_1}{3}\sin 3x$$

15.   Assuming a power series solution of the form $y = \Sigma c_n x^n$, we substitute it into the differential equation $xy' + y = 0$ and find that $(n+1)c_n = 0$ for all $n \geq 0$. This implies that $c_n = 0$ for all $n \geq 0$, which means that the only power series solution of our differential equation is the trivial solution $y(x) \equiv 0$. Therefore the equation has no *non-trivial* power series solution.

17.   Assuming a power series solution of the form $y = \Sigma c_n x^n$, we substitute it into the differential equation $x^2 y' + y = 0$. We find that $c_0 = c_1 = 0$ and that $c_{n+1} = -nc_n$ for $n \geq 1$, so it follows that $c_n = 0$ for all $n \geq 0$. Just as in Problems 15 and 16, this means that the equation has no *non-trivial* power series solution.

In Problems 19–22 we first give the recurrence relation that results upon substitution of an assumed power series solution $y = \Sigma c_n x^n$ into the given second-order differential equation. Then we give the resulting general solution, and finally apply the initial conditions $y(0) = c_0$ and $y'(0) = c_1$ to determine the desired particular solution.

19. $c_{n+2} = -\dfrac{2^2 c_n}{(n+1)(n+2)}$ for $n \geq 0$, so $c_{2k} = \dfrac{(-1)^k 2^{2k} c_0}{(2k)!}$ and $c_{2k+1} = \dfrac{(-1)^k 2^{2k} c_1}{(2k+1)!}$.

$$y(x) = c_0\left(1 - \frac{2^2 x^2}{2!} + \frac{2^4 x^4}{4!} - \frac{2^6 x^6}{6!} + \cdots\right) + c_1\left(x - \frac{2^2 x^3}{3!} + \frac{2^4 x^5}{5!} - \frac{2^6 x^7}{7!} + \cdots\right)$$

$c_0 = y(0) = 0$ and $c_1 = y'(0) = 3$, so

$$y(x) = 3\left(x - \frac{2^2 x^3}{3!} + \frac{2^4 x^5}{5!} - \frac{2^6 x^7}{7!} + \cdots\right)$$

$$= \frac{3}{2}\left[(2x) - \frac{(2x)^3}{3!} + \frac{(2x)^5}{5!} - \frac{(2x)^7}{7!} + \cdots\right] = \frac{3}{2}\sin 2x.$$

21. $c_{n+1} = \dfrac{2nc_n - c_{n-1}}{n(n+1)}$ for $n \geq 1$; with $c_0 = y(0) = 0$ and $c_1 = y'(0) = 1$, we obtain

$$c_2 = 1, \ c_3 = \frac{1}{2}, \ c_4 = \frac{1}{6} = \frac{1}{3!}, \ c_5 = \frac{1}{24} = \frac{1}{4!}, \ c_6 = \frac{1}{120} = \frac{1}{5!}. \text{ Evidently } c_n = \frac{1}{(n-1)!}, \text{ so}$$

$$y(x) = x + x^2 + \frac{x^3}{2!} + \frac{x^4}{3!} + \frac{x^5}{4!} + \cdots = x\left(1 + x + \frac{x^2}{2!} + \frac{x^3}{3!} + \frac{x^4}{4!} + \cdots\right) = xe^x.$$

23. $c_0 = c_1 = 0$ and the recursion relation

$$(n^2 - n + 1)c_n + (n - 1)c_{n-1} = 0$$

for $n \geq 2$ imply that $c_n = 0$ for $n \geq 0$. Thus any assumed power series solution $y = \Sigma c_n x^n$ must reduce to the trivial solution $y(x) \equiv 0$.

25.   Substitution of $\sum_{n=0}^{\infty} c_n x^n$ into the differential equation $y'' = y' + y$ leads routinely — via shifts of summation to exhibit x^n-terms throughout — to the recurrence formula

$$(n+2)(n+1)c_{n+2} = (n+1)c_{n+1} + c_n,$$

and the given initial conditions yield $c_0 = 0 = F_0$ and $c_1 = 1 = F_1$. But instead of proceeding immediately to calculate explicit values of further coefficients, let us first multiply the recurrence relation by $n!$. This trick provides the relation

$$(n+2)!c_{n+2} = (n+1)!c_{n+1} + n!c_n,$$

that is, the Fibonacci-defining relation $F_{n+2} = F_{n+1} + F_n$ where $F_n = n!c_n$, so we see that $c_n = F_n/n!$ as desired.

27.   **(b)** The roots of the characteristic equation $r^3 = 1$ are $r_1 = 1$, $r_2 = \alpha = (-1 + i\sqrt{3})/2$, and $r_3 = \beta = (-1 - i\sqrt{3})/2$. Then the general solution is

$$y(x) = Ae^x + Be^{\alpha x} + Ce^{\beta x}. \qquad\qquad (*)$$

Imposing the initial conditions, we get the equations

$$A + B + C = 1$$
$$A + \alpha B + \beta C = 1$$
$$A + \alpha^2 B + \beta^2 C = -1.$$

The solution of this system is $A = 1/3$, $B = (1 - i\sqrt{3})/3$, $C = (1 + i\sqrt{3})/3$. Substitution of these coefficients in (*) and use of Euler's relation $e^{i\theta} = \cos\theta + i\sin\theta$ finally yields the desired result.

SECTION 11.2

SERIES SOLUTIONS NEAR ORDINARY POINTS

Instead of deriving in detail the recurrence relations and solution series for Problems 1 through 15, we indicate where some of these problems and answers originally came from. Each of the differential equations in Problems 1–10 is of the form

$$(Ax^2 + B)y'' + Cxy' + Dy = 0$$

with selected values of the constants A, B, C, D. When we substitute $y = \Sigma c_n x^n$, shift indices where appropriate, and collect coefficients, we get

$$\sum_{n=0}^{\infty} \left[An(n-1)c_n + B(n+1)(n+2)c_{n+2} + Cnc_n + Dc_n \right] x^n = 0.$$

Thus the recurrence relation is

$$c_{n+2} = -\frac{An^2 + (C-A)n + D}{B(n+1)(n+2)} c_n \quad \text{for } n \geq 0.$$

It yields a solution of the form

$$y = c_0 y_{\text{even}} + c_1 y_{\text{odd}}$$

where y_{even} and y_{odd} denote series with terms of even and odd degrees, respectively. The even-degree series $c_0 + c_2 x^2 + c_4 x^4 + \cdots$ converges (by the ratio test) provided that

$$\lim_{n \to \infty} \left| \frac{c_{n+2} x^{n+2}}{c_n x^n} \right| = \left| \frac{Ax^2}{B} \right| < 1.$$

Hence its radius of convergence is at least $\rho = \sqrt{|B/A|}$, as is that of the odd-degree series $c_1 x + c_3 x^3 + c_5 x^4 + \cdots$. (See Problem 6 for an example in which the radius of convergence is, surprisingly, greater than $\sqrt{|B/A|}$.)

In Problems 1–15 we give first the recurrence relation and the radius of convergence, then the resulting power series solution.

1. $\quad c_{n+2} = c_n; \qquad \rho = 1; \qquad c_0 = c_2 = c_4 = \cdots; \qquad c_1 = c_3 = c_5 = \cdots$

$$y(x) = c_0 \sum_{n=0}^{\infty} x^{2n} + c_1 \sum_{n=0}^{\infty} x^{2n+1} = \frac{c_0 + c_1 x}{1 - x^2}$$

3. $$c_{n+2} = -\frac{c_n}{(n+2)}; \qquad \rho = \infty;$$

$$c_{2n} = \frac{(-1)^n c_0}{(2n)(2n-2)\cdots\cdots 4\cdot 2} = \frac{(-1)^n c_0}{n!\,2^n};$$

$$c_{2n+1} = \frac{(-1)^n c_1}{(2n+1)(2n-1)\cdots\cdots 5\cdot 3} = \frac{(-1)^n c_1}{(2n+1)!!}$$

$$y(x) = c_0 \sum_{n=0}^{\infty} (-1)^n \frac{x^{2n}}{n!\,2^n} + c_1 \sum_{n=0}^{\infty} (-1)^n \frac{x^{2n+1}}{(2n+1)!!}$$

5. $$c_{n+2} = \frac{nc_n}{3(n+2)}; \qquad \rho = \sqrt{3}; \qquad c_2 = c_4 = c_6 = \cdots = 0$$

$$c_{2n+1} = \frac{2n-1}{3(2n+1)}\cdot\frac{2n-3}{3(2n-1)}\cdots\cdots\frac{3}{3(5)}\cdot\frac{1}{3(3)}c_1 = \frac{c_1}{(2n+1)3^n}$$

$$y(x) = c_0 + c_1 \sum_{n=0}^{\infty} \frac{x^{2n+1}}{(2n+1)3^n}$$

7. $$c_{n+2} = -\frac{(n-4)^2}{3(n+1)(n+2)}c_n; \qquad \rho \geq \sqrt{3}$$

The factor $(n-4)$ yields $c_6 = c_8 = c_{10} = \cdots = 0$, so y_{even} is a 4th-degree polynomial.

We find first that $c_3 = -c_1/2$ and $c_5 = c_1/120$, and then for $n \geq 3$ that

$$c_{2n+1} = \left(-\frac{(2n-5)^2}{3(2n)(2n+1)}\right)\left(-\frac{(2n-7)^2}{3(2n-2)(2n-1)}\right)\cdots\cdots\left(-\frac{1^2}{3(6)(7)}\right)c_5 =$$

$$= (-1)^{n-2}\frac{[(2n-5)!!]^2}{3^{n-2}(2n+1)(2n-1)\cdots\cdots 7\cdot 6}\cdot\frac{c_1}{120} = 9\cdot(-1)^n\frac{[(2n-5)!!]^2}{3^n(2n+1)!}c_1$$

$$y(x) = c_0\left(1 - \frac{8}{3}x^2 + \frac{8}{27}x^4\right) + c_1\left[x - \frac{1}{2}x^3 + \frac{1}{120}x^5 + 9\sum_{n=3}^{\infty}\frac{[(2n-5)!!]^2(-1)^n}{(2n+1)!\,3^n}x^{2n+1}\right]$$

9. $$c_{n+2} = \frac{(n+3)(n+4)}{(n+1)(n+2)}c_n; \qquad \rho = 1$$

$$c_{2n} = \frac{(2n+1)(2n+2)}{(2n-1)(2n)}\cdot\frac{(2n-1)(2n)}{(2n-3)(2n-2)}\cdots\cdots\frac{3\cdot 4}{1\cdot 2}c_0 = (n+1)(2n+1)c_0$$

$$c_{2n+1} = \frac{(2n+2)(2n+3)}{(2n)(2n+1)}\cdot\frac{(2n)(2n+1)}{(2n-2)(2n-1)}\cdots\cdots\frac{4\cdot 5}{2\cdot 3}c_1 = \frac{1}{3}(n+1)(2n+3)c_1$$

$$y(x) = c_0 \sum_{n=0}^{\infty} (n+1)(2n+1)x^{2n} + \frac{1}{3}c_1 \sum_{n=0}^{\infty} (n+1)(2n+3)x^{2n+1}$$

11. $\qquad c_{n+2} = \dfrac{2(n-5)}{5(n+1)(n+2)}c_n; \qquad \rho = \infty$

The factor $(n-5)$ yields $c_7 = c_9 = c_{11} = \cdots = 0$, so y_{odd} is a 5th-degree polynomial. We find first that $c_2 = -c_1$, $c_4 = c_0/10$ and $c_6 = c_0/750$, and then for $n \geq 4$ that

$$c_{2n} = \frac{2(2n-7)}{5(2n)(2n-1)} \cdot \frac{2(2n-5)}{5(2n-2)(2n-3)} \cdots \frac{2(1)}{5(8)(7)}c_6$$

$$= \frac{2^{n-3}(2n-7)!!}{5^{n-3}(2n)(2n-1)\cdots(8)(7)} \cdot \frac{c_0}{750} =$$

$$= \frac{5^3 \cdot 6!}{2^3 \cdot 750} \cdot \frac{2^n(2n-7)!!}{5^n(2n)(2n)\cdots(8)(7)\cdot 6!} \cdot c_1 = 15 \cdot \frac{2^n(2n-7)!!}{5^n(2n)!}c_0$$

$$y(x) = c_1\left(x - \frac{4x^3}{15} + \frac{4x^5}{375}\right) + c_0\left[1 - x^2 + \frac{x^4}{10} + \frac{x^6}{750} + 15\sum_{n=4}^{\infty}\frac{(2n-7)!!\,2^n}{(2n)!\,5^n}x^{2n}\right]$$

13. $\qquad c_{n+3} = -\dfrac{c_n}{n+3}; \qquad \rho = \infty$

When we substitute $y = \Sigma c_n x^n$ into the given differential equation, we find first that $c_2 = 0$, so the recurrence relation yields $c_5 = c_8 = c_{11} = \cdots = 0$ also.

$$y(x) = c_0 \sum_{n=0}^{\infty}\frac{(-1)^n x^{3n}}{n!\,3^n} + c_1 \sum_{n=0}^{\infty}\frac{(-1)^n x^{3n+1}}{1\cdot 4\cdots(3n+1)}$$

15. $\qquad c_{n+4} = -\dfrac{c_n}{(n+3)(n+4)}; \qquad \rho = \infty$

When we substitute $y = \Sigma c_n x^n$ into the given differential equation, we find first that $c_2 = c_3 = 0$, so the recurrence relation yields $c_6 = c_{10} = \cdots = 0$ and $c_7 = c_{11} = \cdots = 0$ also. Then

$$c_{4n} = \frac{-1}{(4n)(4n-1)} \cdot \frac{-1}{(4n-4)(4n-5)} \cdots \frac{-1}{4\cdot 3}c_0 = \frac{(-1)^n c_0}{4^n n!\cdot(4n-1)(4n-5)\cdots 5\cdot 3},$$

$$c_{3n+1} = \frac{-1}{(4n+1)(4n)} \cdot \frac{-1}{(4n-3)(4n-4)} \cdots \frac{-1}{5\cdot 4}c_1 = \frac{(-1)^n c_1}{4^n n!\cdot(4n+1)(4n-3)\cdots 9\cdot 5}.$$

$$y(x) = c_0\left[1 + \sum_{n=1}^{\infty}\frac{(-1)^n x^{4n}}{4^n n!\cdot 3\cdot 7\cdots(4n-1)}\right] + c_1\left[x + \sum_{n=1}^{\infty}\frac{(-1)^n x^{4n+1}}{4^n n!\cdot 5\cdot 9\cdots(4n+1)}\right]$$

17.   The recurrence relation

$$c_{n+2} = -\frac{(n-2)c_n}{(n+1)(n+2)}$$

yields $c_2 = c_0 = y(0) = 1$ and $c_4 = c_6 = \cdots = 0$. Because $c_1 = y'(0) = 0$, it follows also that $c_1 = c_3 = c_5 = \cdots = 0$. Thus the desired particular solution is $y(x) = 1 + x^2$.

19.   The substitution $t = x - 1$ yields $(1 - t^2)y'' - 6ty' - 4y = 0$, where primes now denote differentiation with respect to t. When we substitute $y = \Sigma c_n t^n$ we get the recurrence relation

$$c_{n+2} = \frac{n+4}{n+2}c_n.$$

for $n \geq 0$, so the solution series has radius of convergence $\rho = 1$, and therefore converges if $-1 < t < 1$. The initial conditions give $c_0 = 0$ and $c_1 = 1$, so $c_{even} = 0$ and

$$c_{2n+1} = \frac{2n+3}{2n+1} \cdot \frac{2n+1}{2n-1} \cdots \cdot \frac{7}{5} \cdot \frac{5}{3} c_1 = \frac{2n+3}{3}.$$

Thus

$$y = \frac{1}{3}\sum_{n=0}^{\infty}(2n+3)t^{2n+1} = \frac{1}{3}\sum_{n=0}^{\infty}(2n+3)(x-1)^{2n+1},$$

and the x-series converges if $0 < x < 2$.

21.   The substitution $t = x + 2$ yields $(4t^2 + 1)y'' = 8y$, where primes now denote differentiation with respect to t. When we substitute $y = \Sigma c_n t^n$ we get the recurrence relation

$$c_{n+2} = -\frac{4(n-2)}{(n+2)}c_n$$

for $n \geq 0$. The initial conditions give $c_0 = 1$ and $c_1 = 0$. It follows that $c_{odd} = 0$, $c_2 = 4$ and $c_4 = c_6 = \cdots = 0$, so the solution reduces to

$$y = 2 + 4t^2 = 1 + 4(x+2)^2.$$

In Problems 23–26 we first derive the recurrence relation, and then calculate the solution series $y_1(x)$ with $c_0 = 1$ and $c_1 = 0$ as well as the solution series $y_2(x)$ with $c_0 = 0$ and $c_1 = 1$.

23.   Substitution of $y = \Sigma c_n x^n$ yields

$$c_0 + 2c_2 + \sum_{n=1}^{\infty}\left[c_{n-1} + c_n + (n+1)(n+2)c_{n+2}\right]x^n = 0,$$

so

$$c_2 = -\frac{1}{2}c_0, \qquad c_{n+2} = -\frac{c_{n-1}+c_n}{(n+1)(n+2)} \quad \text{for} \quad n \ge 1.$$

$$y_1(x) = 1 - \frac{x^2}{2} - \frac{x^3}{6} + \frac{x^4}{24} + \cdots; \qquad y_2(x) = x - \frac{x^3}{6} - \frac{x^4}{12} + \frac{x^5}{120} + \cdots$$

25. Substitution of $y = \Sigma c_n x^n$ yields

$$2c_2 + 6c_3 x + \sum_{n=2}^{\infty}\left[c_{n-2} + (n-1)c_{n-1} + (n+1)(n+2)c_{n+2}\right]x^n = 0,$$

so

$$c_2 = c_3 = 0, \qquad c_{n+2} = -\frac{c_{n-2} + (n-1)c_{n-1}}{(n+1)(n+2)} \quad \text{for} \quad n \ge 2.$$

$$y_1(x) = 1 - \frac{x^4}{12} + \frac{x^7}{126} + \frac{x^8}{672} + \cdots; \qquad y_2(x) = x - \frac{x^4}{12} - \frac{x^5}{20} + \frac{x^7}{126} + \cdots$$

27. Substitution of $y = \Sigma c_n x^n$ yields

$$c_0 + 2c_2 + (2c_1 + 6c_3)x + \sum_{n=2}^{\infty}\left[2c_{n-2} + (n+1)c_n + (n+1)(n+2)c_{n+2}\right]x^n = 0,$$

so

$$c_2 = -\frac{c_0}{2}, \quad c_3 = -\frac{c_1}{3}, \qquad c_{n+2} = -\frac{2c_{n-2} + (n+1)c_n}{(n+1)(n+2)} \quad \text{for} \quad n \ge 2.$$

With $c_0 = y(0) = 1$ and $c_1 = y'(0) = -1$, we obtain

$$y(x) = 1 - x - \frac{x^2}{2} + \frac{x^3}{3} - \frac{x^4}{24} + \frac{x^5}{30} + \frac{29x^6}{720} - \frac{13x^7}{630} - \frac{143x^8}{40320} + \frac{31x^9}{22680} + \cdots.$$

Finally, $x = 0.5$ gives

$$y(0.5) = 1 - 0.5 - 0.125 + 0.041667 - 0.002604 + 0.001042$$
$$+ 0.000629 - 0.000161 - 0.000014 + 0.000003 + \cdots$$
$$y(0.5) \approx 0.415562 \approx 0.4156.$$

29. When we substitute $y = \Sigma c_n x^n$ and $\cos x = \Sigma(-1)^n x^{2n}/(2n)!$ and then collect coefficients of the terms involving $1, x, x^2, \cdots, x^6$, we obtain the equations

$$c_0 + 2c_2 = 0, \quad c_1 + 6c_3 = 0, \quad 12c_4 = 0, \quad -2c_3 + 20c_5 = 0,$$

$$\frac{1}{12}c_2 - 5c_4 + 30c_6 = 0, \quad \frac{1}{4}c_3 - 9c_5 + 42c_6 = 0,$$

$$-\frac{1}{360}c_2 + \frac{1}{2}c_4 - 14c_6 + 56c_8 = 0.$$

Given c_0 and c_1, we can solve easily for $c_2, c_3, \cdots, c_8$ in turn. With the choices $c_0 = 1$, $c_1 = 0$ and $c_0 = 0$, $c_1 = 1$ we obtain the two series solutions

$$y_1(x) = 1 - \frac{x^2}{2} + \frac{x^6}{720} + \frac{13x^8}{40320} + \cdots \quad \text{and} \quad y_2(x) = x - \frac{x^3}{6} - \frac{x^5}{60} - \frac{13x^7}{5040} + \cdots.$$

33. Substitution of $y = \Sigma c_n x^n$ in Hermite's equation leads in the usual way to the recurrence formula

$$c_{n+2} = -\frac{2(\alpha - n)c_n}{(n+1)(n+2)}.$$

Starting with $c_0 = 1$, this formula yields

$$c_2 = -\frac{2\alpha}{2!}, \quad c_4 = +\frac{2^2\alpha(\alpha-2)}{4!}, \quad c_6 = -\frac{2^3\alpha(\alpha-2)(\alpha-4)}{6!}, \quad \cdots.$$

Starting with $c_1 = 1$, it yields

$$c_3 = -\frac{2(\alpha-1)}{3!}, \quad c_5 = +\frac{2^2(\alpha-1)(\alpha-3)}{5!}, \quad c_7 = -\frac{2^3(\alpha-1)(\alpha-3)(\alpha-5)}{7!}, \quad \cdots.$$

This gives the desired even-term and odd-term series y_1 and y_2. If α is an integer, then obviously one series or the other has only finitely many non-zero terms. For instance, with $\alpha = 4$ we get

$$y_1(x) = 1 - \frac{2 \cdot 4}{2}x^2 + \frac{2^2 \cdot 4 \cdot 2}{24}x^4 = 1 - 4x^2 + \frac{4}{3}x^4 = \frac{1}{12}\left(16x^4 - 48x^2 + 12\right),$$

and with $\alpha = 5$ we get

$$y_2(x) = x - \frac{2 \cdot 4}{6}x^3 + \frac{2^2 \cdot 4 \cdot 2}{120}x^5 = x - \frac{4}{3}x^3 + \frac{4}{15}x^5 = \frac{1}{120}\left(32x^5 - 160x^3 + 120\right).$$

The figure below shows the interlaced zeros of the 4th and 5th Hermite polynomials.

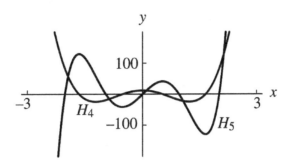

35. **(a)** If

$$y_0 = 1 + \sum_{n=1}^{\infty} \frac{(2n-1)!!}{2^{3n}n!} x^{2n} = 1 + \sum_{n=1}^{\infty} a_n z^n$$

where $a_n = \dfrac{(2n-1)!!}{2^{3n}n!}$, then the radius of convergence of the series in $z = x^2$ is

$$\rho = \lim_{n\to\infty} \left| \frac{a_n}{a_{n+1}} \right| = \lim_{n\to\infty} \frac{(2n-1)!!/2^{3n}n!}{(2n+1)!!/2^{3n+3}(n+1)!} = \lim_{n\to\infty} \frac{2^3(n+1)}{2n+1} = 4.$$

Thus the series in z converges if $-4 < z = x^2 < 4$, so the series $y_0(x)$ converges if $-2 < x < 2$, and thus has radius of convergence equal to 2.

(b) If

$$y_1 = x\left(1 + \sum_{n=1}^{\infty} \frac{n!}{2^n(2n+1)!!} x^{2n}\right) = x\left(1 + \sum_{n=1}^{\infty} b_n z^n\right)$$

where $b_n = \dfrac{n!}{2^n(2n+1)!!}$, then the radius of convergence of the series in z is

$$\rho = \lim_{n\to\infty} \left| \frac{b_n}{b_{n+1}} \right| = \lim_{n\to\infty} \frac{n!/2^n(2n+1)!!}{(n+1)!/2^{n+1}(2n+3)!!} = \lim_{n\to\infty} \frac{2(2n+3)}{n+1} = 4.$$

Hence it follows as in part (a) that the series $y_1(x)$ has radius of convergence equal to 2.

SECTION 11.3

REGULAR SINGULAR POINTS

1. Upon division of the given differential equation by x we see that $P(x) = 1 - x^2$ and $Q(x) = (\sin x)/x$. Because both are analytic at $x = 0$ — in particular, $(\sin x)/x \to 1$ as $x \to 0$ because

$$\frac{\sin x}{x} = \frac{1}{x}\sum_{n=0}^{\infty}\frac{(-1)^n x^{2n+1}}{(2n+1)!} = \sum_{n=1}^{\infty}\frac{(-1)^n x^{2n}}{(2n+1)!} = 1 - \frac{x^2}{3!} + \frac{x^4}{5!} - \frac{x^6}{7!} + \cdots$$

— it follows that $x = 0$ is an ordinary point.

3. When we rewrite the given equation in the standard form of Equation (3) in this section, we see that $p(x) = (\cos x)/x$ and $q(x) = x$. Because $(\cos x)/x \to \infty$ as $x \to 0$ it follows that $p(x)$ is not analytic at $x = 0$, so $x = 0$ is an irregular singular point.

5. In the standard form of Equation (3) we have $p(x) = 2/(1 + x)$ and $q(x) = 3x^2/(1 + x)$. Both are analytic $x = 0$, so $x = 0$ is a regular singular point. The indicial equation is

$$r(r - 1) + 2r = r^2 + r = r(r + 1) = 0,$$

so the exponents are $r_1 = 0$ and $r_2 = -1$.

7. In the standard form of Equation (3) we have $p(x) = (6 \sin x)/x$ and $q(x) = 6$, so $x = 0$ is a regular singular point with $p_0 = q_0 = 6$. The indicial equation is $r^2 + 5r + 6 = 0$, so the exponents are $r_1 = -2$ and $r_2 = -3$.

9. The only singular point of the differential equation $y'' + \dfrac{x}{1-x}y' + \dfrac{x^2}{1-x}y = 0$ is $x = 1$. Upon substituting $t = x - 1$, $x = t + 1$ we get the transformed equation

$y'' - \dfrac{t+1}{t}y' - \dfrac{(t+1)^2}{t}y = 0$, where primes now denote differentiation with respect to t.

In the standard form of Equation (3) we have $p(t) = -(1+t)$ and $q(t) = -t(1+t)^2$. Both these functions are analytic, so it follows that $x = 1$ is a regular singular point of the original equation.

11. The only singular points of the differential equation $y'' - \dfrac{2x}{1-x^2}y' + \dfrac{12}{1-x^2}y = 0$ are $x = +1$ and $x = -1$.

$x = +1$: Upon substituting $t = x - 1$, $x = t + 1$ we get the transformed equation

$$y'' + \frac{2(t+1)}{t(t+2)} y' - \frac{12}{t(t+2)} y = 0,$$ where primes now denote differentiation with respect to

t. In the standard form of Equation (3) we have $p(t) = \frac{2(t+1)}{t+2}$ and $q(t) = -\frac{12t}{t+2}$.

Both these functions are analytic at $t = 0$, so it follows that $x = +1$ is a regular singular point of the original equation.

$x = -1$: Upon substituting $t = x + 1$, $x = t - 1$ we get the transformed equation

$$y'' + \frac{2(t-1)}{t(t-2)} y' - \frac{12}{t(t-2)} y = 0,$$ where primes now denote differentiation with respect to

t. In the standard form of Equation (3) we have $p(t) = \frac{2(t-1)}{t-2}$ and $q(t) = -\frac{12t}{t-2}$.

Both these functions are analytic at $t = 0$, so it follows that $x = -1$ is a regular singular point of the original equation.

13. The only singular points of the differential equation $y'' + \dfrac{1}{x+2} y' + \dfrac{1}{x-2} y = 0$ are

$x = +2$ and $x = -2$.

$x = +2$: Upon substituting $t = x - 2$, $x = t + 2$ we get the transformed equation

$$y'' + \frac{1}{t+4} y' + \frac{1}{t} y = 0,$$ where primes now denote differentiation with respect to t. In the

standard form of Equation (3) we have $p(t) = \dfrac{t}{t+4}$ and $q(t) = t$. Both these

functions are analytic at $t = 0$, so it follows that $x = +2$ is a regular singular point of the original equation.

$x = -2$: Upon substituting $t = x + 2$, $x = t - 2$ we get the transformed equation

$$y'' + \frac{1}{t} y' + \frac{1}{t-4} y = 0,$$ where primes now denote differentiation with respect to t. In the

standard form of Equation (3) we have $p(t) \equiv 1$ and $q(t) = \dfrac{t^2}{t-4}$. Both these

functions are analytic at $t = 0$, so it follows that $x = -2$ is a regular singular point of the original equation.

15. The only singular point of the differential equation $y'' - \dfrac{x^2-4}{(x-2)^2} y' + \dfrac{x+2}{(x-2)^2} y = 0$ is

$x = 2$. Upon substituting $t = x - 2$, $x = t + 2$ we get the transformed equation

$$y'' - \frac{t+4}{t} y' + \frac{t+4}{t^2} y = 0,$$ where primes now denote differentiation with respect to t. In

the standard form of Equation (3) we have $p(t) = -(t+4)$ and $q(t) = t+4$. Both these functions are analytic, so it follows that $x = 2$ is a regular singular point of the original equation.

Each of the differential equations in Problems 17–20 is of the form

$$Axy'' + By' + Cy = 0$$

with indicial equation $Ar^2 + (B - A)r = 0$. Substitution of $y = \Sigma c_n x^{n+r}$ into the differential equation yields the recurrence relation

$$c_n = -\frac{C c_{n-1}}{A(n+r)^2 + (B - A)(n+r)}$$

for $n \geq 1$. In these problems the exponents $r_1 = 0$ and $r_2 = (A - B)/A$ do *not* differ by an integer, so this recurrence relation yields two linearly independent Frobenius series solutions when we apply it separately with $r = r_1$ and with $r = r_2$.

17.   With exponent $r_1 = 0$: $c_n = -\dfrac{c_{n-1}}{4n^2 - 2n}$

$$y_1(x) = x^0 \left(1 - \frac{x}{2} + \frac{x^2}{24} - \frac{x^3}{720} + \cdots \right) = \sum_{n=0}^{\infty} \frac{(-1)^n \left(\sqrt{x}\right)^{2n}}{(2n)!} = \cos\sqrt{x}$$

With exponent $r_2 = \dfrac{1}{2}$: $c_n = -\dfrac{c_{n-1}}{4n^2 + 2n}$

$$y_2(x) = x^{1/2} \left(1 - \frac{x}{6} + \frac{x^2}{120} - \frac{x^3}{5040} + \cdots \right) = \sum_{n=0}^{\infty} \frac{(-1)^n \left(\sqrt{x}\right)^{2n+1}}{(2n+1)!} = \sin\sqrt{x}$$

19.   With exponent $r_1 = 0$: $c_n = \dfrac{c_{n-1}}{2n^2 - 3n}$

$$y_1(x) = x^0 \left(1 - x - \frac{x^2}{2} - \frac{x^3}{18} - \frac{x^4}{360} - \cdots \right) = 1 - x - \sum_{n=2}^{\infty} \frac{x^n}{n!(2n-3)!!}$$

With exponent $r_2 = \dfrac{3}{2}$: $c_n = \dfrac{c_{n-1}}{2n^2 + 3n}$

$$y_2(x) = x^{3/2} \left(1 + \frac{x}{5} + \frac{x^2}{70} + \frac{x^3}{1890} + \frac{x^4}{83160} + \cdots \right) = x^{3/2} \left[1 + 3 \sum_{n=1}^{\infty} \frac{x^n}{n!(2n+3)!!} \right]$$

The differential equations in Problems 21–24 are all of the form

$$Ax^2 y'' + Bxy' + (C + Dx^2)y = 0 \tag{1}$$

with indical equation

$$\phi(r) = Ar^2 + (B - A)r + C = 0. \tag{2}$$

Substitution of $y = \Sigma c_n x^{n+r}$ into the differential equation yields

$$\phi(r)c_0 x^r + \phi(r+1)c_1 x^{r+1} + \sum_{n=2}^{\infty}\left[\phi(r+n)c_n + Dc_{n-2}\right]x^{n+r} = 0. \tag{3}$$

In each of Problems 21–24 the exponents r_1 and r_2 do *not* differ by an integer. Hence when we substitute either $r = r_1$ or $r = r_2$ into Equation (*) above, we find that c_0 is arbitrary because $\phi(r)$ is then zero, that $c_1 = 0$ — because its coefficient $\phi(r+1)$ is then nonzero — and that

$$c_n = -\frac{Dc_{n-2}}{\phi(r+n)} = -\frac{Dc_{n-2}}{A(n+r)^2 + (B-A)(n+r)+C} \tag{4}$$

for $n \geq 2$. Thus this recurrence formula yields two linearly independent Frobenius series solutions when we apply it separately with $r = r_1$ and with $r = r_2$.

21.   With exponent $r_1 = 1$: $c_1 = 0$, $c_n = \dfrac{2c_{n-2}}{n(2n+3)}$

$$y_1(x) = x^1\left(1 + \frac{x^2}{7} + \frac{x^4}{154} + \frac{x^6}{6930} + \cdots\right) = x\left[1 + \sum_{n=1}^{\infty}\frac{x^{2n}}{n!\cdot 7\cdot 11\cdots (4n+3)}\right]$$

With exponent $r_2 = -\dfrac{1}{2}$: $c_1 = 0$, $c_n = \dfrac{2c_{n-2}}{n(2n-3)}$

$$y_2(x) = x^{-1/2}\left(1 + x^2 + \frac{x^4}{10} + \frac{x^6}{270} + \cdots\right) = \frac{1}{\sqrt{x}}\left[1 + \sum_{n=1}^{\infty}\frac{x^{2n}}{n!\cdot 1\cdot 5\cdots (4n-3)}\right]$$

23.   With exponent $r_1 = \dfrac{1}{2}$: $c_1 = 0$, $c_n = \dfrac{c_{n-2}}{n(6n+7)}$

$$y_1(x) = x^{1/2}\left(1 + \frac{x^2}{38} + \frac{x^4}{4712} + \frac{x^6}{1215696} + \cdots\right) = \sqrt{x}\left[1 + \sum_{n=1}^{\infty}\frac{x^{2n}}{2^n n!\cdot 19\cdot 31\cdots (12n+7)}\right]$$

With exponent $r_2 = -\dfrac{2}{3}$: $c_1 = 0$, $c_n = \dfrac{c_{n-2}}{n(6n-7)}$

$$y_2(x) = x^{-2/3}\left(1 + \frac{x^2}{10} + \frac{x^4}{680} + \frac{x^6}{118320} + \cdots\right) = x^{-2/3}\left[1 + \sum_{n=1}^{\infty}\frac{x^{2n}}{2^n n!\cdot 5\cdot 17\cdots (12n-7)}\right]$$

25.   With exponent $r_1 = \dfrac{1}{2}$: $c_n = -\dfrac{c_{n-1}}{2n}$

$$y_1(x) = x^{1/2}\left(1 - \frac{x}{2} + \frac{x^2}{8} - \frac{x^3}{48} + \frac{x^4}{384} - \cdots\right) = \sqrt{x}\sum_{n=0}^{\infty}\frac{(-1)^n x^n}{n!2^n} = \sqrt{x}\,e^{-x/2}$$

With exponent $r_2 = 0$: $c_n = -\dfrac{c_{n-1}}{2n-1}$

$$y_2(x) = x^0\left(1 - x + \frac{x^2}{3} - \frac{x^3}{15} + \frac{x^4}{105} - \cdots\right) = 1 + \sum_{n=1}^{\infty} \frac{(-1)^n x^n}{(2n-1)!!}$$

The differential equations in Problems 27–29 (after multiplication by x) and the one in Problem 31 are of the same form (1) above as those in Problems 21–24. However, now the exponents r_1 and $r_2 = r_1 - 1$ *do* differ by an integer. Hence when we substitute the smaller exponent $r = r_2$ into Equation (3), we find that c_0 and c_1 are *both* arbitrary, and that c_n is given (for $n \geq 2$) by the recurrence relation in (4). Thus the *smaller* exponent r_2 yields the general solution $y(x) = c_0 y_1(x) + c_1 y_2(x)$ in terms of the two linearly independent Frobenius series solutions $y_1(x)$ and $y_2(x)$.

27.   Exponents $r_1 = 0$ and $r_2 = -1$; with $r = -1$: $c_n = -\dfrac{9c_{n-2}}{n(n-1)}$

$$y(x) = \frac{c_0}{x}\left(1 - \frac{9x^2}{2} + \frac{27x^4}{8} - \frac{81x^6}{80} + \cdots\right) + \frac{c_1}{x}\left(x - \frac{3x^3}{2} + \frac{27x^5}{40} - \frac{81x^7}{560} + \cdots\right)$$

$$= \frac{c_0}{x}\left(1 - \frac{9x^2}{2} + \frac{81x^4}{24} - \frac{729x^6}{720} + \cdots\right) + \frac{c_1}{3x}\left(3x - \frac{27x^3}{6} + \frac{243x^5}{120} - \frac{2187x^7}{5040} + \cdots\right)$$

$$y(x) = c_0 \frac{\cos 3x}{x} + \frac{1}{3}c_1 \frac{\sin 3x}{x}$$

The figure below shows the graphs of the independent solutions $y_1(x) = \dfrac{\cos 3x}{x}$ and $y_2(x) = \dfrac{\sin 3x}{x}$.

29. Exponents $r_1 = 0$ and $r_2 = -1$; with $r = -1$: $c_n = -\dfrac{c_{n-2}}{4n(n-1)}$

$$y(x) = \frac{c_0}{x}\left(1 - \frac{x^2}{8} + \frac{x^4}{384} - \frac{x^6}{46080} + \cdots\right) + \frac{c_1}{x}\left(x - \frac{x^3}{24} + \frac{x^5}{1920} - \frac{x^7}{322560} + \cdots\right)$$

$$= \frac{c_0}{x}\left(1 - \frac{x^2}{2^2 \cdot 2} + \frac{x^4}{2^4 \cdot 24} - \frac{x^6}{2^6 \cdot 720} + \cdots\right) + \frac{2c_1}{x}\left(\frac{x}{2} - \frac{x^3}{2^3 \cdot 6} + \frac{x^5}{2^5 \cdot 120} - \frac{x^7}{2^7 \cdot 5040} + \cdots\right)$$

$$y(x) = \frac{c_0}{x}\cos\frac{x}{2} + \frac{2c_1}{x}\sin\frac{x}{2}$$

The figure below shows the graphs of the independent solutions

$$y_1(x) = \frac{\cos x/2}{x} \quad \text{and} \quad y_2(x) = \frac{\sin x/2}{x}.$$

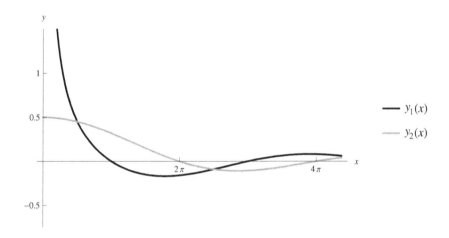

31. The given differential equation $4x^2 y'' - 4xy' + (3 - 4x^2)y = 0$ has indicial equation
$4r^2 - 8r + 3 = (2r - 3)(2r - 1) = 0$, so its exponents are $r_1 = 3/2$ and $r_2 = 1/2$.
With $r = 3/2$, the recurrence relation $c_n = c_{n-2}/n(n-1)$ yields the general solution

$$y(x) = c_0 x^{1/2}\left(1 + \frac{x^2}{2} + \frac{x^4}{24} + \frac{x^6}{720} + \cdots\right) + c_1 x^{1/2}\left(x + \frac{x^3}{6} + \frac{x^5}{120} + \frac{x^7}{5040} + \cdots\right)$$

$$y(x) = c_0 \sqrt{x}\cosh x + c_1 \sqrt{x}\sinh x.$$

The figure below shows the graphs of the independent solutions $y_1(x) = \sqrt{x}\cosh x$
and $y_2(x) = \sqrt{x}\sinh x$.

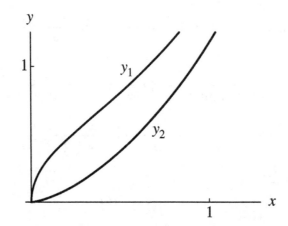

33. Exponents $r_1 = 1/2$ and $r_2 = -1$. With each exponent we find that c_0 is arbitrary and we can solve recursively for c_n in terms of c_{n-1}.

$$y_1(x) = \sqrt{x}\left(1 + \frac{11x}{20} - \frac{11x^2}{224} + \frac{671x^3}{24192} - \frac{9577x^4}{387072} + \cdots\right)$$

$$y_2(x) = \frac{1}{x}\left(1 + 10x + 5x^2 + \frac{10x^3}{9} - \frac{7x^4}{18} + \cdots\right)$$

35. Substitution of $y = x^r \sum c_n x^n$ into the differential equation yields a result of the form

$$-rc_0 x^{r-1} + (\cdots)x^r + (\cdots)x^{r+1} + \cdots = 0,$$

so we see immediately that $c_0 \neq 0$ implies that $r = 0$. Then substitution of the power series $y = \sum c_n x^n$ yields

$$(c_0 - c_1) + (4c_1 - 2c_2)x + (9c_2 - 3c_3)x^2 + (16c_3 - 4c_4)x^4 + \cdots = 0$$

Evidently $c_n = nc_{n-1}$, so if $c_0 = 1$ it follows that $c_n = n!$ for $n \geq 1$. But the series $\sum n! x^n$ has zero radius of convergence, and hence converges only if $x = 0$. We therefore conclude that the given differential equation has *no* nontrivial Frobenius series solution.

37. Substitution of $y = x^r \sum c_n x^n$ into the differential equation $x^3 y'' - xy' + y = 0$ yields a result of the form

$$(r-1)^2 c_0 x^r + (\cdots)x^{r+1} + (\cdots)x^{r+2} + \cdots = 0,$$

so it follows that $r = 1$. But then substitution of $y = x \sum c_n x^n$ into the differential equation yields

$$c_1 x^2 + 4c_2 x^3 + 9c_3 x^4 + 16c_4 x^5 + 25c_5 x^6 + \cdots = 0,$$

so it follows that $c_1 = c_2 = c_3 = c_4 = \cdots = 0$. Hence $y(x) = c_0 x$.

39. Exponents $r_1 = 1$ and $r_2 = -1$; with $r = +1$: $c_1 = 0$, $c_n = -\dfrac{c_{n-2}}{n(n+2)}$

$$y(x) = c_0 x \left(1 - \frac{x^2}{8} + \frac{x^4}{192} - \frac{x^6}{9216} + \frac{x^8}{737280} - \cdots \right)$$

$$= c_0 x \left(1 - \frac{x^2}{2^2 1! 2!} + \frac{x^4}{2^4 2! 3!} - \frac{x^6}{2^6 3! 4!} + \frac{x^8}{2^8 4! 5!} - \cdots \right)$$

If $c_0 = 1/2$, then

$$y(x) = J_1(x) = \frac{x}{2} \sum_{n=0}^{\infty} \frac{(-1)^n}{n!(n+1)} \left(\frac{x}{2} \right)^{2n}.$$

Now, consider the smaller exponent $r_2 = -1$. A Frobenius series with $r = -1$ is of the form $y = x^{-1} \sum_{n=0}^{\infty} c_n x^n$ with $c_0 \neq 0$. However, substitution of this series into Bessel's equation of order 1 gives

$$-c_1 + c_0 x + (c_1 + 3c_3)x^2 + (c_2 + 8c_4)x^3 + (c_3 + 15c_5)x^5 + \cdots = 0,$$

so it follows that $c_0 = 0$, after all. Thus Bessel's equation of order 1 does not have a Frobenius series solution with leading term $c_0 x^{-1}$. However, there is a little more here that meets the eye. We see further that c_2 is arbitrary and that $c_1 = 0$ and $c_n = c_{n-2}/n(n-2)$ for $n > 2$. It follows that our assumed Frobenius series $y = x^{-1} \sum_{n=0}^{\infty} c_n x^n$ actually reduces to

$$y(x) = c_2 x \left(1 - \frac{x^2}{8} + \frac{x^4}{192} - \frac{x^6}{9216} + \frac{x^8}{737280} - \cdots \right).$$

But this is the same as our series solution obtained above using the larger exponent $r = +1$ (calling the arbitrary constant c_2 rather than c_0).

SECTION 11.4

BESSEL FUNCTIONS

Of course Bessel's equation is the most important special ordinary differential equation in mathematics, and every student should be exposed at least to Bessel functions of the first kind.

1. $$J_0'(x) = D_x\left(1 + \sum_{m=1}^{\infty} \frac{(-1)^m x^{2m}}{2^{2m}(m!)^2}\right) = \sum_{m=1}^{\infty} \frac{(-1)^m 2m\, x^{2m-1}}{2^{2m}(m!)^2}$$

$$= \sum_{m=1}^{\infty} \frac{(-1)^m x^{2m-1}}{2^{2m-1}(m-1)!(m!)} = \sum_{m=0}^{\infty} \frac{(-1)^{m+1} x^{2m+1}}{2^{2m+1}(m)!(m+1!)}$$

$$= -\sum_{m=0}^{\infty} \frac{(-1)^m x^{2m+1}}{2^{2m+1}(m)!(m+1!)} = -J_1(x)$$

3. **(a)** $$\Gamma\left(m + \frac{2}{3}\right) = \Gamma\left(\frac{3m+2}{3}\right) = \frac{3m-1}{3}\cdot\frac{3m-4}{3}\cdot\Gamma\left(\frac{3m-4}{3}\right)$$

$$= \frac{3m-1}{3}\cdot\frac{3m-4}{3}\cdots\cdot\frac{5}{3}\cdot\frac{2}{3}\cdot\Gamma\left(\frac{2}{3}\right) = \frac{2\cdot5\cdot8\cdots\cdot(3m-1)}{3^m}\Gamma\left(\frac{2}{3}\right)$$

(b) $$J_{-1/3}(x) = \sum_{m=0}^{\infty} \frac{(-1)^m}{m!\,\Gamma(m+2/3)}\left(\frac{x}{2}\right)^{2m-1/3} = \frac{(x/2)^{-1/3}}{\Gamma(2/3)}\sum_{m=0}^{\infty} \frac{(-1)^m 3^m x^{2m}}{m!\,2\cdot3\cdot8\cdots\cdot(3m-1)}$$

5. Starting with $p = 3$ in Equation (26) we get

$$J_4(x) = \frac{6}{x}J_3(x) - J_2(x) = \frac{6}{x}\left[\frac{4}{x}J_2(x) - J_1(x)\right] - J_2(x)$$

$$= \left(\frac{24}{x^2} - 1\right)\left[\frac{2}{x}J_1(x) - J_0(x)\right] - \frac{6}{x}J_1(x) = \frac{x^2-24}{x^2}J_0(x) + \frac{8(6-x^2)}{x^3}J_1(x)$$

9. $\Gamma(p + m + 1) = (p + m)(p + m - 1)\cdots\cdot(p + 2)(p + 1)\Gamma(p + 1)$, so

$$J_p(x) = \sum_{m=0}^{\infty} \frac{(-1)^m}{m!\,\Gamma(p+m+1)}\left(\frac{x}{2}\right)^{2m+p}$$

$$= \frac{(x/2)^p}{\Gamma(p+1)}\sum_{m=0}^{\infty} \frac{(-1)^m}{m!(p+1)(p+2)\cdots\cdot(p+m)}\left(\frac{x}{2}\right)^{2m}.$$

In Problems 11–18 we use a conspicuous dot • to indicate our choice of u and dv in the integration by parts formula $\int u \cdot dv = uv - \int v\, du$. We use repeatedly the facts (from Example 1) that $\int x J_0(x)\, dx = x J_1(x) + C$ and $\int J_1(x)\, dx = -J_0(x) + C$.

11.
$$
\begin{aligned}
\int x^2 J_0(x)\, dx &= \int x \cdot x J_0(x)\, dx \\
&= x^2 J_1(x) - \int x \cdot J_1(x)\, dx \\
&= x^2 J_1(x) - \left(-x J_0(x) + \int J_0(x)\, dx\right) \\
&= x^2 J_1(x) + x J_0(x) - \int J_0(x)\, dx + C
\end{aligned}
$$

13.
$$
\begin{aligned}
\int x^4 J_0(x)\, dx &= \int x^3 \cdot x J_0(x)\, dx \\
&= x^4 J_1(x) - 3\int x^3 \cdot J_1(x)\, dx \\
&= x^4 J_1(x) - 3\left(-x^3 J_0(x) + 3\int x \cdot x J_0(x)\, dx\right) \\
&= x^4 J_1(x) + 3x^3 J_0(x) - 9\left(x^2 J_1(x) - \int x \cdot J_1(x)\, dx\right) \\
&= x^4 J_1(x) + 3x^3 J_0(x) - 9x^2 J_1(x) + 9\left(-x J_0(x) + \int J_0(x)\, dx\right) \\
&= (x^4 - 9x^2) J_1(x) + (3x^3 - 9x) J_0(x) + 9\int J_0(x)\, dx + C
\end{aligned}
$$

15.
$$
\begin{aligned}
\int x^2 J_1(x)\, dx &= \int x^2 \cdot J_1(x)\, dx \\
&= -x^2 J_0(x) + 2\int x J_0(x)\, dx = -x^2 J_0(x) + 2x J_1(x) + C
\end{aligned}
$$

17.
$$
\begin{aligned}
\int x^4 J_1(x)\, dx &= \int x^4 \cdot J_1(x)\, dx \\
&= -x^4 J_0(x) + 4\int x^2 \cdot x J_0(x)\, dx \\
&= -x^4 J_0(x) + 4\left(x^3 J_1(x) - 2\int x^2 \cdot J_1(x)\, dx\right) \\
&= -x^4 J_0(x) + 4x^3 J_1(x) - 8\left(-x^2 J_0(x) + 2\int x J_0(x)\, dx\right) \\
&= (-x^4 + 8x^2) J_0(x) + (4x^3 - 16x) J_1(x) + C
\end{aligned}
$$

Problems 19–30 are routine applications of the theorem in this section. In each case it is necessary only to identify the coefficients A, B, C and the exponent q in the differential equation

$$
x^2 y'' + Axy' + (B + Cx^q)y = 0. \tag{1}
$$

Then we can calculate the values

$$\alpha = \frac{1-A}{2}, \quad \beta = \frac{q}{2}, \quad k = \frac{2\sqrt{C}}{q}, \quad p = \frac{\sqrt{(1-A)^2 - 4B}}{q} \qquad (2)$$

and finally write the general solution

$$y(x) = x^\alpha \left[c_1 J_p(kx^\beta) + c_2 J_{-p}(kx^\beta) \right] \qquad (3)$$

specified in Theorem 1 on solutions in terms of Bessel functions. This is a "template procedure" that we illustrate only in a couple of problems.

19.   We have $A = -1, B = 1, C = 1, q = 2$ so

$$\alpha = \frac{1-(-1)}{2} = 1, \quad \beta = \frac{2}{2} = 1, \quad k = \frac{2\sqrt{1}}{2} = 1, \quad p = \frac{\sqrt{(1-(-1))^2 - 4(1)}}{2} = 0,$$

so our general solution is $y(x) = x[c_1 J_0(x) + c_2 Y_0(x)]$, using $Y_0(x)$ because $p = 0$ is an integer.

21. $y(x) = x[c_1 J_{1/2}(3x^2) + c_2 J_{-1/2}(3x^2)]$

23.   To match the given equation with Eq. (1) above, we first divide through by the leading coefficient 16 to obtain the equation

$$x^2 y'' + \frac{5}{3} xy' + \left(-\frac{5}{36} + \frac{1}{4} x^3 \right) y = 0$$

with $A = 5/3, B = -5/36, C = 1/4,$ and $q = 3$. Then

$$\alpha = \frac{1 - 5/3}{3} = -\frac{1}{3}, \quad \beta = \frac{3}{2}, \quad k = \frac{2\sqrt{1/4}}{3} = \frac{1}{3}, \quad p = \frac{\sqrt{(1 - 5/3)^2 - 4(-5/36)}}{3} = \frac{1}{3},$$

so our general solution is $y(x) = x^{-1/3}[c_1 J_{1/3}(x^{3/2}/3) + c_2 J_{-1/3}(x^{3/2}/3)]$.

25. $y(x) = x^{-1}[c_1 J_0(x) + c_2 Y_0(x)]$

27. $y(x) = x^{1/2}[c_1 J_{1/2}(2x^{3/2}) + c_2 J_{-1/2}(2x^{3/2})]$

29. $y(x) = x^{1/2}[c_1 J_{1/6}(x^3/3) + c_2 J_{-1/6}(x^3/3)]$

31.   We want to solve the equation $xy'' + 2y' + xy = 0$. If we rewrite it as

$$x^2 y'' + 2xy' + x^2 y = 0$$

then we have the form in Equation (1) with $A = 2$, $B = 0$, $C = 1$, and $q = 2$. Then Equation (2) gives $\alpha = -1/2$, $\beta = 1$, $k = 1$, and $p = 1/2$, so by Equation (3) the general solution is

$$y(x) = x^{-1/2}\left[c_1 J_{1/2}(x) + c_1 J_{-1/2}(x)\right]$$

$$= x^{-1/2}\left[c_1\sqrt{\frac{2}{\pi x}}\cos x + c_2\sqrt{\frac{2}{\pi x}}\sin x\right]$$

$$= \frac{1}{x}\left(a_1\cos x + a_2\sin x\right)$$

(with $a_i = c_i\sqrt{2/\pi}$), using Equations (19) in Section 11.4.

33.   The substitution

$$y = -\frac{u'}{u}, \qquad y' = \frac{(u')^2}{u^2} - \frac{u''}{u}$$

immediately transforms $y' = x^2 + y^2$ to $u'' + x^2 u = 0$. The equivalent equation

$$x^2 u'' + x^4 u = 0$$

is of the form in (1) with $A = B = 0$, $C = 1$, and $q = 4$. Equations (2) give $\alpha = 1/2$, $\beta = 2$, $k = 1/2$, and $p = 1/4$, so the general solution is

$$u(x) = x^{1/2}\left[c_1 J_{1/4}(x^2/2) + c_2 J_{-1/4}(x^2/2)\right].$$

To compute $u'(x)$, let $z = x^2/2$ so $x = 2^{1/2}z^{1/2}$. Then Equation (22) in Section 11.4 with $p = 1/4$ yields

$$\frac{d}{dx}\left(x^{1/2}J_{1/4}(x^2/2)\right) = \frac{d}{dz}\left(2^{1/4}z^{1/4}J_{1/4}(z)\right)\cdot\frac{dz}{dx}$$

$$= 2^{1/4}z^{1/4}J_{-3/4}(z)\cdot\frac{dz}{dx}$$

$$= 2^{1/4}\cdot\frac{x^{1/2}}{2^{1/4}}J_{-3/4}(x^2/2)\cdot x = x^{3/2}J_{-3/4}(x^2/2).$$

Similarly, Equation (23) in Section 11.4 with $p = -1/4$ yields

$$\frac{d}{dx}\left(x^{1/2}J_{-1/4}(x^2/2)\right) = \frac{d}{dz}\left(2^{1/4}z^{1/4}J_{-1/4}(z)\right)\cdot\frac{dz}{dx} = -x^{3/2}J_{3/4}(x^2/2).$$

Therefore

$$u'(x) = x^{3/2}\left[c_1 J_{-3/4}(x^2/2) - c_2 J_{3/4}(x^2/2)\right].$$

It follows finally that the general solution of the Riccati equation $y' = x^2 + y^2$ is

$$y(x) = -\frac{u'}{u} = x \cdot \frac{J_{3/4}\left(\frac{1}{2}x^2\right) - c\,J_{-3/4}\left(\frac{1}{2}x^2\right)}{c\,J_{1/4}\left(\frac{1}{2}x^2\right) + J_{-1/4}\left(\frac{1}{2}x^2\right)}$$

where the arbitrary constant is $c = c_1/c_2$.

34.   Substitution of the series expressions for the Bessel functions in the formula for $y(x)$ in Problem 33 yields

$$y(x) = x \cdot \frac{A\left(\frac{1}{2}x^2\right)^{3/4}(1+\cdots) - c\,B\left(\frac{1}{2}x^2\right)^{-3/4}(1+\cdots)}{c\,C\left(\frac{1}{2}x^2\right)^{1/4}(1+\cdots) + D\left(\frac{1}{2}x^2\right)^{-1/4}(1+\cdots)}$$

where each pair of parentheses encloses a power series in x with constant term 1, and

$$A = 2^{-3/4}/\Gamma(7/4) \qquad\qquad B = 2^{3/4}/\Gamma(1/4)$$
$$C = 2^{-1/4}/\Gamma(5/4) \qquad\qquad D = 2^{1/4}/\Gamma(3/4).$$

Multiplication of numerator and denominator by $x^{1/2}$ and a bit of simplification gives

$$y(x) = \frac{2^{-3/4}Ax^3(1+\cdots) - 2^{3/4}c\,B(1+\cdots)}{2^{-1/4}c\,Cx(1+\cdots) + 2^{1/4}D(1+\cdots)}.$$

It now follows that

$$y(0) = \frac{-2^{3/4}cB}{2^{1/4}D} = \frac{-2^{1/2}\left(2^{3/4}/\Gamma(1/4)\right)}{2^{1/4}/\Gamma(3/4)} = -2c \cdot \frac{\Gamma(3/4)}{\Gamma(1/4)}. \qquad (*)$$

(a)   If $y(0) = 0$ then (*) gives $c = 0$ in the general solution formula of Problem 33.

(b)   If $y(0) = 1$ then (*) gives $c = -\Gamma(1/4)/2\Gamma(3/4)$. More generally, (*) yields the formula

$$y(x) = x \cdot \frac{2\Gamma\left(\frac{3}{4}\right)J_{3/4}\left(\frac{1}{2}x^2\right) + y_0\,\Gamma\left(\frac{1}{4}\right)J_{-3/4}\left(\frac{1}{2}x^2\right)}{2\Gamma\left(\frac{3}{4}\right)J_{-1/4}\left(\frac{1}{2}x^2\right) - y_0\,\Gamma\left(\frac{1}{4}\right)J_{1/4}\left(\frac{1}{2}x^2\right)}$$

for the solution of the initial value problem $y' = x^2 + y^2$, $y(0) = y_0$.

APPENDIX A

EXISTENCE AND UNIQUENESS OF SOLUTIONS

In Problems 1–12 we apply the iterative formula

$$y_{n+1} = b + \int_a^x f(t, y_n(t))\, dt$$

to compute successive approximations $\{y_n(x)\}$ to the solution of the initial value problem

$$y' = f(x, y), \qquad\qquad y(a) = b.$$

starting with $y_0(x) = b$.

1. $y_0(x) = 3$

$y_1(x) = 3 + 3x$

$y_2(x) = 3 + 3x + 3x^2/2$

$y_3(x) = 3 + 3x + 3x^2/2 + x^3/2$

$y_4(x) = 3 + 3x + 3x^2/2 + x^3/2 + x^4/8$

$y(x) = 3 - 3x + 3x^2/2 + x^3/2 + x^4/8 + \cdots = 3e^x$

3. $y_0(x) = 1$

$y_1(x) = 1 - x^2$

$y_2(x) = 1 - x^2 + x^4/2$

$y_3(x) = 1 - x^2 + x^4/2 - x^6/6$

$y_4(x) = 1 - x^2 + x^4/2 - x^6/6 + x^8/24$

$y(x) = 1 - x^2 + x^4/2 - x^6/6 + x^8/24 - \cdots = \exp(-x^2)$

5. $y_0(x) = 0$

$y_1(x) = 2x$

$y_2(x) = 2x + 2x^2$

$y_3(x) = 2x + 2x^2 + 4x^3/3$

$y_4(x) = 2x + 2x^2 + 4x^3/3 + 2x^4/3$

$y(x) = 2x + 2x^2 + 4x^3/3 + 2x^4/3 + \cdots = e^{2x} - 1$

7. $y_0(x) = 0$

$y_1(x) = x^2$

$y_2(x) = x^2 + x^4/2$

$y_3(x) = x^2 + x^4/2 + x^6/6$

$y_4(x) = x^2 + x^4/2 + x^6/6 + x^8/24$

$y(x) = x^2 + x^4/2 + x^6/6 + x^8/24 + \cdots = \exp(x^2) - 1$

9. $y_0(x) = 1$

$y_1(x) = (1 + x) + x^2/2$

$y_2(x) = (1 + x + x^2) + x^3/6$

$y_3(x) = (1 + x + x^2 + x^3/3) + x^4/24$

$y(x) = 1 + x + x^2 + x^3/3 + x^4/12 + \cdots = 2e^x - 1 - x$

11. $y_0(x) = 1$

$y_1(x) = 1 + x$

$y_2(x) = (1 + x + x^2) + x^3/3$

$y_3(x) = (1 + x + x^2 + x^3) + 2x^4/3 + x^5/3 + x^6/9 + x^7/63$

$y(x) = 1 + x + x^2 + x^3 + x^4 + \cdots = 1/(1 - x)$

13.

$$\begin{bmatrix} x_0(t) \\ y_0(t) \end{bmatrix} = \begin{bmatrix} 1 \\ -1 \end{bmatrix}$$

$$\begin{bmatrix} x_1(t) \\ y_1(t) \end{bmatrix} = \begin{bmatrix} 1 + 3t \\ -1 + 5t \end{bmatrix}$$

$$\begin{bmatrix} x_2(t) \\ y_2(t) \end{bmatrix} = \begin{bmatrix} 1 + 3t + \frac{1}{2}t^2 \\ -1 + 5t - \frac{1}{2}t^2 \end{bmatrix}$$

$$\begin{bmatrix} x_3(t) \\ y_3(t) \end{bmatrix} = \begin{bmatrix} 1 + 3t + \frac{1}{2}t^2 + \frac{1}{3}t^3 \\ -1 + 5t - \frac{1}{2}t^2 + \frac{5}{6}t^3 \end{bmatrix}$$